A FESTSCHRIFT HONORING THE CONTRIBUTIONS OF CALIFORNIA ARCHAEOLOGIST JAY VON WERLHOF

Edited by:
Russell L. Kaldenberg

Fall 2006

TABLE OF CONTENTS

Page

DEDICATION

This work is dedicated to Sherilee von Werlhof, Jay's devoted wife and dedicated field partner for over thirty years. With the floors of her home full of archaeologists, Sherilee always knew what was happening.

PREFACE

This publication is the result of the work of many of Jay von Werlhof's dedicated friends and colleagues who in 2004 at the Society for California Archeology Annual Meeting held a symposium dedicated to Jay. The contents of the volume represent a sampling of papers which were delivered at the symposium and others which were contributed by colleagues who were influenced in one way or another by Jay over their professional or personal lives. Still other papers will assist in helping the reader understand the eclectic interests that Jay has, including his assessment of the "Battle of the Bulge," a critical fight during World War II that Jay participated in as a young man.

We received permission to reprint his outstanding study of the Rocky Hill Rock Art site near Exeter, California, a paper that has been unavailable since 1958, yet is still relevant. Recent photographs of the exquisite rock art from Rocky Hill have been reproduced for this volume. And papers explaining the Malpais by Morlin Childers and "Ancient Man" in San Diego County by George Carter are included to provide a background for the reader to understand the milieu in which Jay worked when he arrived at the Imperial Valley College to begin his teaching career there in 1972. These papers are reprinted by permission of the Imperial Valley Desert Museum.

The outstanding contribution on "Tufa-Encrusted Trees from the Late Pleistocene Searles Lake, California" by Diane M. Erwin and Howard E. Shorn, was originally scheduled to be published by the University of California Museum of Paleontology. It may also be cited as UCMP Contribution No. 1876. Finally, several heart-felt testimonials from Jay's friends about him and his influence on them are included.

Many thanks to all of the above and to Archaeological Systems Management, Inc., and PARC for their contribution to the editing and publishing of this work. Marcia Sandusky worked many hours trying to resolve complicated issues related to the placement of figures and to the figures themselves that came into ASM in various formats. Without her vigilance this would have never been accomplished. Also, many thanks to John Cook for reaching deep into his pocket to cover the publishing expense of this volume. The good folks at both the Friends of China Lake Archaeology and the Searles Valley Historical Association are recognized for their support while I neglected many projects to get this done. Particularly noted are folks who waited and waited for me to get back to them including Bill Wight, Jerry Grimsley, Richard Stewart, Virginia Bickford, Vanessa Zelko, Lit Brush, Jim and Bonnie Fairchild, Sharon Hartley and Bill and Barbara Gossett. Thanks are owed also to Carolyn Shepherd for her uplifting comments and support and to Judyth Reed for her continued encouragement. The last thanks go to the Maturango Museum Publications Committee and volunteers and employees including Jane Burbank-Larsen, Liz Babcock, and Alexander "Sandy" Rogers for assisting in ensuring that this became a volume in their series on the Archeology of the Desert.

Russell L. Kaldenberg
October 2006

INTRODUCTION

It is only fitting that we commemorate the life and contributions of Jay von Werlhof with this collection of papers. Mentor to many, friend to all, and archaeologist's archaeologist, he is honored in this volume for his scholarly pursuits and influence which have been and remain many and varied throughout his lifetime. These cover the gamut from ethnohistoric research of the Tachi on Santa Rosa Rancheria in the 1950s to his most recent dedication to and passion for the geoglyphs of the far southwestern deserts. The breadth and variety of tributes contained herein illustrate the influence Jay has effected in the archaeological community, and even more so amply illustrates his unique and highly valued acumen in the role as a mentor to many.

Jay has always been a true gentleman and an avant-garde scholar interested in every aspect of our world. His main focus has been on the understanding of cosmological questions regarding the unknowns of prehistory and particularly how Native Peoples dealt with such questions through their spiritual beliefs as manifest archaeologically on the ground for us to decode.

My friendship and admiration of Jay go back to some of my earliest experiences in archaeology when volunteering with Paul Ezell and Jay during excavation of prehistoric sites in the sand dunes of Westmoreland, in the Imperial Valley. Even then, in the early 1970s Jay was thought of as the "grand old man," the protector and the wisdom seeker of geoglyphs and petroglyphs, the true seeker of knowledge, not only to understand it but to preserve and protect icons and their significance for future generations. He has spent most of his life protecting the sites and features of the "old ones" in the Colorado Desert and Imperial Valley. Those that know his history and even those who don't should be sure to read Kaldenberg's short biography of von Werlhof in this volume, followed by Jay's cogent account of lessons learned from his experience at the Battle of the Bulge.

Jay trained many professionals, many of whom are nearing the apex of their careers presently, and they left an astounding volume of data for the future in what we refer to as the gray literature. Jay has produced thousands of pages that will be useful for decades to come. His students are all grateful for his guidance and the standards he set. All of us remember his detailed sketches and notebooks; they are second to none in the profession.

One of his most important contributions has been the conceptualization of the Imperial Valley Desert Museum in Ocotillo. His ability to obtain funding to get the museum started and his ability to weave a selling story to local donors has been the major reason that the beautiful museum was conceived and awaits its grand opening to the public in late 2006.

It is an honor and privilege for ASM PARC to fund the publication of this monograph memorializing Jay's lifelong contribution to history, prehistory, Native American studies, and intellectualism. That so many colleagues contributed to this festschrift is a testament to his lifetime in archaeology and to the gentleman who has left his mark on American archaeology. I hope you find the monograph a fitting tribute to Jay and his legacy and I hope you join me and all of the contributors in wishing Jay and his wife, Sherilee, the best in the days ahead.

John R. Cook

TUFA-ENCRUSTED TREES FROM LATE PLEISTOCENE SEARLES LAKE, CALIFORNIA

Diane M. Erwin, Ph.D. and Howard E. Schorn, M.A.
Museum of Paleontology, University of California, Berkeley, California 94720
Key words: Pleistocene, Searles Lake, paleoclimate, fossil tufa, *Juniperus osteosperma*
email: dmerwin@berkeley.edu, 4270how@comcast.net

ABSTRACT

Remnants of woody plants encrusted with a rind of tufa, 2.0-17.0 cm thick, have been found in a small Pleistocene paleoembayment along the western shore of Searles Lake, near the Point of Rocks (POR) area of the Argus Range in southern California. The tufa-encrusted stems occur at 17 sites with the highest number of sites occurring in an area extending ~125 m north-south along the 672 m (2040 ft.) contour. Most are preserved as imprints 1.0 – 30.0 cm wide, showing long, narrow, parallel ridges and shallow grooves that run lengthwise along the stem, a pattern most similar to the strip-like bark of *Juniperus* (juniper). Wood anatomy from one specimen is also consistent with juniper showing short, narrow, uniseriate rays 1-9 cells high; the majority are five cells or less and they lack horizontal resin ducts. Of the two junipers in the region today, the POR juniper is most likely *Juniperus osteosperma* (Utah juniper), which does grow on Argus Peak at 1402 m (4600 ft.) northwest of and roughly five miles upslope from the fossil site. The next closest occurrences of Utah juniper are on Maturango Peak about 23 km (14 mi.) northeast of the POR embayment and to the east in the Panamint Range, but at elevations of 610 to 2073 m (2000 - 6800 ft.) higher than the fossil locality. Many of the fossil sites appear to be *in situ*, therefore indicating that juniper became established and grew within 204 m (670 ft.) of the present floor of Searles Lake basin. Presence of fossil juniper further suggests these plants were growing under Pleistocene climatic conditions that exist during major glacial (pluvial) episodes that depress the altitudinal ranges of plants and promote shifts in their latitudinal ranges.

INTRODUCTION

Today Searles Lake, California is a playa bound by the Slate Range to the east and the Argus Range to the west. The present vegetation in the area is part of the Desert Province characterized by low desert-scrub with Creosote bush (*Larrea*) dominant (Hickman, 1993). Searles Lake is probably best known for "the Pinnacles," massive calcareous tufa towers and mounds that formed through algal mediated precipitation of calcium-rich springs mixing with alkaline lake water along faults (joints?) running through the lake bottom (Packard, 1948; Scholl, 1960; Rieger, 1992; Figures 1a, b). Currently Searles Lake near Trona and Pioneer Point is the site of commercial mining operations extracting the valuable evaporite minerals that were laid down as the lake dried up over the last 100,000+ years (Smith, 1979; Smith et al., 1980; Benson et al., 1990). Because of its economic value, Searles Lake was the focus of early geological studies (e.g., Gale, 1914; Flint and Gale, 1958; Benson et al., 1990). However, later research has focused more on understanding the lake's Late Cenozoic history and its hydrological relationship to other Quaternary paleolakes in the western Great Basin [viz., Lakes Russell (Mono Lake), Owens, China, Panamint, Manly (Death Valley), Lahontan and Bonneville) and in the reconstruction of regional climate and vegetation change through lake core analysis (Smith, 1979, 1984, 1987, 2000; Benson et. al., 1990; Dorn et al., 1990; Woolfenden, 1996; Lin et al., 1998; Litwin et al., 1999).

In this paper, we report on the presence of tufa-encrusted conifer remains found near the Point of Rocks (POR) area of the Argus Range near the towns of Trona and Pioneer Point (Fig 1a, b). Although coniferous pollen and woody plant fragments have been found in Searles Lake cores (Leopold, 1964; Spaulding et al., 1983; Woolfenden, 1996; Litwin et al., 1999), the POR plants are the first record of woody plant stem macrofossils preserved in tufa found along Searles Lake Pleistocene shorelines. Currently seventeen plant-bearing sites have been recorded on the east-facing slope of the Argus Range between 609 and 658 m (2000 - 2160 ft.). Here we describe the remains of *Juniperus* (juniper) recovered thus far and discuss the implied environmental and climatic conditions for this area at the time these plants were growing. From our investigation it is clear that the POR fossil tufa deposits are scientifically significant, deserving future detailed study by those interested in understanding the biological and physico-chemical processes of tufa (carbonate) deposition, Quaternary climate, glaciation, paleohydrology, and vegetation history of this region of the western United States.

MATERIALS AND METHODS

The POR fossil plants occur within a small embayment on the northwestern shore of Searles Lake (Figure 1a, UCMP PA1065). Although tufa blankets the entire embayment encrusting exposed bedrock above ground and at the surface from elevations of 537 m to over 671 m (1760 - 2200 ft.) (e.g. Figures 3, 4, 7-10), the tufa-encrusted plant fossils occur in discreet areas of concentration (Figures 2, 11, 12, 23a, 25a). The plant-bearing sites are located in a roughly 125-m-long area that runs north-south along the 622 m (2040 ft.) contour (Figure 2). At most sites the stems are broken apart into segments along fractures running perpendicular and parallel to the long axis (Figures 13-17, 23a-d). Stem segments at some sites have separated close to one another, whereas at other sites some distance separates the pieces suggesting they have undergone some transport. Only four specimens were collected for lab examination because removing specimens would take them out of context with associated pieces and the landscape.

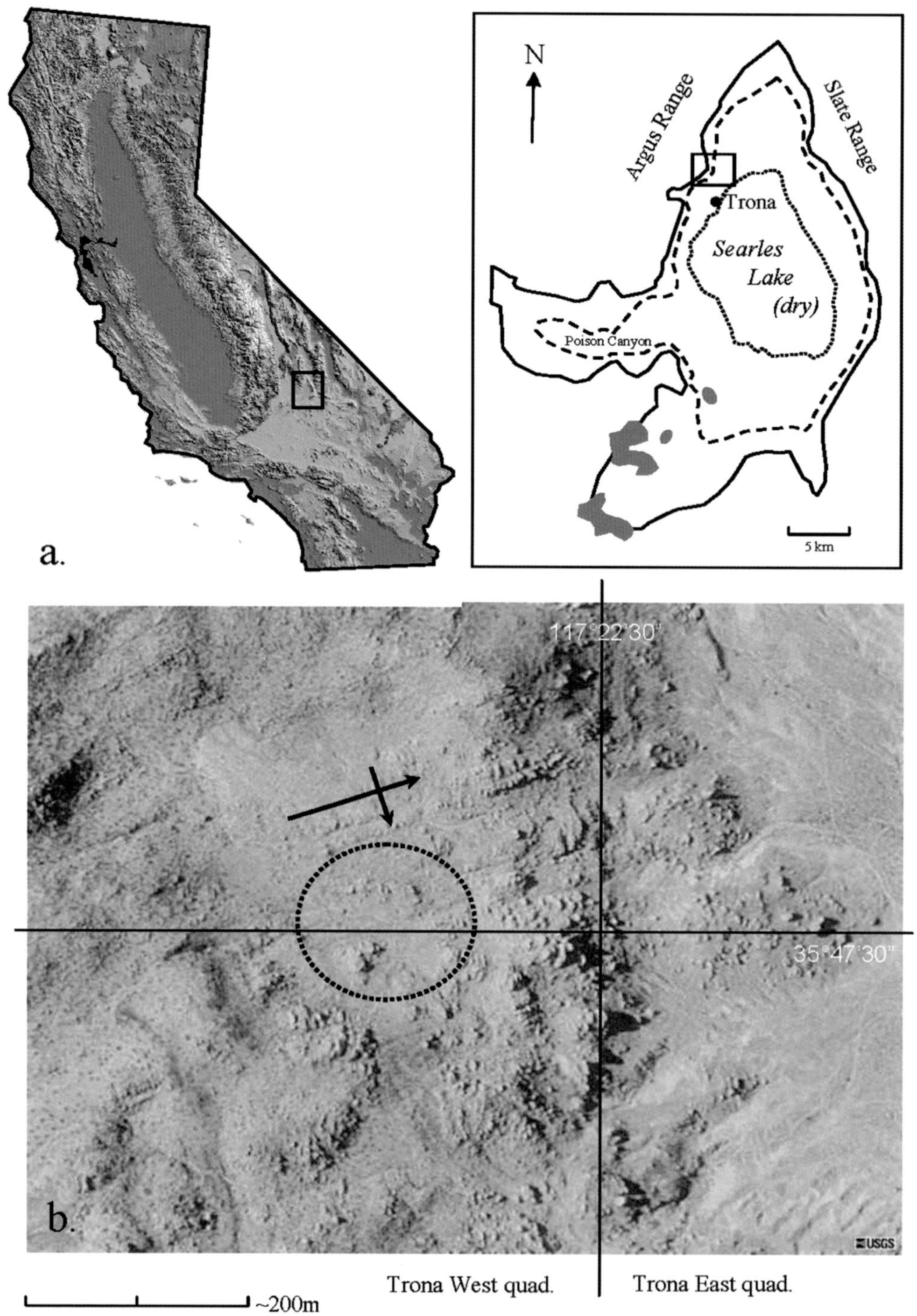

Figure 1. (a) Relief map of California showing location of Searles Lake (small box) and enlargement of area showing the lake outline with location of the POR tufa-bearing paleoembayment marked by the rectangle.
Fine dotted line is present dry lakebed, dashed line marks Tioga shoreline, the solid line marks the Tahoe shoreline at 689 m (2260 ft.), and gray areas mark location of the "Pinnacles." Searles Lake map modified from Joel Michaelsen http://www.geog.ucsb.edu/~joel/g148_f03/lecture_notes/basin_range/searles_lake.jpg (b) Aerial photo of the POR embayment showing the direction and jointing pattern of the granodiorite (arrows) and location of tufa-encrusted plant concentrations marked by dotted line.

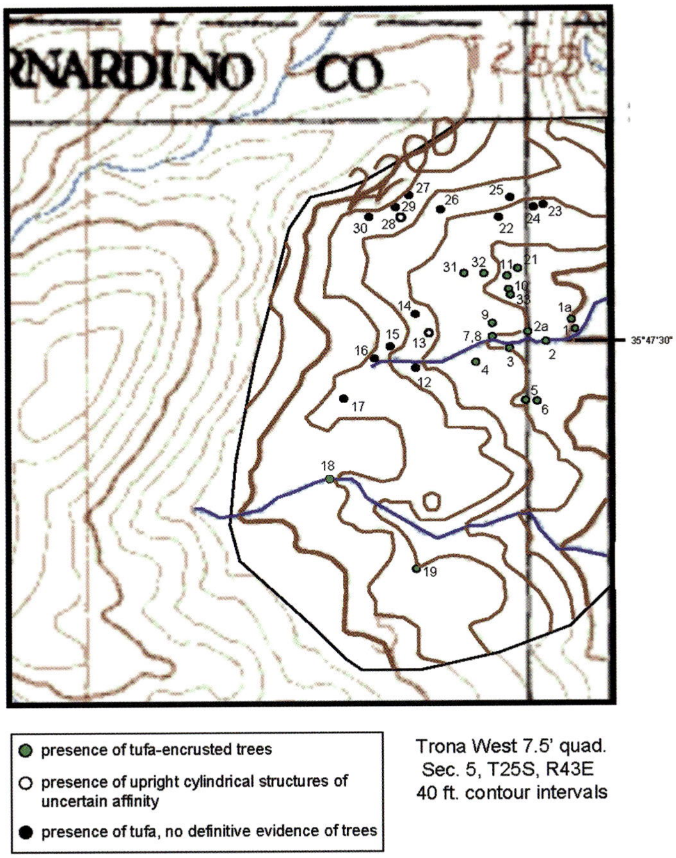

Figure 2. Distribution of sites of tufa concentration within the Point of Rocks study area. Sites are plots from coordinates taken with a Garmin Map 76 GPS unit using the 1927 datum. Note Map Site 20 not shown. Plot is on the Trona East USGS quad.

Figure 3. Tufa-encrusted granodiorite near pathway leading up to the study area. Note tufa in crevice at arrow, tufa talus at lower left and spheroidal weathering of the granodiorite.

Figure 4. View of Argus Range looking west from near 658 m (2160 ft.) elevation. Tufa at arrow with occurrences higher upslope. Paleoshoreline at dashed yellow line.

Figure 5. View looking east across the POR paleoembayment showing typical low, desert scrub vegetation. Creosote bushes in foreground.

Figure 6. View looking south over POR paleoembayment. Arrows indicate outcropping intrusions of calcium-rich gabbro.

Figure 7. Granodiorite boulder completely encrusted with tufa at 579 m (1900 ft.) near Map Site 14. Some tufa has sloughed off from weathering.

Figure 8. Weathered tufa cracking and sloughing off granodiorite at 579 m (1900 ft.) near Map Site 26.

Figure 9. Weathered and cracked tufa covering the ground at 597 m (1960 ft.) near Map Site 27

Figure 10. Granodiorite covering the ground at 567 m (1860 ft.) near Map Site 4. Tufa remaining is that in the crevices.

Figure 11.

Figure 12. View looking southeast toward Searles Lake from Map Site 5. 11. This site has sandy tufa (T) and large, disarticulated pieces of tufa-encrusted trees lying on the granodiorite (arrows). 12. Close-up of tufa-encrusted trees shown at arrows in Figure 11.

Figures. 13. Remnants of disarticulated tufa-encrusted trees at Map Site 5. 13. Shows how specimens fracture perpendicular and parallel to the long axis of the stem. Scale bar = 15 cm. 14. Close-up of branched stem. Scale bar = 10 cm.

Figures. 14. Remnants of disarticulated tufa-encrusted trees at Map Site 5. 13. Shows how specimens fracture perpendicular and parallel to the long axis of the stem. Scale bar = 15 cm. 14. Close-up of branched stem. Scale bar = 10 cm.

Figure 15a. Close-up of specimen at Map Site 5 showing surface pattern of the stem. Scale bar = 4 cm.

Figure 15b. Detail of the internal surface of specimen we interpret as the internal view of the bark. Striate, strip-like pattern is similar to *Juniperus*. Scale bar = 1 cm.

Figure 16. Mold of tufa-encrusted tree at Map Site 5 showing five branches, impression of the internal stem surface, and thick coating of tufa (T). Scale bar = 4 cm.

Figure 17. a. One of the most complete specimens. No stem surface pattern preserved but branching consistent with that of a woody plant. Associated pieces probably part of one individual. b. Closer view of same specimen in a. showing branching. Map Site 11. Scale bar = 15 cm.

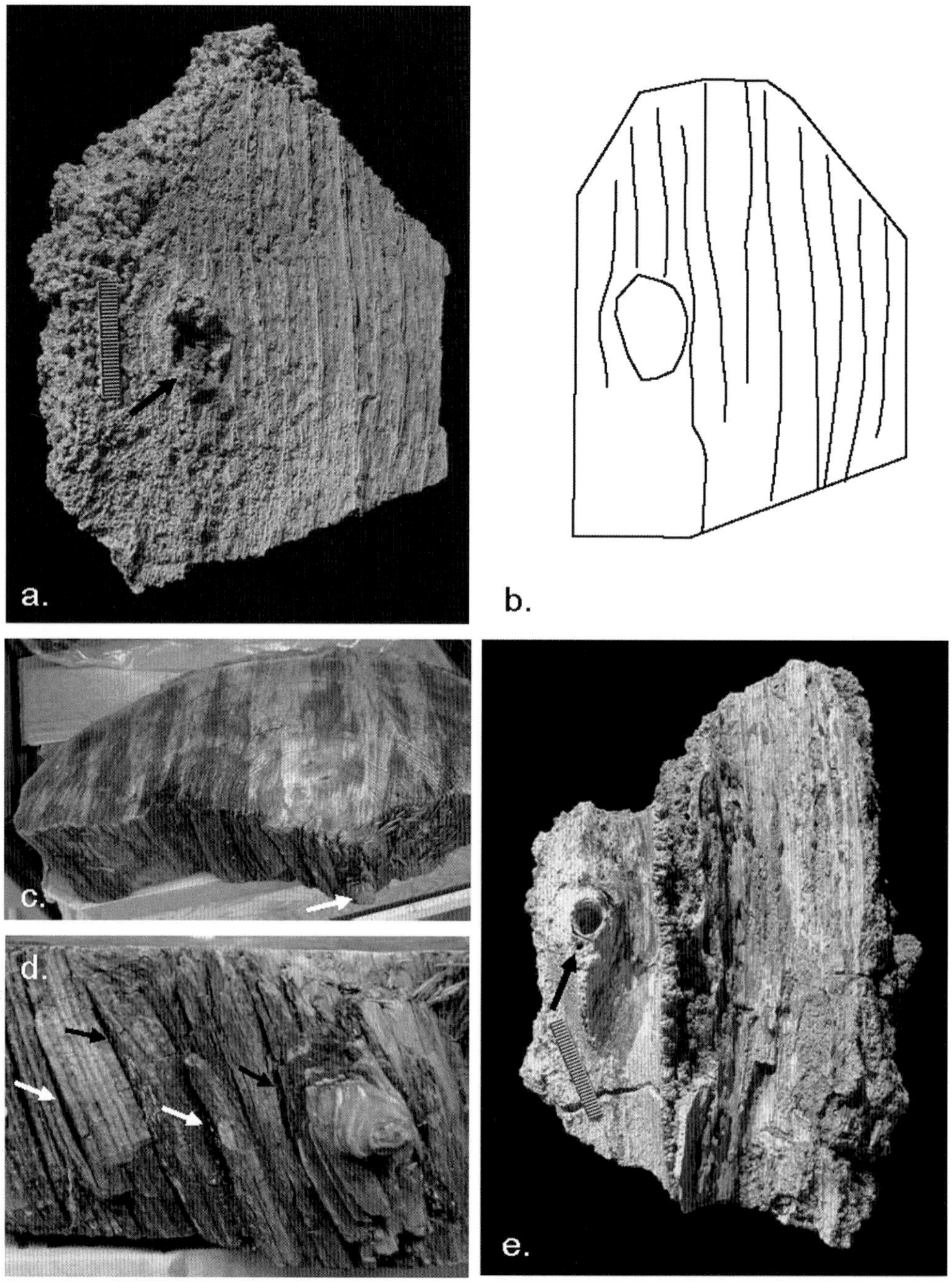

Figure 18. a. UCMP 151928. Tufa fragment showing impression of a stem surface. Branch scar at arrow. Scale = 3 cm. b. Line drawing illustrating striate pattern of internal view of stem surface. c. For comparison, section of Holocene *Pseudotsuga* (Douglas fir) trunk (UCMP 151930) broken in half revealing branch that was embedded by later growth of the wood (arrow). View of growth rings in cross section and side view. d. Close-up of the fracture surface of the specimen in "c" as seen in side view showing the branch and similar striated pattern produced by separation of the growth rings (arrows). e. UCMP 151927. Mold of tufa-encrusted stem from vicinity of Map Site 19 showing a similar view and striate surface pattern and branch at left. Specimen has remnants of wood preserved. Scale = 3 cm.

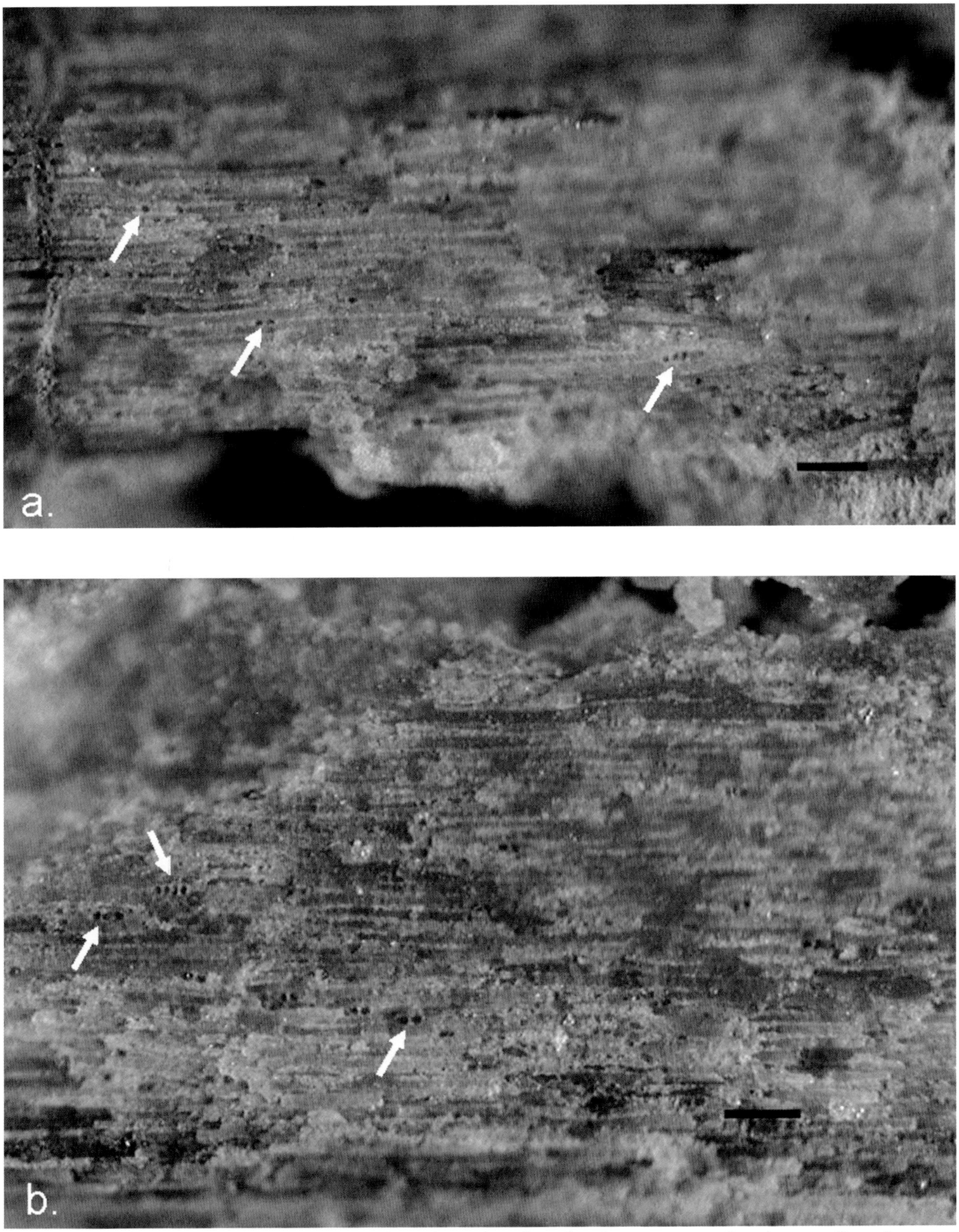

Figure 19. UCMP 151927. a, b. Tangential view of wood surface showing remnants of tracheary cells and narrow, short, uniseriate rays lacking horizontal resin canals (arrows). Scale bar = 100 microns.

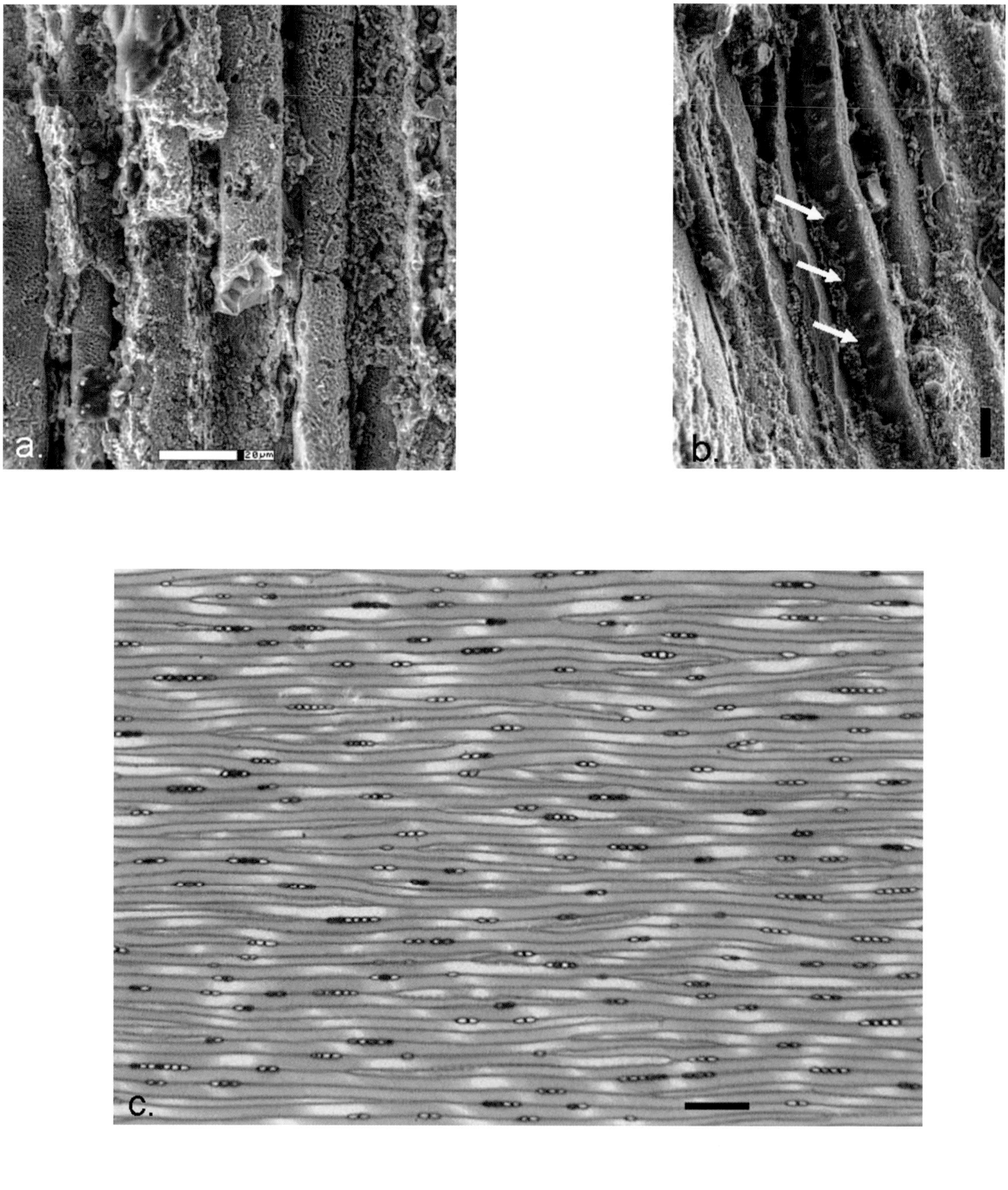

Figure 20. a, b. UCMP 151927. a. ESEM micrograph of wood showing degraded nature of the cells. b. Tracheid showing circular bordered pits with remnants of in-filled pit cavities (arrows). Scale = 20 microns. c. Tangential section of modern juniper wood showing short, narrow, uniseriate rays lacking horizontal resin canals similar to the POR fossil wood.

Figure 21. a, b. Mold of large tree fragment lying in drainage at Map Site 18. c. This is a sandy tufa just above specimen and may be site specimen fell from. The specimen in the drainage is preserved in conglomeratic sandstone. Large clast at arrow, some remnants of tufa adhering to sandstone. d. Granodiorite encrusted by conglomeratic sandstone at Map Site 17 which is along similar elevation as Site 18 (see Figure 2). Arrow points to area where tufa covers lower left corner of sandstone. Rock hammer = 28 cm long for scale.

Figure 22. a-d. *Juniperus chinensis.* a. Shrub growing along south side of the Valley Life Sciences Building, UC Berkeley campus. b. Close-up of the multi-stemmed base of *J. chinensis.* c, d. Close-up of the bark of *J. chinensis* characterized by more or less even, parallel strips. e. Similar strip-like pattern on *J. bermudiana.*

Figure 23. a-d. Large, complete specimen at Map Site 3. a. Pieces labeled 1-3 may be part of same individual. Smaller associated pieces also may belong to this tree. View looking east. Scale bar = 15 cm. Figure 33. b. Close-up of curved branch segment labeled "1" in "a." This specimen has one lateral branch (arrow). c. Segments "2" and "3" of large specimen in "a" that appear to have been broken apart at arrow. d. Close-up of segment "3" showing branching. Internal stem pattern not well preserved. Scale bar = 15 cm.

Future work that is not within the scope of this paper will provide detailed maps and depositional analysis of the principal plant-bearing sites in order to determine their *in situ* status. Preliminary observations suggest that some of the sites are in place. Future data collected in the field from each site will include images of stem fragments, their length and width measurements, tufa thickness, and orientation of the stems relative to an east - west transect.

The POR plant fossils are preserved primarily as three-dimensional molds (e.g. Figures 13-17, 23a-d, 25a) with little to no wood structure preserved. In some specimens the surface pattern of the stem is clearly visible (Figures 15, 18a, e), while others show only the surface pattern of the tufa (Figures 23a-d, 24a-d). One tufa fragment found at Map Site 19 has remnants of petrified wood attached (Figure 18e). The wood was imaged using a Leica 10 binocular microscope with an attached Sony DKC-5000 digital camera. Small pieces were also mounted on a standard SEM stub with carbon sticky tape and viewed with a Phillips Environmental Scanning Electron Microscope at 10 kv. Comparisons to extant conifers were made using wood thin-sections and the softwoods database with the GUESS Version 1.1 computer-aided wood identification program of Wheeler et al. (1986). All images were processed using Adobe Photoshop 6.0.

For those cylindrical tufa structures lacking wood, we used the following criteria to recognize them as plants. Specimens had to exhibit at least a combination of two features: surface pattern corresponding to bark or wood and branch scars and/or a branching pattern consistent with plants. Also if there was present at least one tree fragment at a site, we considered other pieces likely to be plant as well. This inference was used especially at sites where cylindrical pieces were still intact and not split open. The POR specimens are housed in the paleobotanical collection at the University of California Museum of Paleontology (UCMP).

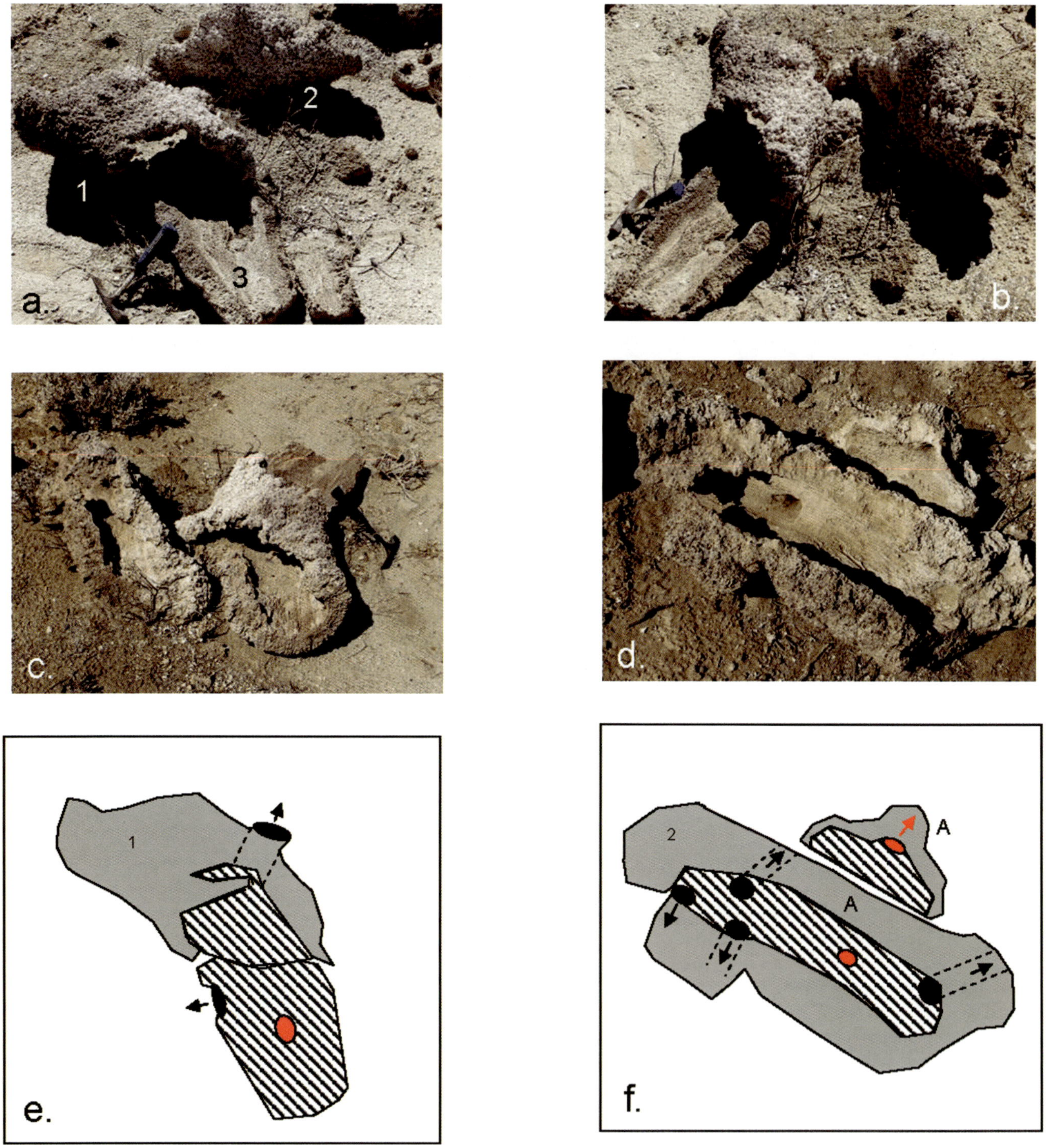

Figure 24. Large tree specimen at Map Site 31. Specimen in three pieces with other pieces in the vicinity. a – c. Different views of the specimen. Note piece "3" was reunited with "1" by DME. d. Close-up of piece "2" showing branching. Small piece at "A" broke off in handling. e, f. Line illustrations showing branching. Solid ovals mark branches the one in red in "e" is going in toward the ground the one in "f" is coming out toward the reader. Gray is encrusting tufa, stripped pattern the stem impression. Rock hammer = 28 cm long for scale.

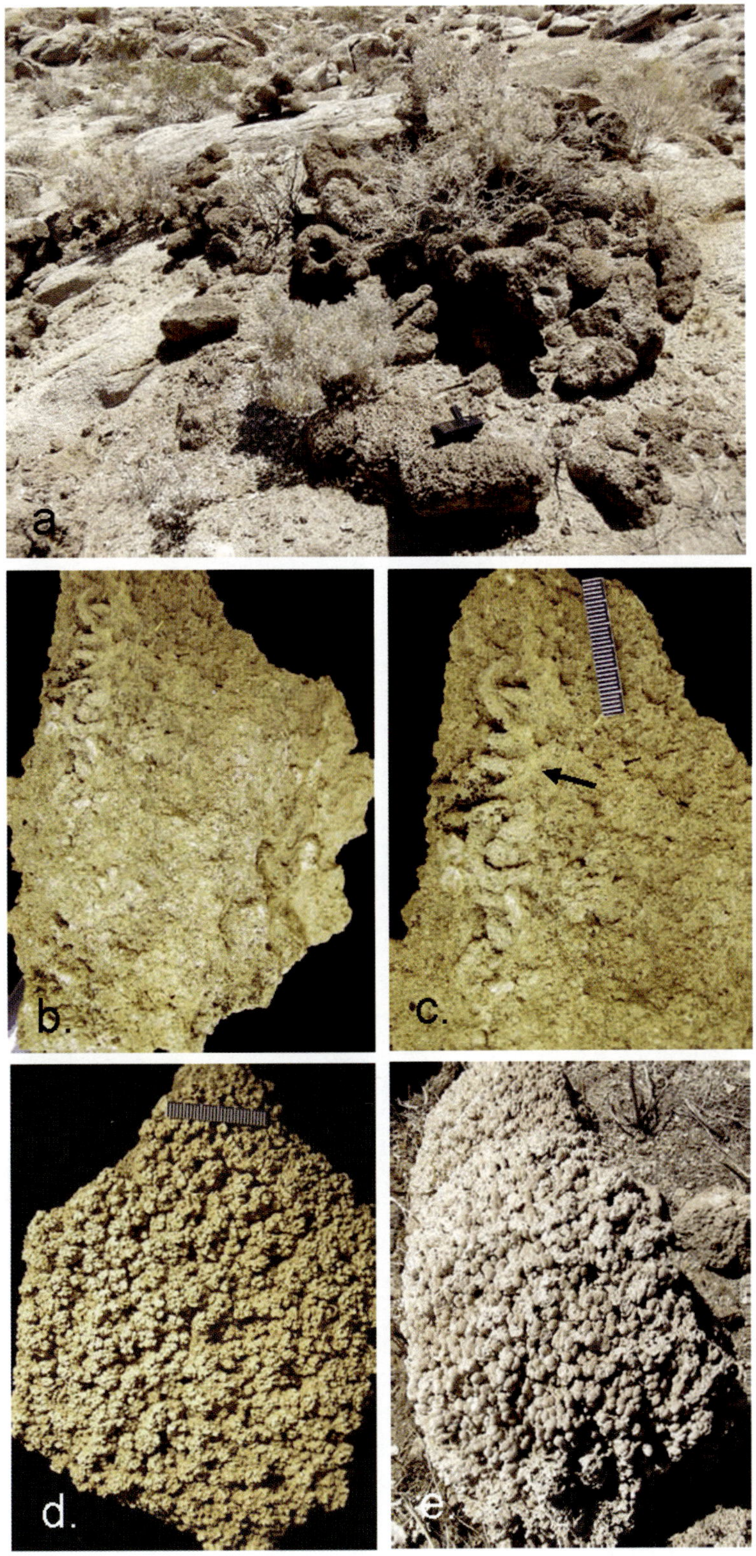

Figure 25. a. Shrub–like appearance of tufa-encrusted stems at Map Site 10 (=33). a-d. Types of plant encasing tufas. a, b. Sandy tufa. Note burrows at arrow in "b." Scale = 3 cm. c, d. Primarily calcium carbonate. Finer texture in "c," coarser texture in "d." Scale = 3 cm in "c."

Figure 26. a – e. Cylindrical tufa structures of uncertain affinity. These lack features to definitively identify them as plant remains. a, b. Upright specimens at Map Site 28. Note presence of at least two layers of tufa growth on this specimen. Scale bars = 15 cm. c, d. Two upright structures in vicinity of Map Site 28 with tufa-filled centers. Scale in cm. e. Cylindrical tufa structures lying in vicinity of Map Site 32 also with in-filled centers. Scale bar = 15 cm.

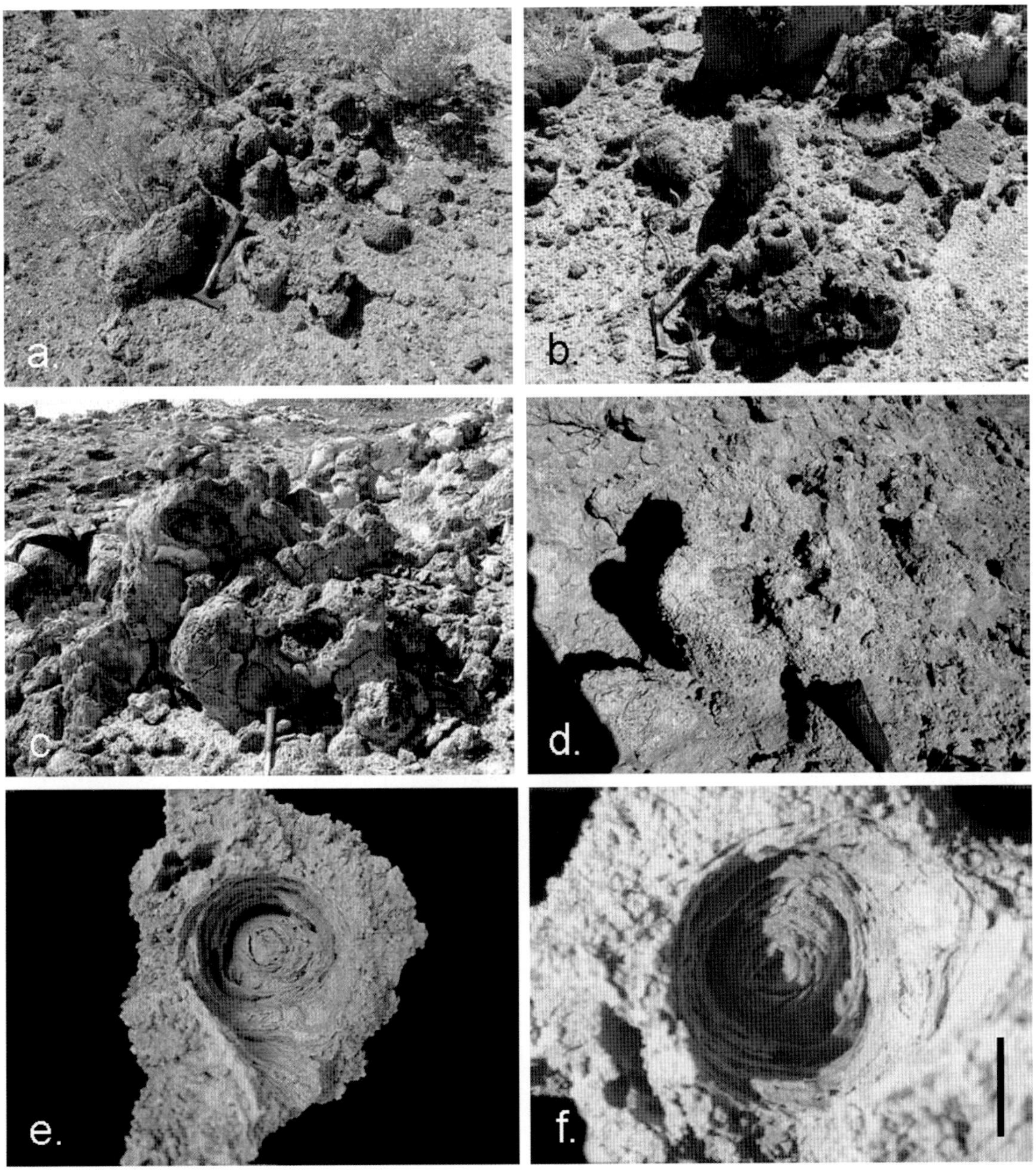

Figure 27. a-f. Cylindrical tufa structures of uncertain affinity. These lack features to definitively identify them as plant remains. a. Map Site 4, b. Map Site 32, c. Map Site 16, d. Map Site 30. e, f. UCMP 151929. This specimen has concentric rings that look like growth rings but no cellular structure to indicate this specimen is a woody plant. f. Scale bar = 2 cm.

Geological and Depositional Setting

The tufa-bearing paleoembayment is located in the southern portion of the Argus Range. The Argus Range is part of the greater Basin and Range Geologic Province and as such is oriented north-south (north northwest- south southeast). The ranges of the Basin and Range Geologic Province are fundamentally blocks of Earth's crust that have remained elevated while the crust has extended and other portions have collapsed to form the basins, such as the Searles Lake Basin. The ranges have served as both latitudinal and elevational corridors for Pleistocene plant dispersal. This topic is more fully discussed in the later section devoted to paleoclimatic interpretations.

The rocky nature of the area (e.g., Figures 3-6) is largely a result of the jointing (Figure 1b) and subsequent weathering of the bedrock. In the POR study area, the bedrock of the Argus Range is igneous intrusives of granodiorite composition. As the granodiorite is exposed to the near surface and the overlying confining pressure is lessened, the rock body expands along a jointing pattern that develops essentially quadrangle blocks as seen in aerial view (Figure 1b). The blocks in this area are oriented approximately E 15° N, S 15° E. The subsequent spheroidal weathering of these blocks is typical of uniform-grained granitic rocks, and this type of weathering forms the rounded and erect, pillar-like rocky landscape (Figures 3-6, 17a, 29b).

In the study area, the east-northeast and northwest orientations of the joints form a checkerboard pattern (Figure 1b). The joints are directed downslope toward the east, and this serves to direct the flow of both the ground and surface waters. Note how the drainage courses are largely a consequence of the original jointing pattern (Figure 1b). It is also inferred that the jointing pattern directly influenced the distribution of the calcium-bearing groundwater that precipitated with the CO_3 radical in the lake waters to form the abundant tufa ($CaCO_3$) that encrusted the rocks/fossil trees of the area (e.g. Figures 3, 7-10, 13-17). The calcium supplied to the system must have come, at least in part, from the weathering of the calcium-rich feldspar in the granodiorite bedrock. It also would have come from more localized intrusions of a calcium-rich gabbro (Figure 6, arrows).

Point of Rocks Fossil Remains

Fragments with wood preserved – Stem fragment, UCMP 151927, is the most informative. This specimen found near Map Site 19 (Figure 2) has remnants of petrified wood still adhering to the tufa surface (Figures 18e). The specimen is 17.5 cm long and 11.0 cm wide and shows one branch (Figure 18e). The branch base is 6.5 cm long and 2.0 cm wide with a 9.0 cm diameter circular vascular trace (Figure 18e). Next to the larger trace is a smaller oval-shaped trace 2.0 mm wide (Figure 18e, arrow).

Figures 19a and b show close-ups of the wood still attached to the tufa surface. Visible are the tracheary elements and rays as seen in tangential section viewed with light microscopy at a magnification of 80X. Although it was not possible to prepare transverse and radial sections from the specimen, the tangential view does show that the rays are uniseriate and lack horizontal resin canals. The rays are short and narrow, 1-9 cells high, with the majority being five cells or less. Looking at the wood with the Environmental Scanning Electron Microscope clearly shows just how much the cell structure is degraded (Figures 20a, b). Although Figure 20 b clearly shows remnants of bordered pits diagnostic of conifer wood, with the light-colored raised areas representing the in filled pit cavities (Figure 20b, at arrows), they reveal no additional features of taxonomic significance. Based on the anatomical features and cell size, this specimen is most similar to the wood structure of *Juniperus* (juniper) (Figure 20c). We do note that there is some resemblance to *Calocedrus* in wood anatomy, but the cells of *Calocedrus* are larger and the bark pattern is somewhat different from the impression specimens examined thus far.

A mold of another large tree fragment lying in the drainage at Map Site 18 may also have remnants of wood preserved (Figures 21a, b) and should be investigated further. It is one of the largest specimens and has a width of 30 cm. It may represent portions of a trunk, but could possibly be a portion of a rooting base. Unlike most of the other POR specimens, the encasing tufa appears to have a higher sand content. The specimen does lie just below a ledge of sandy tufa (Figure 21c) and the specimen may have fallen from this area. This tree fragment is preserved in conglomeratic sandstone with some remnants of tufa adhering to the sandstone (Figure 21b). The arrow in Figure 21b marks one of the larger clasts. At Map Site 17, which is at nearly the same elevation as Map Site 18 (see Figure 2) a portion of exposed granodiorite is also encrusted by a similar conglomeratic sandstone (Figure 21d). The arrow in Figure 21d points to an area where tufa covers the lower left corner of the sandstone.

Specimens with surface pattern preserved – We stress here that interpreting specimens with only a stem surface imprint is difficult. This is because woody plant stems have three main parts: the harder, more resistant vascular cylinder (wood), the soft, fleshy vascular cambium, and the bark. The vascular cambium lies between the bark and vascular cylinder (Raven et al., 1981). In life, cells of the cambial layer are actively dividing to produce cells of the wood and bark. Because cambial cells are in an embryonic stage, their walls are thin and the layer itself is soft and "slippery/slimy" to the touch. Consequently, it is a plane of weakness. Furthermore, cells that make up the wood for the most part are fibrous; however, the parenchyma and ray cells can be softer and less resistant to decay, or disruption than the adjacent thicker-walled tracheary cells. The taphonomic factors that must be considered in interpreting the stem surfaces therefore include: (1) whether or not the bark was attached or (2) whether both the bark and portions of the wood had been stripped away prior to being tufa-encrusted and fossilized. Figure 18c is a section of a Holocene *Pseudotsuga* (Douglas fir) trunk collected from Marin County,

California in which the bark had sloughed off prior to burial and the outermost surface is that of the wood.

Obviously different grades of preservation may yield disparate or in some cases similar surface patterns. For example, the section of the Marin Douglas fir trunk, that has been broken in half reveals a fracture surface that seen in side view shows a similar striate pattern that, could be produced by the imprint from striated bark (e.g. Figures 22c-e). In the case of the Douglas fir, the resulting striate pattern is from the separation of the growth rings of the wood (Figure 18d, arrows). As seen in Figure 18e, the mold of the tufa-encrusted tree fragment UCMP 151927 shows a similar view and striate surface pattern. Considering these taphonomic aspects is especially important at the POR site where pre-encrustation abrasion from transport down slope into the lake via streams or through lake level changes would bring plants in contact with the lakeshore subjecting them to wave action and their incorporation as driftwood.

With that said, several specimens found do show the well-preserved impression of a stem surface (Figures 15, 16, 18a, b, e) and branch scars (e.g. 18a, e). Although we interpret the pattern as that of the bark where the linear ridges correspond to the narrow grooves formed between the adjacent strips of bark, we do so with some caution. This pattern is seen in the specimens shown in Figures 18a and e and several at Map Site 5 (Figures 15a, b, 16) where long, thin ridges that are relatively straight do run parallel to one another along the length of the stem. This type of pattern is very similar to that found in species of juniper. Figures 22a-d are images taken of *Juniperus chinensis* that is currently growing along the south side of the Valley Life Sciences Building on the UC Berkeley campus. The bark pattern of this species as well as of other junipers (e.g. *J. bermudiana*; Figure 22e) is typically characterized by long strips more or less equal in width that run parallel along the length of the stem. The Chinese juniper also illustrates the shrubby, highly branched habit typical of some juniper species (Figures 22a, b).

Among the junipers currently growing near the POR study area, the fossils are most likely the Utah juniper, *Juniperus osteosperma*. Based on the relatively smaller branch sizes, the highly branched stems demonstrated by several of the larger POR specimens, what may be the basal remnants of a shrub (Figure 25a), suggest the POR junipers were shrubby, and so in this feature also similar to Utah juniper.

Branched specimens – The specimens range from pieces 15 -18 cm long and 11-12-cm-wide you can hold in your hand (Figures 18a, e) to larger, multi-branched specimens (Figures 13, 14, 16, 17a, b, 23a, 24, 25a). The specimen in Figure 17 represents one of the larger more impressive ones found thus far. It appears that it has broken into at least four pieces. The length of the largest section is about 112 cm the width of the largest branch is 10 cm. Another large branched specimen occurs at Map Site 3 (Figures 23a-d). Again this specimen consists of a number of associated, disarticulated pieces that appear to have broken apart over time based on their orientation and so are likely to fit together as one individual. The limb section marked "1" in Figure 23a and illustrated at higher magnification in Figure 23b is about 84 cm long and 7 cm wide. It is curved and has one lateral branch. The section marked "2" is about 100-cm-long and 18-cm-wide and consists of two main branches (Figure 23a). Section "3" is about 130-cm-long and 13-cm-wide, and branches at least twice. Although it is suggested here that these pieces belong to the same individual, this is where detailed mapping and measurement of the pieces at this site and others needs to be done to determine how the pieces relate to each other. This would undoubtedly help determine if the trees at Map Site 3 and others were in place or if they represent driftwood in possibly a channel-lag-type deposit or left beached after lake levels dropped. One other aspect that needs to be recorded is the thickness of the tufa. If the tufa is the same thickness around the stem we can assume that the stem was not lying on its side either floating or lying submerged on the lake bottom or shore.

Another large, branched specimen is located at Map Site 31. This specimen too has at least four pieces associated with it (Figure 24a-d). The piece labeled "3" in Figure 24a was actually lying several feet away and was recognized by DME as being part of this specimen. Collectively the larger piece (1 + 3) is 106 cm long and 21.0 cm at its widest point. This specimen branches at least three times (Figure 24e), and the branches are 4.0, 5.0, and 6.5 cm wide. The piece labeled "2" may also be part of "1." This specimen is 66 cm long and branches at least five times over this distance (Figure 20f). The branch widths are 2.5 (at A), 4.5, 5, and 5.5 cm. Based on this specimen and others described. it is clear that, unlike today, relatively large trees or shrubs were growing in the POR area in the past.

Types of Tufa-Encrusting the Stems

We noted that among the plant specimens found there were generally three different types or morphologies of the encrusting tufa. Two specimens were found in which the tufa had more of a sandy texture or content (Figures 18e, 21b). Figures 25b and c are the backside of UCMP 151927 illustrated in Figure 18e. The sandy nature of this specimen and that of the other suggests the tufa was deposited on these specimens as they were lying on or near the lake shore or possibly near a stream channel. Also this specimen shows the remains of tubular burrows (Figures 25c, arrow). The last two types of tufa are primarily calcium carbonate with little sand. The tufa encrusting UCMP 151928 (Figure 18a) is of this type. This specimen's tufa growths are finer, shorter and appear denser (Figure 25d) than those where they are coarser, longer, and terminate in larger spheroidal heads (Figure 25e).

CYLINDRICAL TUFA STRUCTURES OF UNCERTAIN AFFINITY

Figures 26a-e and 27a-d illustrate several examples of cylindrical tufa structures that lack features to identify them as plants. Those in Figure 26 are plant-like only in the sense that they are upright or were at one time and have a cylindrical shape with a central core (Figure 26b-e). These structures should be looked at in more detail by a carbonate sedimentologist as they may be some type of vent structure. If they are not such structures then perhaps one could argue they are plant remains. The structures in Figures 27a-d are also suggestive of possibly being the remnants of the basal portions of shrubs, but again there are no conclusive botanical features to identify them as plants. The specimen UCMP 151929 illustrated in Figure 27e and f has concentric rings that look like the growth rings of a woody plant but no cellular detail remains and there is no surface pattern to the stem.

PALEOCLIMATIC INTERPRETATIONS

Plants are directly linked to their environments. Present-day vegetation in the POR area is one of low desert scrub. Creosote bush (*Larrea tridentata*), with stem diameters of less than 2 cm, is the largest woody plant at lower elevations and in the POR area. Plants are tied to their environment and therefore they are excellent indicators of past climate (Bailey and Sinnot, 1915, 1916; Wolfe, 1995; Wilf, 1997). This is especially true for woody plants (i.e., trees and shrubs) from the last few thousand years of the Pleistocene because these plants are considered direct analogs of their living counterparts, and so temperature and precipitation requirements of the living (Thompson et al., 1999, 2000) can be inferred for their fossil counterparts with a relatively high degree of confidence.

Wood anatomy suggests at least some of the tufa-encrusted trees from the Searles Lake area are most similar to juniper (*Juniperus*). This identification is used to estimate the difference between the Pleistocene and present climate of the POR area. Two species of juniper presently live in this region of California, the Utah juniper (*Juniperus osteosperma*) and the Western juniper (*Juniperus occidentalis*) (Figure 28). The Utah juniper is present in the Argus Range in the areas of Maturango (Figure 29) and Argus peaks (Figures 30a, b). The Argus Peak occurrence was recently documented by Russell L. Kaldenberg (written communication, 1/20/2005) (Figure 30b). In this area, the plants occur down to 4600 ft. (1400 m) in an area dominated by springs and seeps and approximately 5 miles (8 km) from the POR fossil sites. Observations indicate that in the pinyon-juniper woodlands so typical of our southwest (Grayson1993; Charlet, 1996) juniper trees always occur first at the lower elevations before pinyon. In checking areas other than Argus Peak, we applied this general principle to infer that in areas not actually visited the lowest elevation trees visible on the aerial photographs are juniper.

Utah juniper has a shrubby habit and today is found at elevations of 1300 to 2600 m in Arizona, Colorado, Idaho, Montana, Nevada, New Mexico, Utah, Wyoming and California growing in areas that are dry with rocky soils and slopes (Adams, 1993). The occurrence of fossil Utah juniper at the POR sites implies that local climatic conditions at the time of tufa deposition were such that the distribution of this species was extended ~5 miles (8 km) southeast and down slope in elevation by a minimum of 2400 feet (~730 m). Using climate data directly associated with living plants of the Utah juniper, such as Carson City, Nevada, ~215 miles (345 km) and ~ 3.5 degrees latitude north of POR, a comparison between the Pleistocene fossil sites and present-day Point of Rocks provides an estimate of the amount of climate change that has occurred since the time of deposition (Table 1).

We emphasize that the data given in the Pleistocene column of Table 1 should not be taken as exact. They are more an estimate of the local Pleistocene conditions at POR. At this general level of comparisons, these local Pleistocene estimates are in fair agreement with the Pleistocene climate trends already understood for the region (Wells and Woodcock, 1985; McCarten and Van Devender, 1988; Woolfenden, 1996; Thompson *et al.*, 1999). During the time of maximum lake level and the deposition of the tufa-encrusted junipers: 1) available moisture was about 2.5 times that of present; 2) the mean annual temperature was lower than present by about 14-16°F; 3) summers were cooler by about 17-19°F; and 4) therefore there was less temperature difference between summers and winters, *i.e.*, temperatures were less extreme than today. Pleistocene climatic conditions that existed during major glacial (pluvial) episodes depressed the altitudinal ranges of plants, and generally shifted their latitudinal ranges. In the POR area, junipers grew at least as low as 670 m (2200 ft.) in elevation, or about 183m (600 ft.) above the present floor of Searles Lake basin.

Table 1. Comparison of Estimated Pleistocene Climate at POR, Present-day POR and Pleistocene Yucca Mountain, Nevada

	Pleistocene	Yucca Mountain	Present
Mean Annul Ppt	10.0'' (254mm)	10.2'' (260mm)	4.0'' (101 mm)
Mean Annual Temp	50°F (10°C)	45.5-47.3°F (7.5 -8.5°C)	67°F (19.4°C)
Warm Month (July) Mean Temp	70°F (21.1°C)	69.8-71.6°F (21-22°C)	90°F (32.2°C)
Cold Month (Dec.) Mean Temp	33°F (0.6°C)	24.8-26.6°F (-4 to -3°C)	46°F (7.8°C)
Mean Annual Range Temp	36.7°F (13.9°C)	45°F (25°C)	44°F (23.86°C)

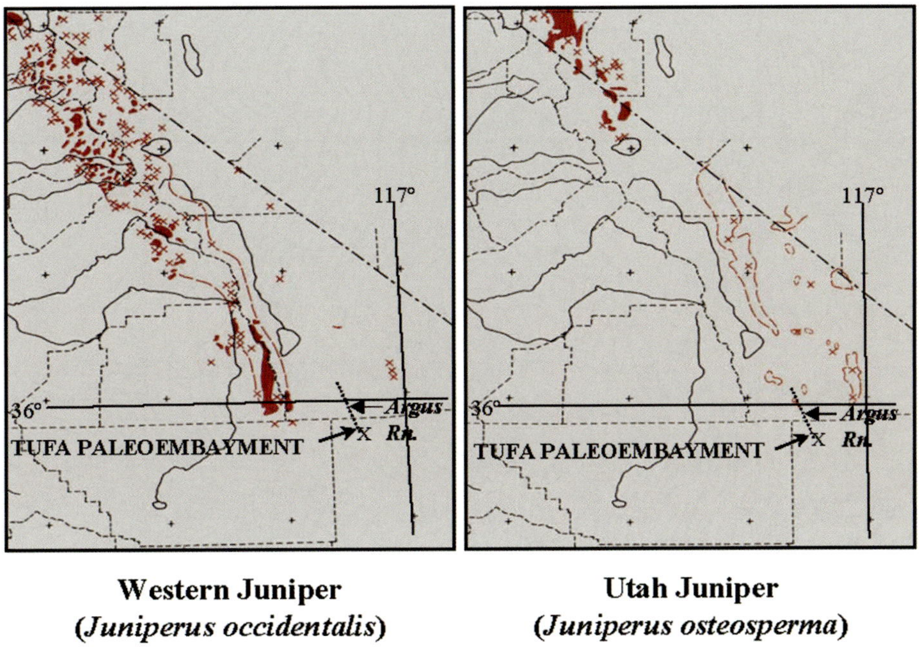

Figure 28. Distribution of extant junipers in the area of the Point of the Rocks paleoembayment. Distribution maps from: Griffin, J.R. and Critchfield, W.B. 1972. The distribution of forest trees in California. U.S.D.A. Forest Service Research Paper PSW- 82/1972. (Reprinted with Supplement, 1976.)

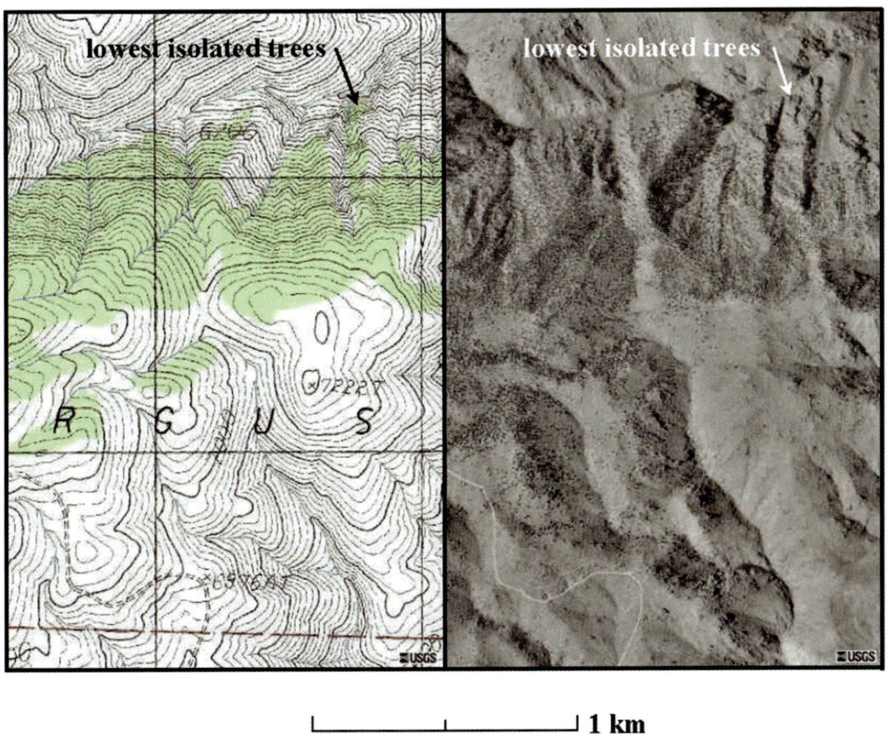

Figure 29. Portion of the Maturango Peak topographic quadrangle and aerial photograph. This area is just north of Shepard Canyon at ~117°29' W, 36°03'N, and is used to illustrate the fairly high degree of correspondence between the actual distribution of living trees in the field and the tree overlay print expressed on topographic maps. The living trees are assumed to be junipers of the species found at the fossil tufa locality. This area of living trees is ~14 miles (23 km) north and some 3500+ feet (1066m) higher than the Point of Rocks fossil tufa deposits.

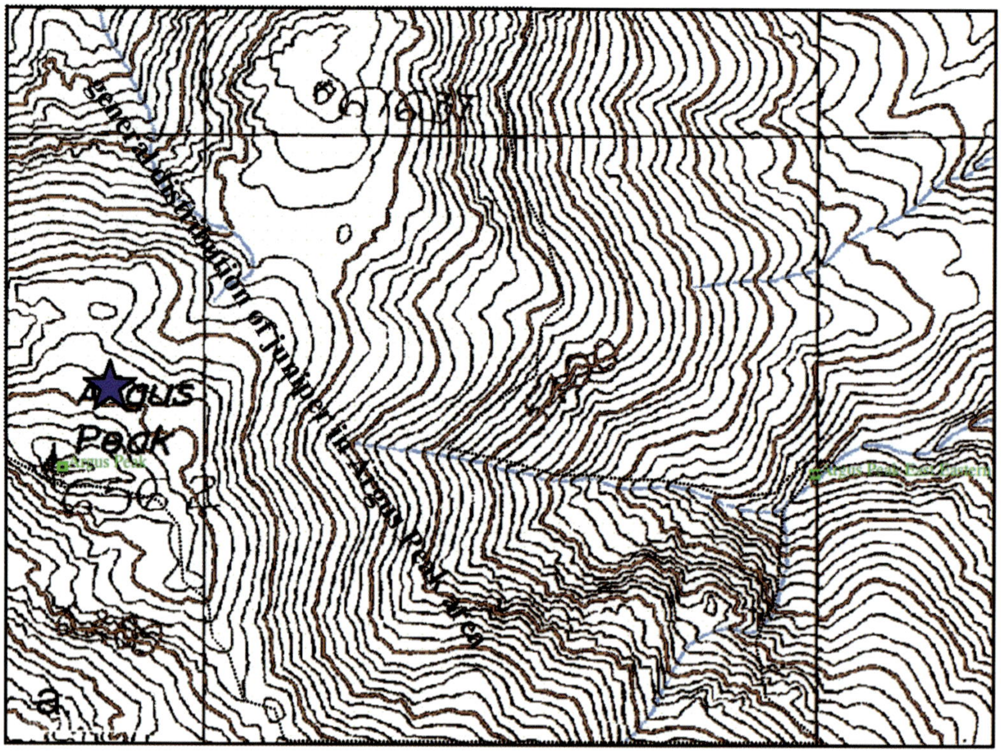

Figure 30. Portion of Trona West Quad. showing general distribution of juniper in the Argus Range. Star represents area visited by Russell L. Kaldenberg (see b) confirming the presence of juniper at this elevation. Site is about 5 miles from POR paleoembayment. b. Photo showing ruggedness of the Argus Peak terrain and presence of juniper (arrow) at about 4600 ft. Photo courtesy of R.L.K.

Conclusions

There are reports of tufa and marl throughout the Pleistocene paleo-lake system in southern California, but as far as we know, there have been no reports of unequivocal macrofossil remains of woody plants encased by tufa. Moreover, some of the sites of tufa concentration in the POR area appear to represent sites where the trees were possibly growing in place (e.g. Figures 11, 12, 23a, 25a). These sites in particular need additional detailed study and are especially significant as they may give a definite fixed position (elevation) at which these plants were growing. Their occurrence obviously implies that the level of Searles Lake was lower at the time the trees were growing. It therefore follows that the lake level then later rose completely submerging the trees for some period of time long enough for thick rinds of tufa 17 to 20 cm on some specimens to grow (viz. Map Site 5; Figures11-15a, 16).

Based on the specimens studied, we interpret some of the fossil stems to be most similar to *Juniperus* (juniper) based on fossil wood anatomy in combination with the interpretation that the striate surface pattern closely resembles the bark of juniper. We further suggest that the POR fossils most likely represent the Utah juniper, *J. osteosperma*. This is based on wood anatomy, inferred habit, and the current presence of this species in the Argus and Panamint Ranges, with the nearest trees being on Argus Peak roughly 8 km (5 mi.) upslope from the POR fossil site (Figure 30a, b).

The presence of the POR fossils indicates that in the later Pleistocene junipers grew at least as low as 670 m (2200 ft.) elevation, or about 183 m (600 ft.) above the present floor of Searles Lake basin implying climatic conditions at that time were conducive to the growth of these trees on the lower rocky slopes overlooking Searles Lake. Using the present-day climate tolerances for Utah juniper, the inferred Late Pleistocene climatic conditions in the POR area were characterized by: 1) available moisture about 2.5 times that of present; 2) the mean annual temperature lower than present by about 14-16°F; and 3) summers cooler by about 17-19°F; with less temperature difference between summers and winters, *i.e.*, less extreme than today.

Acknowledgments

The authors wish to thank Russell L. Kaldenberg, David Lawler, and Judyth Reed for bringing this site to our attention, assistance in the field, and helpful discussions. Partial support for this project was provided by the Miss Annie Alexander Endowment to the University of California Museum of Paleontology. This is UCMP Contribution No. 1876.

References

Adams, Robert P.
1993 *Juniperus*. Flora of North America Editorial Committee (eds.): Flora of North America North of Mexico, Vol. 2. Oxford University Press.

Bailey, I.W. and E. W. Sinnott
1915 A botanical index of Cretaceous and Tertiary climates. Science 41: 831-834.
1916 The climate distribution of certain types of angiosperms leaves. American Journal of Botany 3: 24-39.

Benson, L., D. R. Currey, R. I. Dorn, K. R. Lajoie, C.G. Oviatt, S. W. Robinson, G.I. Smith, and S. Stine
1990 Chronology of expansion and contraction of four Great Basin lake systems during the past 35,000 years. Palaeogeography, Palaeoclimatology, Palaeoecology 78:241-286.

Charlet, D. A. 1996. Atlas of Nevada Conifers. University of Nevada Press, Reno and Las Vegas. 320 pp.

Dorn, R. I., A.J.T. Jull, D.J. Donahue, T.W. Linick, and L.J. Toolin
1990 Latest Pleistocene lake shorelines and glacial chronology in the Western Basin and Range province, U.S.A.: insights from AMS radiocarbon dating of rock varnish and paleoclimatic implications. Palaeogeography, Palaeoclimatology, Palaeoecology 78:315-331.

Flint, R., F. and Gale, W.A.
1958 Stratigraphy and radiocarbon dates at Searles Lake, California. American Journal of Science 256:689-714.

Gale, H.S.,
1914 Salines in the Owens, Searles, and Panamint Basins, southeastern California: U.S. Geological Survey Bulletin 580-L, p.251-323.

Hickman, J.C.
1993 The Jepson Manual Higher Plants of California. J.C. Hickman (ed.), University of California Press, Berkeley and Los Angeles, 1400 pp.

Leopold, E.B.
1967 Summary of palynological data from Searles Lake. In Pleistocene ecology and palynology, Searles Valley, California. Guidebook for Friends of the Pleistocene, Pacific Coast Section, Sept, 23-24, 1967, p52-56.

Lin, J. C., Wallace S. Broecker, Sidney R. Hemming, Irena Hajdas, Robert F. Anderson, George I. Smith, Maxwell Kelley, Georges Bonani
1998 A Reassessment of U-Th and 14C Ages for Late-Glacial High-Frequency Hydrological Events at Searles Lake, California. Quaternary Research 49:11-23.

Litwin, R.J., Smoot, J.P., Durika, N.J., Smith, G.I.
1999 Calibrating late Quaternary terrestrial climate signals; radiometrically dated pollen evidence from the southern Sierra Nevada, USA. Quaternary Science Reviews 18:1151-1171.

McCarten, N. and T. R. Van Devender
1988 Late Wisconsin vegetation of Robber's Roost in the western Mojave Desert, California. Madroño, 35(3):226-237.

Morrison, R. B.
1991 Quaternary stratigraphic, hydrologic, and climatic history of the Great Basin, with emphasis on Lakes Lahontan, Bonneville, and Tecopa in: Morrison, R.B., ed., Quaternary Nonglacial Geology: Conterminous U.S., Geological Society of America, Boulder, Colorado, p. 283-320.

Packard, C.
1948 Mysterious giants of the desert [tufa pinnacles, California]. Rocks and Minerals 23:99-104.

Raven, P. H., R. F. Evert, and H. Curtis
1981 Biology of Plants. 3rd edition. Worth Publishers, Inc., New York, New York.

Reiger, T.
1992 Calcareous tufa formations, Searles Lake and Mono Lake, California Geology, 99-109.

Scholl, D. W.
1960 Pleistocene algal pinnacles at Searles Lake, California. Journal of Sedimentary Petrology, 30(3):414-431.

Smith, G. I.
1979 Subsurface stratigraphy and geochemistry of Late Quaternary evaporites, Searles Lake, California. U.S. Geological Survey Professional Paper 1043:130 pp.

Smith, George I.
1984 Paleohydrologic regimes in the southwestern Great Basin, 0-3.2 my ago, compared with other long records of "global" climate. Quaternary Research 22:1-17.

Smith, G. I.
1987 Searles Valley, California; outcrop evidence of a Pleistocene lake and its fluctuations, limnology, and climatic significance. Cordilleran Section of the Geological Society of America, Geol. Soc. Am., Boulder, CO, United States (USA), pp. 137-142.

Smith, G. I.
2000 Late Pliocene and Pleistocene Searles Lake, California, U.S.A. AAPG Studies in Geology, 46:591-596.

Spaulding, W. G.
1983 Late Wisconsin macrofossil records of desert vegetation in the American Southwest. Quaternary Research 19, 256-264.

Thompson, R. S., K. H. Anderson , and P. J. Bartlein
1999 Quantitative paleoclimatic reconstructions from Late Pleistocene plant macrofossils of the Yucca Mountain region. U.S. Geological Survey Open-File Report 99-338.

Thompson, R.S., Anderson, K.H. and Bartlein, P.J.
2000 Atlas of relations between climatic parameters and distributions of important trees and shrubs in North America—Introduction and Conifers. U.S. Geological Survey Professional Paper 1650-A.

Wells, P. V. and D. Woodcock
1985 Full-glacial vegetation of Death Valley, California: Juniper woodland opening to *Yucca* semi-desert. Madroño, 32(1):11-23.

Wilf, P.
1997 When are leaves good thermometers? A new case for leaf margin analysis. *Paleobiology* 23(3): 373-390.

Wolfe, J.A.
1995 Paleoclimatic estimates from Tertiary leaf assemblages. Annual Review of Earth and Planetary Sciences, 23:119-142.

Woolfenden, W. B.
1996 Quaternary vegetation history, p. 47-70, Sierra Nevada Ecosystem Project, Final Report to Congress, Assessments and Scientific Basis for Management Strategies. Volume II. University of California, Center for Water and Wildland Resources, Davis.

ONE MAN'S DECISION: WHY SHAEF FAILED TO HALT THE BATTLE OF THE BULGE

Jay von Werlhof
114th Signal Radio Intelligence Company
U.S. Army

It was 5:30 in the morning of 16 December 1944 when the distant crash of artillery and the rumble of tanks ended the military stalemate along the northwestern front. The Battle of the Bulge was on. The Germans began this last and most brutal attack in World War II against a divided and weak Allied line. The sudden offense over ran the front line sending our Allied troops into bold retreat and the surprised Supreme Headquarters of Allied Expeditionary Forces (SHAEF) command into hastily drawn but ineffective defenses. The Germans rolled west unopposed except in the Elsenborn sector.

Since the St. Malo breakthrough in July, SHAEF headquarters and tactical field commanders had relied heavily on air reconnaissance and G2 (Intelligence) ground units for detecting enemy troop movements. But in early December, bad weather swept in from the North Sea grounding our air force and shielding German movements from other visible spies. For the first time, the Allied commanders were forced to depend almost entirely on the G2 reports of the Signal Radio Intelligence (SRI) Companies for tactical information along the Western front. Sporadic data from prisoners and espionage resources continued to filter through the line, but for continuity and reliability the G2 daily reports from the SRI Companies operating near the front became the eyes of SHAEF. By the end of 1944, there were eight SRI Companies operating along the West Wall, including the 113th, 114th, 116th, 117th, 118th, 129th, 135th, and 137th. In addition, there was the 113th SRI Battalion attached to Lt. Gen. Omar Bradley's 12th Army Group.

If the Allies immediately and obviously responded to intercepted information that the Germans were planing a particular move, the enemy would have suspected correctly [that] the Allies were on to their communications system. Even so, the Germans had to assume that a certain proportion of their messages were being intercepted, and they devised various means to at least slow down the process by which Allies received these communications.

Following the Avranches breakthrough, the Germans revised the rather cumbersome way in which call signs and frequencies were changed. They published a book that showed for each day what call signs and radio frequencies were to be used over a six month period.

The changes were to take place as each day's operations began. The books were distributed to all German division headquarters allowing a more rapid and smoother transition in message exchanges as a new day opened. Unknown to the Germans, however, at least two of the books fell into the hands of the advancing First Army which turned one over to the 114th SRI Company, and perhaps later, the Third Army's 118th SRI Company received the other. We nicknamed our book "The Mike Book," since the German label for the book was prefixed with a capital M. Our TA Section never told the Intercept Platoon how we knew what the changes were. That was one of our own top secrets.

Sometime after we bypassed Paris and were around Soisson or into Luxembourg, one of the American crews at Army headquarters "lost" an Ultra decoding machine. Despite General Eisenhower's direct orders to "find that truck" containing the machine, it was never located. Probably it had fallen into German hands during some military move, for soon after that all intercepted messages appeared from random call signs and radio frequencies rather than the orderly changes the Mike Book had provided. The only clue our intercept operators now had was from their familiarity with the "fist" of a particular sender, that is, his style of sending. The Intercept Operator immediately alerted his Trick Chief who passed this important point along to the TA Section, which then urged the operator to stay on the frequency and listen for additional clues or messages. Even that continuity with the past was lost as new replacement operators entered the field. The continuity we still had going, however, was the map that our Order of Battle Section kept on the wall of its hutch. The DF Platoon gave us a five-point fix on the location of any sender who stayed on the air long enough for the crews to "shoot it" in. Though a frequency and call sign changed, the position of the sender might remain fixed over a few days. In itself this would have been a valuable piece of information on the static condition of the front and possibly clues for the change in transmission modes.

Linguists in the TA unit could handle any written European language or dialect. Sgt. "Jeeter" McPheeter, for example, had a Ph.D. in European languages, and taught Romance languages at University of North Carolina. Joseph B. Rounds, who occupied the second half of my pup tent, had set up a library for the League of Nations, was on leave as Head Librarian of Grovsner Library in Buffalo, and spoke fluent French and German. Others in the unit covered the Slavonic languages. One of the members of our TA Platoon was a personal friend of Walt Disney who agreed to design the logo for our reorganized company. The original is in the National Archives.

Our southernmost DF unit was at the edge of the Ardennes, and our northernmost unit was near Eupen, giving us more than a twenty-mile coverage of the northwestern front. With this placement we were west of the central sector of V Corps. The 9th Infantry Division was to our north, the 2nd and 99th Infantry Divisions were to our east, and the 106th Infantry Division to our southeast. The V and VII Corps were fighting to the northeast of Eupen towards Aachen and the Roer River dams as the main elements of General Dwight D. Eisenhower's code-named Operation Queen. They were still within our sector but extended our range to forty miles. All of this was soon to change.

Gasoline and ammunition were always in short supply. To prevent one fuel dump from German capture, the Americans poured 125,000 gallons into a ditch and torched it. At the time near Metz, when General Patton was complaining about gasoline shortages for his tanks, the 114th SRI Company intercepted a clear-text message from a panicky soldier that he and two others were guarding a 50,000-gallon gasoline dump and needed help. We contacted headquarters which saw to it that several airplanes supplied the help until a patrol could seize the site from the hapless defenders.

By the 1st of December, the V and VII Corps had fought beyond Aachen, the ancient capitol of Charlemagne, and were along the Roer River facing the 272nd Volksgrenadier Division. This drive under Operation Queen, aimed to prevent Germans from opening the Roer dams and flooding the Lower Roer Valley which would have stopped General Hodge's drive towards Cologne. Successes at the Roer River brought General Gerow's V Corps back to Monschau for regrouping and rest. This was just prior to the German counter-offense.

As early as October, Hitler began pulling his best divisions from the West Front for refitting in Eastern Germany. His favorite, the Sixth SS Panzer Army, was brought up to full strength while the Fifth SS Panzer Army and the Seventh Army remained understaffed.

Maj. Gen. Leonard T. Gerow's V Corps was holding the north pocket of the Ardennes front. Within it, the 2nd Infantry Division under Maj. Gen. Walter M. Robertson held the right flank while Maj. Gen. Walter Lauer's 99th Infantry Division commanded the left. Adjoining the right flank of the V Corps was Maj. Gen. Troy H. Middleton's VIII Corps composed of a cavalry regiment, three infantry divisions, and one armored division, all stretched out over a sixty-five-mile course, with but 500 men per mile.

Even considering the activity along the Roer River, a military slowdown had been gradually evolving along the West Wall. As mid-December approached, the front was relatively quiet. In the north, General Hodges was anxious to continue Operation Queen to Duren on the east bank of the Roer River, while in the south General George S. Patton was in the final stage of preparing his Third Army for the planned 19 December all-out offense into Saarland, southwestern Germany. In between these separated salients American troops were at rest or refitting while unknown to their Higher Command, the Germans were amassing for a major assault through this quiet sector of the Ardennes. Hitler's plan, that he dubbed Wacht Am Rhein, would supposedly propel three armies of the Aryan super-race to Brussels and Antwerp and, in some mystic way bring a quick end to the war. Readied, on 15 December they commanded 250,000 troops - 3,100 men per mile.

General der Panzertruppen Hasso-Ecard von Manteuffel's Fifth Panzer Army struck at the north edge of VIII Corps, and Seventh Army General Panzertruppen Erich von Brandenberger attacked to the southwest to stop General Patton's advance. With the Sixth SS Panzer Army were other units that recently arrived from the east, notably the 12th Volksgrenadier Division, the 2nd SS Panzer Division, and the 246th Volksgrenadier Division. In the Fifth SS Panzer Army was the Panzer Lehr Division which recently had arrived by rail from Bulgaria. Also from the Eastern Front were the 26th Volksgrenadier Division and the 62nd Volksgrenadier Division. From the Italian front was the tough 15th Panzer Grenadier Division. Specially trained were the new 560th Volksgrenadier Division and the 18th Volksgrenadier Division, both ready to be battle tempered.

General Patton is not known for graciousness, but it would be hard for any man to emulate his military penchant for daring and risk. However, he did appreciate meritoriousness in others and awarded the 118th SRI Company, attached to the Third Army, a special plaque for outstanding service as a frontline unit. Military historians have highly complimented the 114th SRI Company for the "excellent" coverage we had of German units during Operation Queen. Of greater importance to us, however, was the order of battle that gradually emerged on our map of the Sixth SS Panzer Army. Even though Hitler had ordered strict radio silence of the three armies being stretched between Monschau and Eckternach, the Sixth SS Panzer Army seemed to have trouble getting its complex unit lined up in the few days before the attack. This army was the spearhead, and it was imperative that all units participating in the drive be fully organized. While the 114th SRI Company received such data from Sixth SS Panzer Army divisions, SRI Companies covering other German armies adhering to the blackout did not receive similar messages. The effective tracking zone of our 114th SRI Company was northeast of Eupen and southeast of St. Vith in General Hodge's First Army sector. Our area overlapped V, VII and VIII Corps. Though messages were usually very short, they showed over a few days a growing pattern of amassed power.

Our DF Platoon and Cryptoanalysis Section detailed a concentration of armed might not known before, behind the West Wall. Some of the units could be identified, and our DF Platoon "shot in" many more. Operators recognizing "fists" of senders helped the Order of Battle Section to keep track of ground movements. Between 9-11 December, the 114th SRI Company constructed an impressive Order of Battle Map.

While the daily reports we sent to First Army Headquarters and SHAEF showed increased activity east of the front, the lieutenant of our TA Platoon, 1st Lt. "D," refused to send our consolidated report and map of 11 December. He compared G2 records from First Army Headquarters with our data and cited records that showed some of our identified units to be on the Eastern Front, destroyed, or in rest areas. Roy Duke, technical sergeant in charge of the operations of the TA Platoon, discussed with Lt. "D" the difference between other

G2 data and ours, but to no avail. To us, these G2 data were obviously obsolete in a fast-changing tactical situation. Capt. Alfred R. Braddock, 114th SRI Company Commander did not involve himself in intersectional matters which gave each senior platoon lieutenant the final word. We immediately set to work re-doing our report from additional field data. By the 14th of December we had added several units to the Order of Battle Map, and incorporated these data in a new report. Again, our lieutenant refused to except our report as credible military intelligence. Traffic all but ceased on the afternoon of the 15th, apparently as the Sixth SS Panzer Army under General Dietrich and the Kampfgruppe Peiper Armored Regiment under SS Colonel Joachim Peiper had the main elements of their attack force in line and ready to move. They now waited for zero hour.

G2 at Army or SHAEF levels was very slow in sending its accumulated data to its field companies. Not all of its reported resources were reliable. Too, intelligence officers were not above political squabbles, as the intelligence head of First Army was not on good terms with the intelligence head of General Bradley's 12th Army Group—the 113th SRI Battalion—resulting in poor exchanges of data. On 14 December Eisenhower's G2 officer's Order of Battle Map showed only four divisions facing the VIII Corps front line.

Up to the time of the attack at 5:30 [a.m.] on Saturday, 16 December, Eisenhower's ULTRA provided no real clues that anything unusual was happening on the German side of the West Wall. The 118th and 137th SRI Companies did not report the kind of activity in their sectors that the Sixth SS Panzer Army passed on to the 114th. The other resources that G2 Army and SHAEF Headquarters relied upon were not working due to bad weather, bad assumptions, and radio silence in all but the Sixth SS Army sector. While other reports were based on rumor or partial truths, it seems the 114th SRI was the only resource collecting hard data.

But it was one man's decision that kept the 114th SRI Company TA Section from sending the only reliable and current reports available to First Army and SHAEF Headquarters. In the ensuing battle, the American total casualties were 80,987; German totals were 81,834. That is 162,821 men killed, wounded and missing. Had First Army and SHAEF received and accepted the 114th report, there was time from the 11th to the 16th of December to blunt the attack. But one man's decision blocked that possibility.

IN THE FOOTSTEPS OF MALCOLM ROGERS: ARCHAEOLOGICAL SITE RECORDATION AT INDIAN SPRING, SOUTH RANGE, CHINA LAKE NAVAL AIR WEAPONS STATION

Mark W. Allen, Ph.D.
Department of Geography and Anthropology
California State Polytechnic University
3801 W. Temple Avenue
Pomona, California 91768

This paper is a summary of a brief site recordation project conducted at Indian Spring in the western Mojave Desert. This is an extant and reliable spring some 10-12 kilometers north of the Superior Valley playas, near the southeastern corner of the South Range of China Lake Naval Air Weapons Station (Figure 1). It lies at an elevation of roughly 1000 meters and it is fed by a major drainage that flows south between two ridges separated by only a few hundred meters. Despite substantial use of the area over the past 100-150 years for stock and military tests, a brief survey of the Indian Spring vicinity confirms a wealth of archaeological resources dating over the past several thousand years. This preliminary work is only a beginning. Moreover, this project provides the opportunity to marvel once again at the tremendous accomplishments of the pioneers of California desert archaeology.

Previous Archaeological Investigations at Indian Spring

The Indian Spring area was apparently first subjected to archaeological investigation by Malcolm Rogers of the San Diego Museum of Man from October 29-31, 1929. His notes (Rogers n.d.), while brief, do include some detail and interpretations concerning several archaeological sites within a kilometer or two of Indian Spring. The two most prominent sites that he mentions are Petroglyph Cave west of the spring and Zoo Cave east of the spring. Because of the importance of these observations, his handwritten notes are transcribed here in full:

M-99 Indian Spring

> Comes out of the center of a great lava boulder fan and is well hidden. Considerable dacite (some pre-pottery) scattered about. Queer location as ground is so rocky only a few attempts to clear space for house sites seen. Two of the larger boulders have old geometric petroglyphs on them. A strong trail comes up from the East Superior Lake to the south and then bends off to the northwest where it passes by a cliff with caves in it. This is one mile from the spring. The big shelter is partitioned off into rooms with low rough masonry laid up will a brush fill, these appear to be late Shoshonean shelters...1/4 mile east of spring is another very low shelter in a clift [sic]. In the tuff floor are six or eight shallow mortar holes. These would have to be worked in from a reclining position as the roof is so low. On the hill above this cave are a few lava blocks covered with early broad type Mohave glyphs. There is also some indefinite ceremonial work on the hill top. It takes the form of an east and west alignment. Two broken flat type metates found in front of this cave.

The fan glomerate [sic] surface in the area about the spring has old erosional channels which are more or less sandy and filled with fine gravels. In them are series of boulder rimmed house sites on a large scale. Some are 10' in diameter. No metates about them only Playa or Pinto-Gypsum flaking and blades in black dacite. There is [sic] some ceremonial alignments here which are difficult to decipher because of the thickness of the natural distribution of the lava boulders. One circle 15' in diameter was stacked full of large boulders. This ceremonial lies 300' north east of the spring.

Petroglyph Cave ¾ west of spring has 3 Shoshonean boulder rimmed beds in it, and 5 broken lava metates. It was also occupied by Mohaves and Pinto Gypsum or Playa peoples. The latter industries show on the talus....Shelter 40' long- 12' deep and the debris has a max. depth of 5'. In front of the cave is one large slab of very old broad lined petros of the early Mohave type ... a few smaller caves have been slightly occupied.

Petroglyph Cave (1 mile west of M-99)

Talus Material and
Surface objects from cave Floor= M-99A

Paiute Level (0"-12") = M-99-B
Amargosa? Level (12"-30") = M-99-C

To the left of 2 great rocks which have fallen from the face is a vertical slab (3' x 4') entirely covered with the oldest style glyphs of the Mohave desert. They are so greatly eroded and oxidized that few of the lines can be traced.

Back [?] of the 2 big rocks an avalanche of rocks and dirt have covered much of the oldest cave occupation. To the north under the deepest overhang are 2 Shoshonean boulder walled rooms, circular in shape.

In front of one are 4 broken Shoshonean metates of lava and 3 crude lava manos. The total length of the shelter was originally 160' long. The overhang varies greatly, 10' being the maximum at the present time. It was all excavated by wind scour working on a volcanic ash stratum. On the talus Mono type sherds, Early Mohave and some Amargosa stone work is present. Also dacite flaking of Playa or Pinto age.

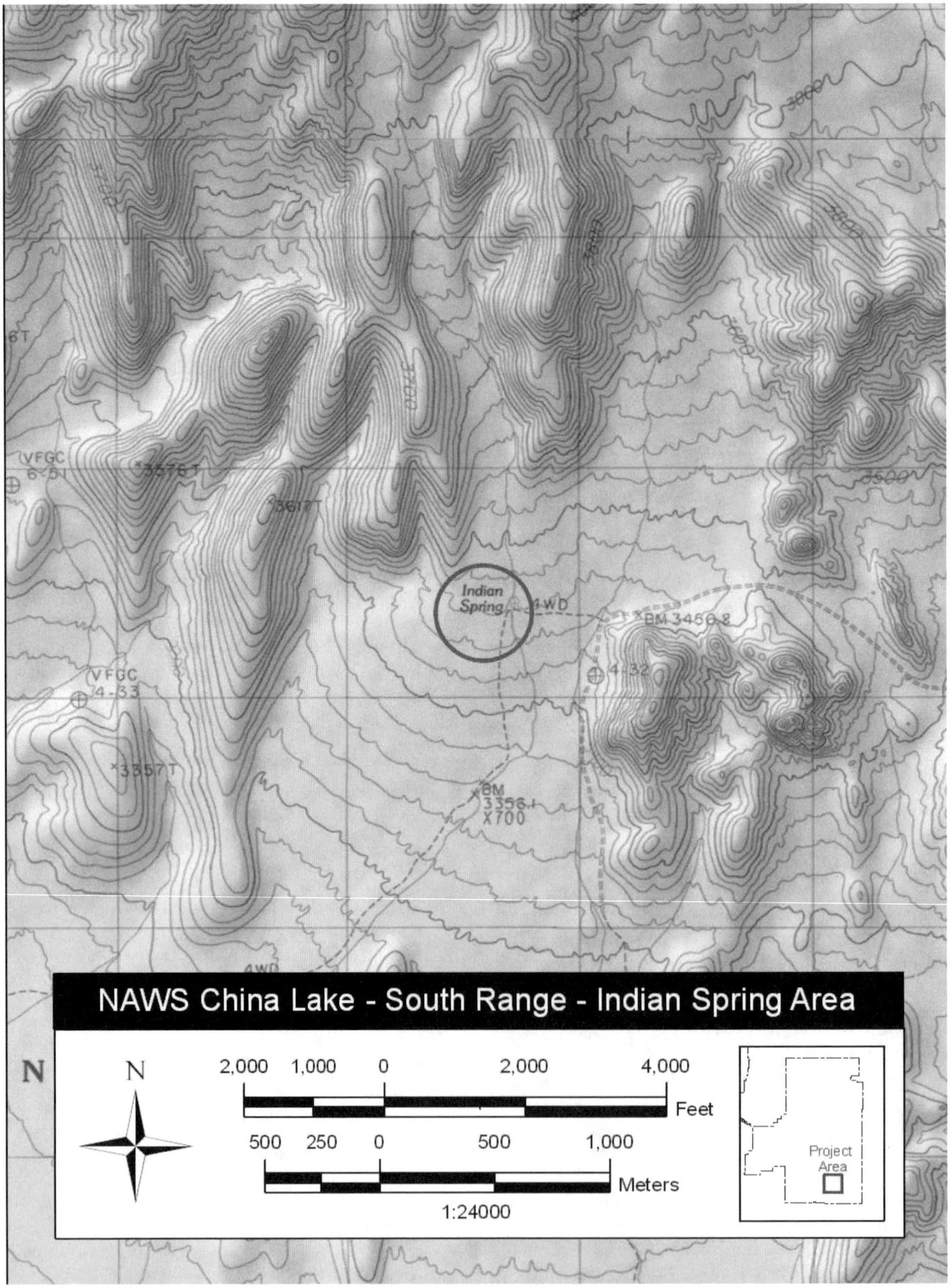

Figure 1. Map of South Range, China Lake Naval Air Weapons Station showing location of Indian Spring.

In addition, some additional typed notes in the Rogers (n.d.) papers provide a bit more detail about a test trench that he excavated inside one of the stone wall features at Petroglyph Cave:

> At Petro Cave which is 40' wide x 12' long, are 3 Panamint rooms set apart by low boulder walls. These are characteristic of this region. They have fiber beds in them. On the floors or in front of them were 5 broken lava metates. The petro rock in the cave front has 6 punctates 2" wide placed vertically on one side. In 1929 we ran a test trench to the floor encountering two distinct levels. The top was Panamint and the level Amargosa (not enough material being obtained to determine the period—but late). As the talus produced Amar. I, this cave would pay to excavate [sic] for a deeper section which seems to lie under a big cave-in of the overhang of the shelter. Test trench max. depth 30."

Finally, he included two quick sketches of the geoglyphs or rock alignments: the large one at Zoo Cave and the smaller one on the alluvial fan closer to the spring. In sum, Rogers noted some highly intact Late Prehistoric cultural features, as well as lithics which apparently dated to as far back as the Pinto or even Lake Mohave cultural periods.

Of the sites noted by Rogers, only two were subsequently recorded for the California Office of Historic Preservation prior to the present investigation. Apparently, Gerald Smith (n.d.) visited the area in 1941 and made some quick notes. At some point, site trinomial CA-SBR-723 was assigned to the spring area itself, while CA-SBR-724 was assigned to Rogers' Petroglyph Cave west of the spring. Smith recorded SBR-723 as "Shelter Cave, several caves in rocks near springs, mortars, pottery, trade beads, points." Clearly he intended this site to include the archaeological evidence from the spring itself eastwards to at least as far as Zoo Cave, but the site location map places only a point at Indian Spring itself (a midden which was located but not re-recorded by this project).

A few years later, in January of 1948, Agnes Bierman and A. Mohr (n.d.) recorded a cluster of four small rock shelter sites a kilometer or so southeast of Indian Spring (CA-SBR-54 through 57). Two of these rock shelters were test excavated by Bierman and Mohr (see site records). SBR-55 was found to be 30 inches deep and yielded lithics, ceramics (a sherd which was "black, coarse, punctate decoration, inner roughly smooth"), and at least one perishable artifact (arrow cane fragment). Tools included a chert point base ("willow-leaf," plano-convex cross-section), an obsidian point mid-section, a metate fragment, and several "scrapers" and a "chopper." Surface artifacts in and around the shelter included lithics (jasper, chalcedony, obsidian, rhyolite) and at least a few metates. Bierman and Mohr recommended further excavation of this site.

SBR-57 was "completely excavated" and yielded some biface fragments, debitage, and a schist metate fragment. Bierman and Mohr noted, "deposit contained charcoal throughout, some ash, and a few chips." The other two sites, SBR-54 and 57 were each pronounced, "very poor site. No excavation recommended." These sites were relocated and briefly visited during the present project, but were not re-recorded owing to time constraints. There appear to be a large number of similar sites on the slopes of the ridges around Indian Spring.

Site Recordation at Indian Spring, July 2004

The first goal of this project was to re-locate and better document the two main sites first noted by Malcolm Rogers in 1929: Petroglyph Cave (CA-SBR-724) and Zoo Cave. A second goal was to identify other archaeological sites in the Indian Spring area, including some noted briefly by Rogers and others identified by NAWS Command Archaeologist Russell Kaldenberg. It should be noted here that a large number of sites have been identified, but not yet recorded. These include several other small rock shelters in the vicinity of Petroglyph Cave, numerous rock circle features around Indian Spring itself, and several possible trails that cross the area (some of which were noted by Rogers).

Over the course of three days in the field at Indian Spring, a detailed site update was completed for the Petroglyph Cave Site (CA-SBR-724) and nine sites were recorded for the first time: 3 stone circle sites, 1 trail, 1 large stone enclosure which may be a historic cooking facility, 1 petroglyph site, 1 geoglyph site, 1 bedrock milling station, and the Zoo Cave complex described by Rogers, but inappropriately lumped in by Smith with CA-SBR-723 (Indian Spring proper). The location of these sites is shown in Figure 2. Clearly, this brief recordation effort only scratches the surface of the Indian Spring area.

All sites were located with GPS (Trimble Geoexplorer III, differentially corrected for maximum accuracy), photographed with a digital camera, and mapped in plan with measurements derived from tapes. Full descriptions, photographs, and maps of each site can be found in Allen (2004). Table 1 presents the temporary site numbers, California trinomial numbers, and a brief description of each site recorded. Here, I hold myself to describing only three of the sites with the most important data potential: Zoo Cave, Petroglyph Cave, and the geoglyph first noted by Rogers in 1929, which I have named the Rogers Geoglyph Site.

Zoo Cave (CA-SBR-11689)

We can only guess why Malcolm Rogers called this site Zoo Cave. Although he noted the majority of the site's features during his work in 1929, our work identified a few more. Based on Rogers' notes and a few photographs he took, the site appears to have suffered very little damage or erosion over the past 75 years.

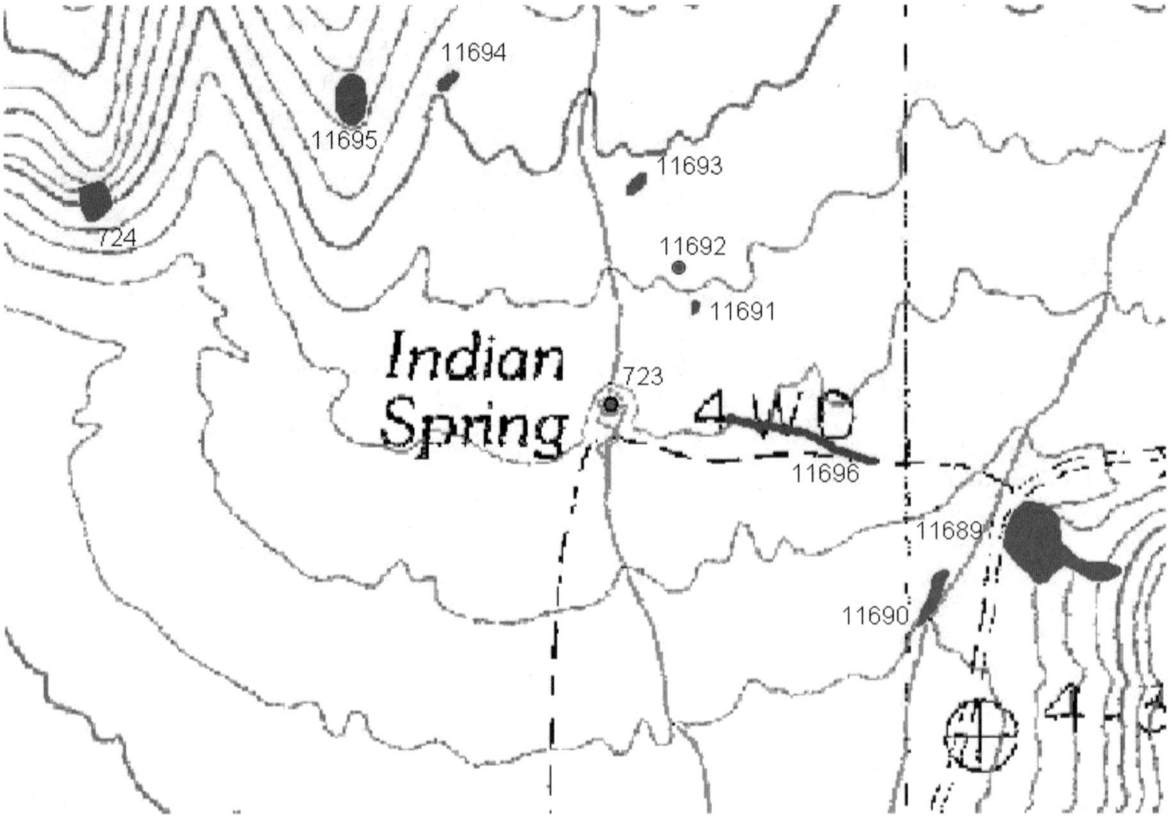

Figure 2. Map showing location of archaeological sites at Indian Spring.

Table 1. Information for Sites Near Indian Spring

Trinomial	Temporary No.	Site Name	Description
CA-SBR-11688	T-8		stone circle
CA-SBR-11689	T-9	Zoo Cave	rock shelters, petroglyphs, 2 geoglyph/rock alignments, mortars, possible cupules, some habitation debris
CA-SBR-724	T-10	Petroglyph Cave	rock shelter, 3 stone wall structures, habitation debris, petroglyphs
CA-SBR-11690	T-11		at least 8 stone circles, groundstone artifacts on surface
CA-SBR-11691	T-12	Rogers Geoglyph	geoglyph
CA-SBR-11692	T-13		bedrock milling feature
CA-SBR-11693	T-14		stone circle, cairn, bedrock milling feature.
CA-SBR-11694	T-15		double enclosure with stone walls
CA-SBR-11695	T-16		petroglyph panels
CA-SBR-11696	T-17	The Richard and Russ Trail	trail

Zoo Cave consists of six distinct loci located on different parts of a prominent volcanic tuff promontory that juts northwest from the hill towards Indian Spring about 500 meters away (Figure 3). Loci A and B are located directly at the base of the promontory on adjoining sides. Locus A is located on the east face and consists of a rockshelter with several bedrock milling features, some groundstone artifacts, and some historic etchings on the tuff. Locus B is at the base of the north face of the promontory and consists of historic etchings, some lithic debitage on the surface, a metate, and some rough stone walls which run perpendicular from the promontory. Locus C is a linear rock feature consisting of 20 separate cairns or lines of volcanic breccia on top of the promontory (Figure 4). It runs roughly 57m. in a gentle S-shaped curve from the center of the site uphill to the east. It also contains a few groundstone artifacts. Locus D is a concentration of 13-14 petroglyph panels on basalt boulders in the southwestern part of the site. Locus E is a set of 12 bedrock features (likely cupules or mortars) that run in a north-south line on top of the promontory. Locus F is another linear rock cairn feature that extends 33.5 m. north-south along the eastern side of the site, nearly perpendicular to Locus C.

Rogers did take a few photographs of some of the petroglyphs at Zoo Cave, and these indicate that these particular panels have suffered very little damage or deterioration over the past three-quarters of a century. The rock art, groundstone, bedrock features, and two rock features constructed of individual cairns are clearly prehistoric in age. Historic artifacts, writing etched into the tuff at Locus A and Locus B, and the nearby presence of Indian Spring also indicate substantial historic use of the site. The periods of this use are not yet identifiable, though it is likely to have been utilized as a place of shade and shelter near the spring since the 1860's.

Figure 3. Plan of Zoo Cave (CA-SBR-11689).

Figure 4. Photograph of linear rock feature (Locus C) at Zoo Cave (CA-SBR-11689).

Petroglyph Cave (CA-SBR-274)

Rogers' Petroglyph Cave appears to be in nearly the same condition that he noted in 1929, with the exception of the site's namesake petroglyph panel, which receives full sun exposure much of the day. The main part of the site is a terrace 6-7 m. wide (nw-se) adjacent to a natural rockshelter or shallow cave in an exposed volcanic tuff face. The terrace is about 28 m. long (ne-sw), but the southern end has been disturbed by a major collapse of the tuff wall and subsequent erosion through this gap. Three architectural features (1-3), probably sleeping areas or shelters, lie in a row along this terrace, each partially sheltered by the natural cave (Figure 5). Features 2 and 3 are much more elaborate than Feature 1. Feature 4 is a highly eroded petroglyph on a basalt boulder (see Figures 6a and 6b for comparison between 1929 and 2004) to the south and slightly down slope from the 3 structures. Feature 5 is another less elaborate petroglyph panel southwest of the terrace and around the curve of the volcanic tuff exposure. A large number of groundstone artifacts are concentrated around the three structures, particularly Feature 2. There is also a fairly continuous scatter of artifacts along the terrace, including several bifacial tool fragments (including likely projectile points), a green slate pendant fragment, and debitage. Materials include chert, obsidian, basalt, and chalcedony. There is also a substantial amount of burnt large mammal bone on the surface. The site then extends down slope to the east: a number of artifacts were identified on the surface of this talus slope, just as Rogers noted. Much of this material has likely washed downhill from the terrace, but there may have been aboriginal use of this area as well.

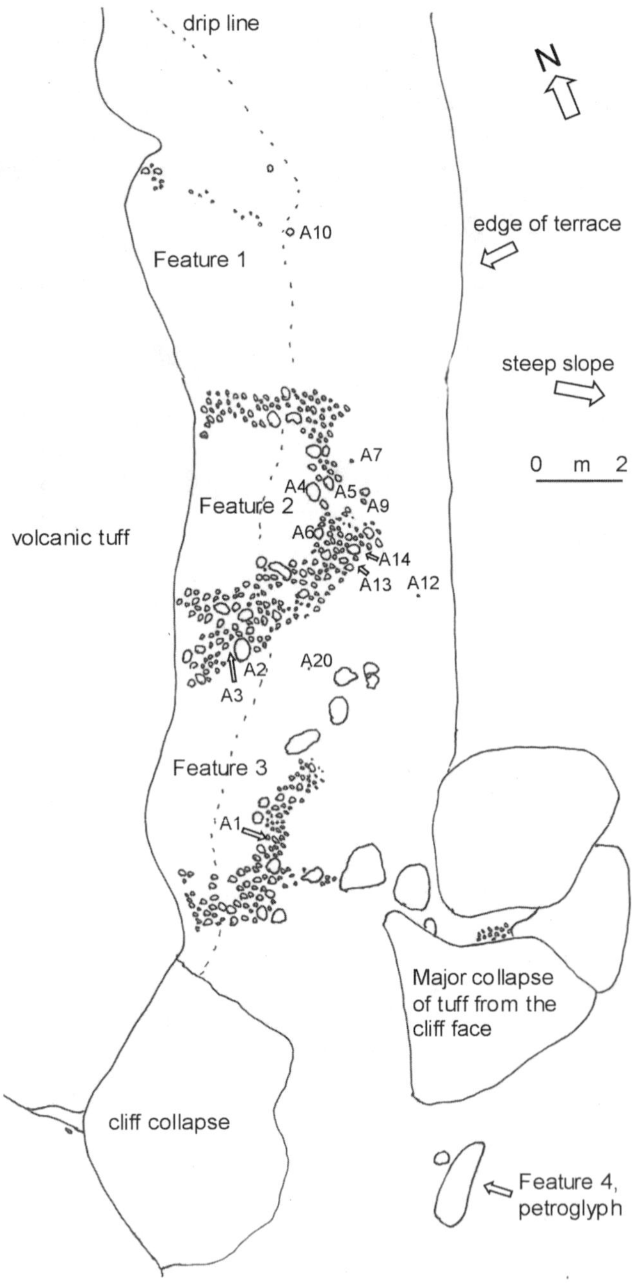

Figure 5. Plan of Petroglyph Cave (CA-SBR-724).

a)

b)

Figure 6a. Photograph of Petroglyph taken during Roger's site recordation in 1929, petroglyph cave (CA-SBR-724)
b. Same petroglyh as Figure 6a., taken in 2004.

Rogers believed that the site had several different occupations based on his test excavations which yielded two stratigraphic layers (he termed them Paiute and Amargosa), the presence of silacious and volcanic (basalt, he calls it dacite) cultural material, and some diagnostic artifacts. The 2004 inspection noted an obsidian Cottonwood Triangular point. It and the relatively intact stone architectural features strongly argue for a Late Prehistoric component. Gypsum and Pinto age deposits appear likely as well. This site promises a great deal of information about at least the Late Prehistoric period (700-200 B.P.). The site is in excellent shape, appears to be undisturbed by humans (other than Rogers' 1929 limited excavations—no sign of which were visible), and contains architectural features which are essentially intact. The presence of burnt large mammal bone and groundstone artifacts indicates that the site is likely to provide information on subsistence practices.

The Rogers Geoglyph Site (CA-SBR-11691)

This site is a small geoglyph located a few hundred meters from Indian Spring. It is constructed with lines of basalt rocks about 10 m. north-south and 7.5 m. east-west (the main axis is 28 degrees east of magnetic north). The geoglyph appears to consist of a half-circle or oval on the northern end connected to a line of rocks which has a single curve halfway along its course (see Figures 7 and 8). This feature appears to be undamaged since Malcolm Rogers quickly sketched it in his field notes. One metate is present at the northern end of the site. Some possible scratching is present on a rock on the west edge. Such features, though not widely noted, are fairly common between the Searles Lake basin region and the Avawatz Mountains at the north edge of Fort Irwin (Kaldenberg 1981.

Figure 7. Photograph of Site CA-SBR-11691.

DISCUSSION

The pioneering survey work of Malcolm Rogers, Agnes Bierman, and A. Mohr, together with the results of this brief three-day project, indicate that the Indian Spring area is a remarkably well-preserved archaeological landscape. Although chronometic data are limited until such time as excavations are conducted, it is certain that the area was utilized during the Late Prehistoric Period (from 700 to 150 B.P.). Moreover, the presence of basalt artifacts and apparently some projectile points led Rogers to infer occupations back to at least the Mid-Holocene (i.e. the Pinto Period, ca. 8,000-4000 BP) and possibly the Lake Mohave Period (ca. 11,000-7,000 B.P.). For detailed discussions of Mojave Desert culture history see Sutton (1986), Sutton et al. (i.b.), and Warren (1984).

This set of resources provides an outstanding opportunity to examine changes in subsistence patterns, technology, resource extraction, social organization, and evidently ideology over at least several thousand years. Detailed survey and site recordation potentially will reveal sufficient context to better understand how the petroglyphs, geoglyphs, rock circles, aligments, trails, and middens fit into a landscape. Likewise, the undisturbed middens and unique adjoining stone circles within the Petroglyph Cave rockshelter have the potential to provide information on subsistence practices and chronology. Finally, Zoo Cave and the Rogers Geoglyph seem to offer the opportunity to further investigate the relationship between ideology and rock art in this part of the Mojave Desert (von Werlhof 1987, 2004; Whitley 2000).

I would also argue that this project is important on another level. The opportunity to be the first archaeologists recording resources that were first examined nearly four generations ago by a figure like Malcolm Rogers is one to savor and reflect on. Too often, and armed with hindsight, we of the present tend to highlight the errors and erroneous conclusions of pioneers like Rogers. As usual, at Indian Spring, Rogers evidently got much of it right.

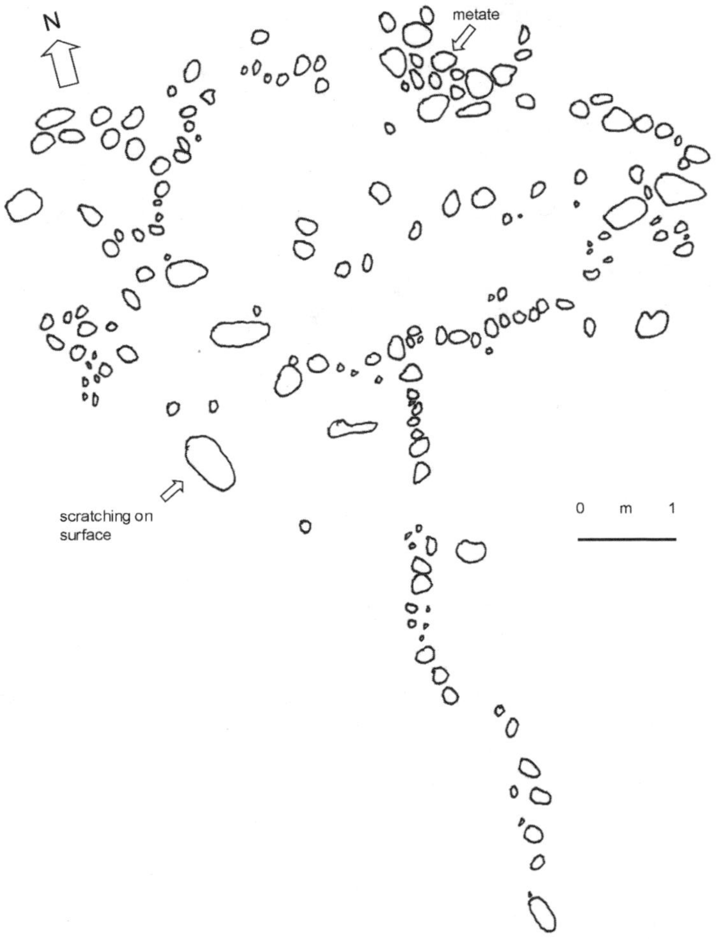

Figure 8. Plan of Site CA-SBR-11691.

Acknowledgments

This project was conducted through Epsilon Systems Solution. The people of the Ridgecrest office are thanked for their extraordinary assistance both in and out of the field. Special thanks are owed to Greg Halsey and Sheila Lloyd. The field crew consisted of Josine Jones, Victoria Harvey, Jose Alvarez, Russ Kaldenberg, Rachel Steiner, Richard Stewart, Judyth Reed, Sandy Rogers, Fran Rogers, Al and Joan Woodman and Elva Younkin. The installation's Command Archaeologist, Russ Kaldenberg, was instrumental in planning the fieldwork, identifying sites to be recorded, and securing substantial volunteer assistance.

References

Allen, Mark W.

2004 Archaeological Investigations at Pilot Knob and Indian Spring, South Range, Naval Air Weapons Station, China Lake, California. Draft report submitted to Naval AirWeapons Station, China Lake

Bierman, Agnes, and A. Mohr

n.d. Unpublished site records for CA-SBR-54 through 57, held at the San Bernardino County Archaeological Information Center, Redlands, California.

Kaldenberg, Russell

1981 The Archaeology of Selected Springs and Playas on Fort Irwin and in Portions of the Avawatz Mountains. San Bernardino Museum Quarterly 27(1 & 2).

2003-2005 Personal communications.

Rogers, Malcolm

n.d. Field notes on file at the San Diego Museum of Man, San Diego, California.

Smith, Gerald A.

n.d. Unpublished site records for CA-SBR 723 and 724, held at the San Bernardino County Archaeological Information Center, Redlands, California

Sutton, Mark Q

1996 The Current Status of Archaeological Research in the Mojave Desert. Journal of California and Great Basin Anthropology 18(2):221-257.

Sutton, Mark Q., Mark E. Basgall, Mark W. Allen, and Jill K. Gardner

In prep. Advances in Understanding Mojave Desert Prehistory.

von Werlhof, Jay

1987 Spirits of the Earth: A Study of Earthen Art in the North American Deserts, Vol. 1, the Northern Deserts. El Centro, California: Imperial Valley College Museum Society.

2004 Spirits of the Earth: That They May Know and Remember, Vol. 2. Ocotillo, California: Imperial Valley College Museum Society.

Warren, Claude

1984 The Desert Region. In Michael Moratto California Archaeology, p. 339-430. Orlando, Florida: Academic Press.

"HI, JAY"

Boma Johnson, M.A.
Archaeology Plus
185 West Center Street
Ivins, Utah 84738

First and foremost, I want to say how much I appreciate the friendship that you and Sherilee have extended to me over the years. I have always admired your balance and wisdom as a professional in a field where these traits are not always in abundance. Thank you for your encouragement to me over the 25 years that I worked for the BLM. It was your vision to me that helped set my research off in the right direction.

And thank you for your kind words about that direction in the preface of Vol. ll of your *Spirits of the Earth*. I appreciate your efforts to gather the primary Yuman Creation stories under one cover, coupled with interfaces of these stories to the geoglyphs in Yuman Territory. Well done!

Jay, this report is formatted as a less than formal or fully documented research paper. What I hope to accomplish in this report is a continuation of the ongoing conversations between us that we have had so many times in the past years. As you can guess, I have been very busy researching and learning from Native Peoples concerning their views about earth figures and rock art as an integral part of their ancient and even modern cultures. As is obvious, my 1985 Earth Figures publication was only the beginning. Since then I have acquired a considerable volume of additional information and insights into the cultural continuum from the archaeological record to the more recent ethnographic and ethnological records. Since it is not likely that I can do a complete review of all this info in this short letter/report, what I hope to accomplish is to do an abbreviated overview about some of my favorite research topics, mainly in your honor and for your benefit. (However, I am OK if the rest of you wish to peer over my shoulder as I talk to Jay.) I am still working on some type of publication on all this material, but as you know, finding funds for a major work is no small feat.

In this report I hope to briefly address the following topics:

1. Review my research style and direction as a preface to the rest of what I say.
2. Build a generalized Native American world view and cosmological backdrop by use of their own stories, myths, metaphors and symbols as seen in their various art displays, imagery and ceremony, to gain greater insight into their cultural ways and beliefs.
3. Use a few of my research projects as typical examples of the above scenario.
4. Reference several of my post-1985 research papers, which I will send directly to you at a later time.
5. Discuss certain topics in your *Spirits of the Earth*, Vol. ll, where I have info to add or share.

My Research Approach

Jay, as you well know, researchers have used many approaches to help understand Native American rock and earth art. Most of these approaches commonly use the time-tested techniques of inductive/deductive logic centered on testable hypothesis, which have emerged out of the scientific method. Logic is a wonderful and practical "sacred cow" to western culture, but often is less useful in understanding the "hows and whys" of nonwestern culture. Most cultures of the world, including our own in many ways, do not function entirely on the same premises of logic. "Reality" is not the same for all peoples. Much of nonwestern cultural experience is not logic-based, but is based on a concept we call myth. A major aspect of Native culture, often overlooked by many researchers, is centered on how these people see and explain their own cultural experiences and not on how we see them through our own cultural eyes.

My primary goal is to blend or overlap the archaeological, ethnographical and ethnological records with the knowledge, beliefs and wisdom of living Native Peoples.

A primary premise of my research is that the study of past culture must include at least three major levels of investigation. The first of these is the traditional archaeological method of studying physical or material cultural remains (artifacts, features, pollen, etc.) in order to delineate past subsistence patterns and environmental settings. This avenue concentrates on using extant material remains to explain what people do to survive in various environments. Knowing what people do is very important, but often overlooks what people believe.

The above material system approach must be balanced by a second major level of investigation that I call the belief system approach. This approach is the study of what people believe and how beliefs become a worldview that highly influences what people do. Since archaeologists are usually not trained in cultural belief systems, this task has traditionally been the work of art historians, ethnographers and ethnologists. In recent years a growing trend has coupled the work of both approaches, leading to much greater cultural process clarity, Figure 1. A multi-variable research design includes far more than artifact analysis and bean counts, and leads to what is sometimes called a holistic approach to Native culture. Under this lofty banner we must look beyond the archaeological and ethnographic records to more sacred concepts and beliefs, ritual and ceremony as held by any particular Native cultural people.

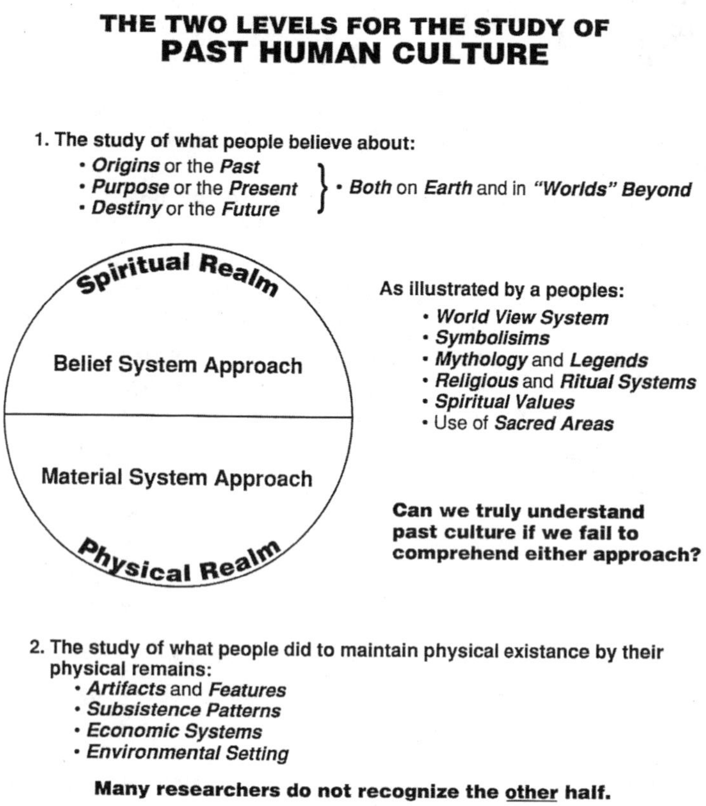

Figure 1. The two levels for the study of past human culture.

The belief systems of most peoples include concepts that range from the profane to the mundane to the sacred. Profane concepts are usually applied to taboo or unproductive activities to be avoided. Mundane concepts are associated with everyday run-of-the-mill activities necessary for maintaining ordinary daily life. Sacred concepts in culture are all about what people consider being of greatest long-range value. In this category are the concepts of (1) "Place." meaning a more than an ordinary place, rather a sacred location, a spiritual landscape or sacred geography, as in Spirit Mountain and the Xam Kwatcan Trail; (2) "Myth" or mythology, meaning sacred stories, legends, events or beliefs which are not always "real," but do serve to explain the unexplainable and account for the otherwise unaccountable; (3) "Ceremony," which means acting out of sacred stories or events on a staging area; (4) "Participants," meaning those persons, mortal or not, who act out sacred dramas of life in ceremony; and (5) "Image," meaning all types of sacred communicative display, including all perceptions of the five senses, especially visual imagery (Figure 2).

Under the category of visual imagery are included all types of designs, symbols, or pictorial displays, and their intended symbolic or metaphoric meanings. The term "communicative display" is an "umbrella term" referencing any type of surface (on wood, stone, earth, leather, cloth, paper, ceramics, etc.), and any type of surface treatment technique (painting, etching, scraping, carving, inlay, piling, etc.) intended to communicate information between individuals or groups. One of the most significant insights we have arrived at in recent decades is that earth figures and rock art are not idle doodling or graffiti as some have tried to tell us, but rather highly intelligent and unique modes of communication by Indian peoples. Gradually over the years we have learned from the Indian people that the communicative displays we call earth figures or geoglyphs and rock art are none other than visual representations of key aspects from their sacred stories, especially their Creation and Great Journey stories.

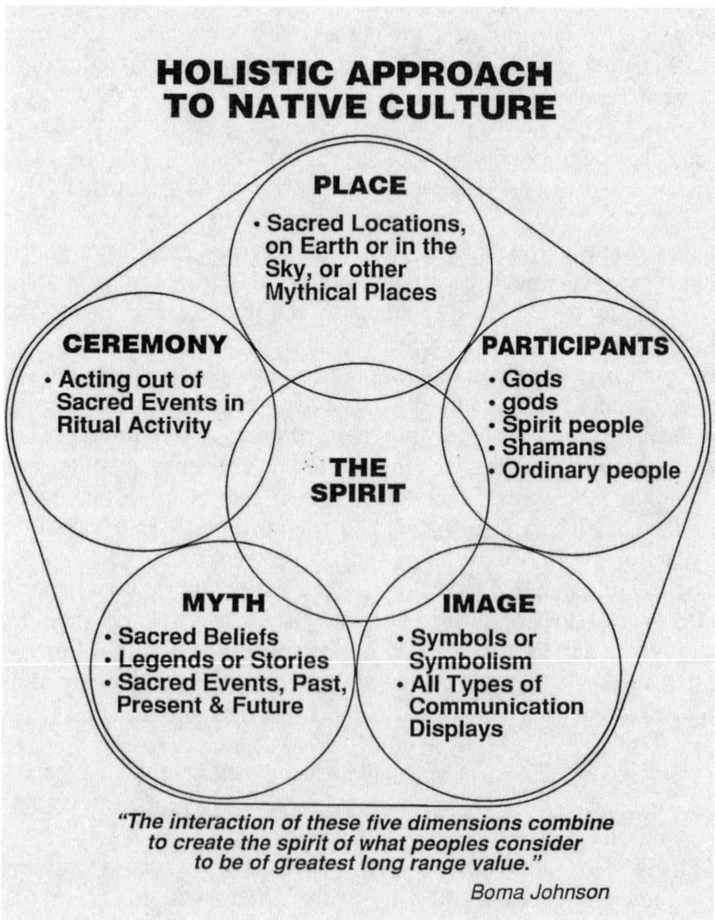

Figure 2. Holistic approach to Native culture.

Native American people have over their long cultural existence developed a communication system based primarily on the use of visual signs and symbols. Hand signs and body gestures allowed people with otherwise unintelligible languages to reach, when face to face, a reasonable degree of understanding. Beyond sign language (which I don't deal with in this report), the primary system for communication was the use of a great variety of visual images in the form of symbols. Various symbols each held one or more meaning or intention. Any single symbol could communicate an independent meaning; however, the more common means of visual communication was the use of multiple symbols grouped together to form a unit. In rock art such a cluster of images are usually called a "panel." A clustering of symbols in concert, whether on rock, the earth, or any medium, provided the "educated viewer" a comprehendible body of information. Like any communication system, local and regional variation of symbol meaning and interpretation clearly did develop over time and space. More importantly, however, is the recognition that many symbols, or clusters of symbols, hold very widespread recognition and meaning.

Levels of Indian Sacred Stories

Not all Indian stories are the same or carry the same weight or function. Much of my research on the relationships between Indian stories and their visual counterparts on the earth or rock has been at the local or regional tribal level. Over the years I have noted, as we all have, many differences in various details in their stories both in time and in space. Yet at the same time I have also noted strong common threads, themes and patterns throughout most local and regional stories. I find that these story variations are mostly local tribal adaptations and reflections of a far greater heritage common to all Native peoples over all the Americas. I find that on a general Pan-Indian level, Native Americans collectively hold an amazingly consistent worldview over most of the western hemisphere. Thus because of a direct relationship between local or regional tribal stories and a Pan American Indian worldview, I find all local tribal stories and their visual counterparts equally valid. One story is not more "correct" or "better" than another because we may think so or want it to be so. Both levels of research are equally productive for understanding the content, purposes, and meanings of earth and rock arts.

Now to the paradox of double, even multiple, meanings or interpretations applied to a particular story, geoglyph, or rock art panel by different persons or groups. Again there are those who ask sincerely which meaning or interpretation is more correct. As non-Indians, we would like it very much if some authority would step up and declare once and for all that Story X or Interpretation B is correct. But alas, it is not to be. My Native teachers over the years have consistently taught me that all stories and their interpretations are "correct" to someone. What are they trying to do to our sense of order? This is what I have been taught by them. The primary meaning of any story or visual representation thereof lies with the original author or maker. However, a person or group, who may or may not know about the original meaning, can develop a secondary, or new meaning. This new meaning is related to a construct in most cultures known as "the eye of the beholder," which allows any person to offer a new interpretation to any preexisting subject. Whoa! My pet theories are in trouble? So what's new? Have fun while you can.

Then comes another twist to the problem of meaning. Not only can almost any person advance new interpretations to old subjects, but also within a single tribal group or clan we find significant changes in story content and meaning, depending on who the listening audience may be. The root of this situation is who is prepared to learn and be responsible for higher levels of sacred knowledge. One teacher put it this way: Knowledge of higher sacred truth can be very dangerous to one's well being if used wrongly or shared with the wrong persons. Indian cultures also have an institutionalized standard for progressive revelation of sacred knowledge to serious initiates. The rest of us, referring to all younger children and white people (non Indians) get only the less-than-sacred stories. But it is not all hopeless. Even non-Indians can learn some midlevel sacred knowledge if they can earn the trust of Native teachers.

So, why the above discussion concerning the presence of various story levels in Indian culture? It boils out like this. Over the years we have worked more directly with local tribal stories as they relate to earth and rock arts, mainly because these stories have been available to us. Yet due to the variety of stories available to us, we have often sought for clarity in understanding by searching for a better "fit" between the often-conflicting story elements and the visual images on the ground or rocks. Out of this conflict I have learned two separate principles. First, it is culturally normal and acceptable for tribal groups to have very different stories about the same subjects. And it is OK. Varity is the spice of life. We can guess what it would be like if we were all alike with no differences in our stories or our abilities.

The second principle that I have learned is perhaps a key to our quandary. Beyond the variation of local stories is a backdrop of collective Pan Indian thought that contributes to most Indian stories. Local stories contribute variety on a "small-picture" level. Pan Indian worldview thought and knowledge contribute unity on a "big-picture" level. The polarization of variety verse unity is a yin/yang type paradox common to most cultural peoples. And that is a good thing. As we look to the Pan Indian "big-picture" worldview, we find a fairly clear backdrop of story commonalities found in local stories across the whole of our continent, indeed, even the world (Bierlein, 1994). Once we can see the unity of the big picture, we can also better understand the source and rational for the often conflicting local stories. This paradoxical relationship applies directly to stories about and interpretations of both earth figures and rock art. We need not get hung up on the fact that stories about these visual images do not always agree. A major key to the dilemma is recognizing a greater overriding unity and harmony at the big-picture level.

The Great Journey in Indian World View

At the local or regional level, and at the mundane level, Indians stories about life adventures are usually told about a more common "here and now " social context. These lower level stories are mostly about people and events that are centered here on earth with only limited reference to a world beyond. However Indian stories told at a big picture or Pan Indian level are typically at a higher sacred level, and are more often involve people and events that reach to worlds beyond this world in past, present and especially future times. Indian people, like people the world over, share similar concerns about life, i.e., who are we? Where did we come from? Why are we here? Where do we go from here, and how do we prepare what will come? We should never forget that most peoples in prehistoric times expressed major interest in the greater cosmological journey from here to the beyond. A common concern by living people for departed ancestor spirits is about their survival in the next world, and that we the living, will soon join them in the next stages of life, see Figure 3.

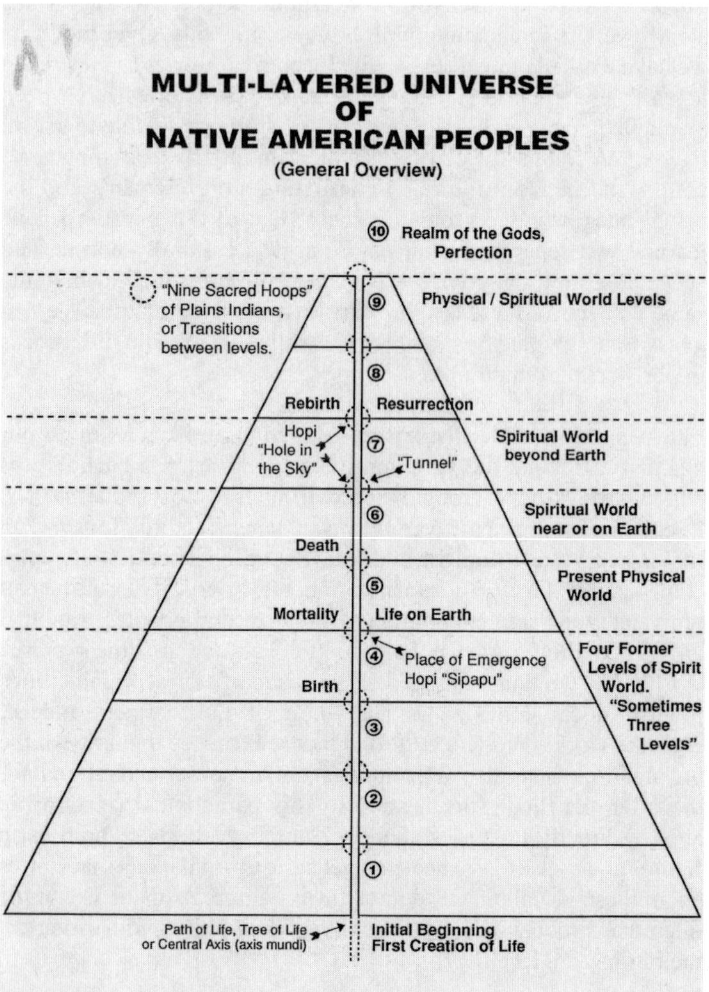

Figure 3. Multi-layered universe of Native American people.

Native American worldviews are complex cosmological scenarios (stories, myths and legends) designed to provide acceptable answers to the above questions. At the heart of many sacred stories are references to places and people in an under world, or in a sky world beyond earth. A primary scenario in sacred stories involves a "great journey" of persons or groups in migration from various past worlds or levels of life to this current world, then on thru several more life experiences in other worlds beyond earth. These migrations or the great journey of our total life experience, here and beyond, are the primary elements of Indian sacred stories and ceremony, and are well represented in the archaeological record, and especially in visual manifestations of rock art and earth figures.

Typically Indian sacred stories are cyclical rather than linear as in western culture. This means that the greater life cycle has no real "beginning" or "end" as we would expect. Rather, Indian sacred stories continually move in an unending series of cycles, divided into life units by portals or passageways between each unit. According to a common Indian worldview scenario, the present cycle which the human family is experiencing had its "beginning" beyond our own island galaxy, the Milky Way, in another galaxy known as "Seven Sisters" or Pleiades in modern language. In many Native stories, a Deity couple known as "First Parents" send two sons, usually twins, from Seven Sisters to a celestial void for the creation of our galaxy. In time with the completion of the Milky Way, and our solar system, the earth would become a temporary learning residency for the human family. After leaving Seven Sisters we entered the Milky Way via a portal at the North Star, then traveled south eventually descending to earth, our current residence. At some point in our great journey we will leave earth, ascending back to the sky world, then travel south again along the Milky Way eventually leaving this galaxy via another portal located at a "J-shaped" void near the galaxy Sagittarius. Our current cycle will "end" with our return back to Seven Sisters and a new cycle of learning and creation in other parts of the greater Universe.

Symbolism and Concepts along the Great Journey

Native American sacred stories and their visual representations begin to make much more sense as we overlay them with a "big picture" of their cosmological journey. Visual representations of their creation and great journey stories are replete with variety of basic symbols still understood by many Indian People over most of the Americas. Among these are the primary symbol for the planet Venus (Figure 4; Johnson 1992) and the Great Cross in its vertical, horizontal and rotating forms. A few other selected symbols that are almost Pan American in both use and understanding are listed below with brief "big picture" definitions or primary meaning/s. These symbols were selected because all are found in the geoglyphs and rock art in greater Southwest.

For many years rock art researchers of the American Southwest have been familiar with the symbol known as the Outlined Cross, representative of Venus, the early morning star of the southeastern sky. In recent years we have learned the meaning of the Double Venus Star, not well known outside of the Lower Colorado River Basin. Local Native Americans tell us that both symbols represent the Creator God Kumastamho and the cycles of birth/death/rebirth.

The descent of the evening star Venus to the Underground (beneath the western horizon) represents the death of Kumastamho, while the rising of the morning star Venus in the eastern sky eight days later represents his rebirth or victory over the underground world. This same interpretation of a similar Venus star symbol is traceable back to Mesoamerica and the birth, death, and rebirth of Kukulcan, the Feathered Serpent and Maya Sky God of Creation, later known as Quetzalcoatl of the Toltec and Aztec.

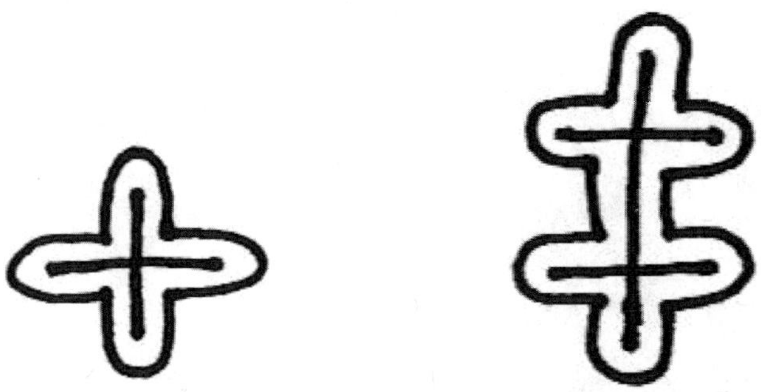

Outlined Cross/Venus Star **Double Venus Star**

Figure 4. The double star symbol in the American Southwest.

The Great Cross: a symbol with many expressions or designs, yet most are equilateral. Perhaps no other symbol is more worldwide in use than the cross. Most cultures recognize the two poles of the earth as the origin for the North-South (NS) line, and the equatorial path of the sun for the East- West (EW) line. These two lines form the Great Cross and the position of the four cardinal directions. A second cross with a 45 degree variance to the Great Cross is the semi-directional cross, also called the lesser cross. For Indian Peoples, both crosses are an integral part of their visual imagery, stories and worldview. The Great Cross in its horizontal position provides for these people a primary image for building of their cosmology. The EW line is usually associated things on a physical plane, including our mortal life on this earth, our birth being identified with the rising Sun and our death with the setting Sun. The NS line is related to the position of the Milky Way in the night sky, and is considered to be on a spiritual plane, and is the path of travel for spirit people before and after mortal life. The point of intersection for the two lines is considered a "center place," or place of balance and harmony between the otherwise opposites of the physical and spiritual planes. To find one's personal "center place" or any number of other uses of this principle (e.g., locating a special place for a village, etc.) is a primary quest in Native worldview.

Celestial bird (eagle/thunderbird, quetzal, condor, etc.): flight, ability to ascend above earth with great power and beauty, associated with the Creator in the beyond world. Also a metaphor for the ultimate transformation from a terrestrial earth being or serpent to a celestial sky being or bird during the ascent through a series of portals enroute, or the reverse if the journey is a descent from the sky world to Earth. This concept is well known in Mesoamerica as the story of the Feathered Serpent Quetzalcoatl (Aztec) or Kukulcan (Mayan). The feathered-serpent idea is midway point of metaphorical transformation of the Creator during His journeys between the Beyond World and Earth.

Serpent (usually rattlesnake): mostly earthbound, often going into the underworld with power and cunning. Serpent is illustrated in story and visual imagery in several formats. Most common: (1) Double format, similar to the caduceus symbol of the Old World. In the New World, the two serpents usually represent the two sons of First Parents in the Sky World, but in their earth mode as mortals. (2) Single format, either the Creator in his role on earth, or sometimes the Creator's not so good twin in his role as the Un-creator, who tries to foil the work of the Creator. (3) Two headed format, usually the Creator represented with his ability to go to any place without restriction, ascend or descend between underworld and sky world. (4) Coiled format, similar to the double headed serpent, the spiral or coiled serpent also indicates descending to earth by a clockwise, head-in-the-center coil. In contrast, a counter-clockwise coil is the Creator/Serpent ascending back into the sky world.

Big Cat (mountain lion or jaguar): "top of the food chain" type power and leadership authority. The Big Cat is also a metaphor for an assistant to the Creator as a guide to spirit people. Tail down format: guide to spirit people in the underworld. Tail over the back format: guide to spirit people in the upper world. This formula is also known in other parts of the ancient world, including ancient Egypt and China.

World Tree, Great Tree or Tree of Life (ceiba tree in Mesoamerica, and cornstalk in the Southwest): a metaphor for the underworld and upper world, and for physical and spiritual realms of life. Tree roots represent this physical Earth where the tree gains moisture and nutrients while the branches and leaves reach toward the sky to gain "spiritual food" (photosynthesis). The concept is that the tree is a symbol of life and represents a required balance of both physical and spiritual "food" for maintaining a balanced and quality life. Often tribal peoples would plant a tree at what they call the "Center Place," or a sacred spot representing a balance between the four directions and the upper and lower worlds.

Ladders, temple stairways, summit paths and vertical lines in rock art: all symbols with a common communicative value; each serving as paths or connectors between the lower and upper worlds. Typically, vertical pathways will contain one or more symbols or features that represent portals that, in turn, separate different levels of our great journey. In contrast, horizontal lines or features usually represent specific places of residence along our journey. For example, on a Pueblo kiva ladder the rungs are symbolic of levels of this life, or different places where we have lived before this world, or after this world, while the ladder pole/s are the vertical pathways connecting these various life units.

Portals or passageways. (holes, circles, rings, openings, pits, etc. (Figure 5): by their nature, usually relatively small and sometimes difficult to pass through, e.g., our birth and death portals in this life unit. The Pueblo kiva portals best illustrate this concept. The sipapuni, or small hole in the kiva floor, represents our emergence place from prior worlds, while the ladder hole in the roof represents the portals we have yet to experience in the future.

The most common term in Indian stories about the portal to the above world is "sky hole" or "hole in the sky." (Figure 6). Visual manifestations of portals are numerous, including (1) "donut hole" type circles, or fully pecked round areas on rock, often referencing portals prior to this world; (2) "donut" or ring type circles with a center hole intact, often referencing portals in the journey beyond this world; (3) the "V"-shape, or a "side view" of a circle, often having a vertical line connecting a "V" at the top and an inverted "V" at the bottom (a common variation of this symbol is the use of several steps or rungs in a single pole ladder with a "V" at each end. Some people call this image a centipede. Both meanings are correct, the first on a sacred level, and the second on a mundane level);and (4) the "T" shape, or three arms of a cross, a widespread symbol for a portal over most of ancient America. In the Southwest the "T" shape is most common in rock art and pueblo doorways. In Mesoamerica it is found in many visual forms, especially murals and architecture.

Figure 5. The Kiva Journey.

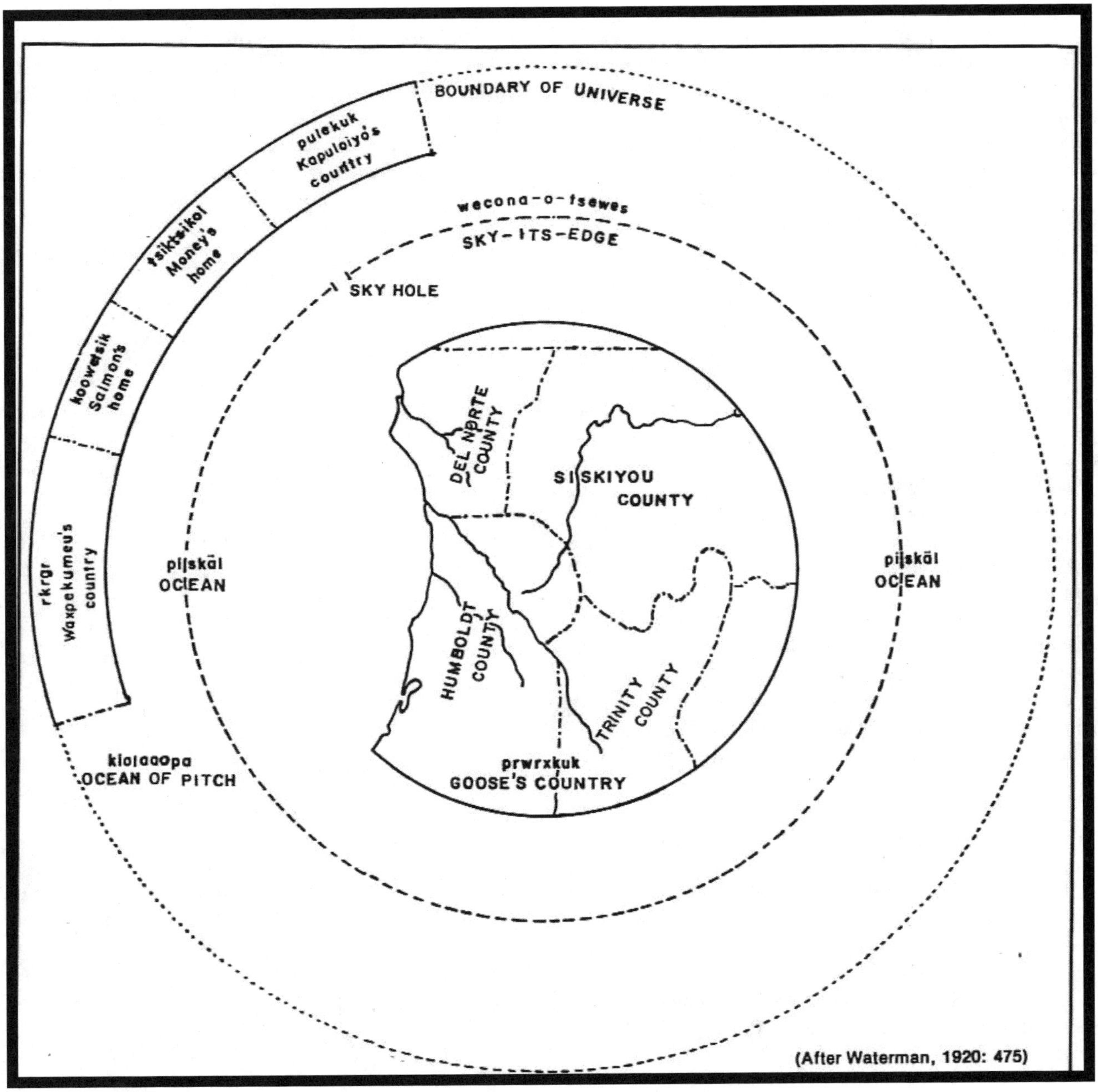

Figure 6. The sky circle.

Two Great Journey symbols, the "S" and the "J" shapes: shapes associated with water. The "S" shape is associated with water, serpents, and movement. In the latter case a vertical "S" usually is related to a pathway and journey between the underworld and the upper world, while a horizontal "S" is related to a pathway or journey on a horizontal and physical plane, such as the one we now reside on. A regular "S" is often a single journey, while several unconnected "S" shapes in a row are either several related journeys or one journey made up of several parts, and when several connected "S" shapes with many extended curves, as in an undulating serpent or river, the reference is often related to an extended, multiphase or difficult journey. However we must continually keep in mind that most symbols like the "S" may hold more than one meaning and the context of usage can offer some clarity or indicate meaning/s.

The same is probably true for the "J" shape; however, in this discussion the reference is specifically to a dark void in the southern Milky Way near the constellation Sagittarius. Near the "J"-shaped void is another circular void that is identified in many great-journey stories as the portal that separates earth's blue sky from black sky (space) beyond the Milky Way. In the Great Journey scenario entry into our world at birth is passage through a forth portal to a fifth world. At "death," or another "birth," we pass through a fifth portal into sixth world located in the blue sky around the earth where ancestor spirits spend some time serving as guardians for relatives still on earth. At some time these spirit people will through pass yet another portal into a seventh world or place. Often passage to the seventh place is encouraged by living relatives on earth during a Keruk-type ceremony where ancestor spirits are invited to leave the blue sky, enter the black sky realm, and get on with their journey back to Seven Sisters. In many stories this portal is most important because it is the great rebirth where all things physical and spiritual, once together in distant times, then separated, are now rejoined again by the power of the Creator.

The Milky Way Trails and the Great Journey

For the Yuman speaking Pai People of the Lower Colorado and Gila River Valleys (Western Arizona, Southeastern California, and parts of Mexico around the Northern part of the Sea of Cortez), no place is more important in terms of sacred geography than the Xam Kwatcan Trail and its numerous sacred component parts (the rivers themselves, the trails, many ceremonial geoglyph and rock art sites, and several sacred mountains) (Figure 7).

The Xam Kwatcan Trail. The primary NS trail or the Xam Kwatcan Trail, along with the nearby Colorado River, extends from Spirit Mountain at the north end of the trail near modern Laughlin, Nevada, south to Pilot Knob just west of Yuma, Arizona. The NS orientation of the trail mostly paralleling the river is no accident. The Trail and River in combination represent in Pai stories an earthly manifestation and mirroring of the Milky Way of the night sky. As such, the River and Trail represent a physical manifestation of the Spirit Pathway (also spiritual) along the Milky Way. Ancestor Spirits are said to often walk the Trail on the earth in memory of events in the time of Creation followed by the Creator leading the first "coming down" migration along the Trail. Quechan people at the Fort Yuma Reservation have informed me that in times past their people would make a special pilgrimage to Spirit Mountain to remember the Creator and His gifts to the human family. To traverse the Trail was/is considered akin to on an earth level to traveling along the Milky Way on a spirit level. An ideal for the Quechan people was to make the journey at least once in a lifetime. The walk was usually accomplished on foot over a period of several weeks; however, for those who knew how, it was sometimes accomplished by an out-of-body experience or in dreamtime in a matter of a few hours. To walk the Trail today can be a blessing or a curse, depending on the frame of mind of the walker. One is never sure who one might meet along the way.

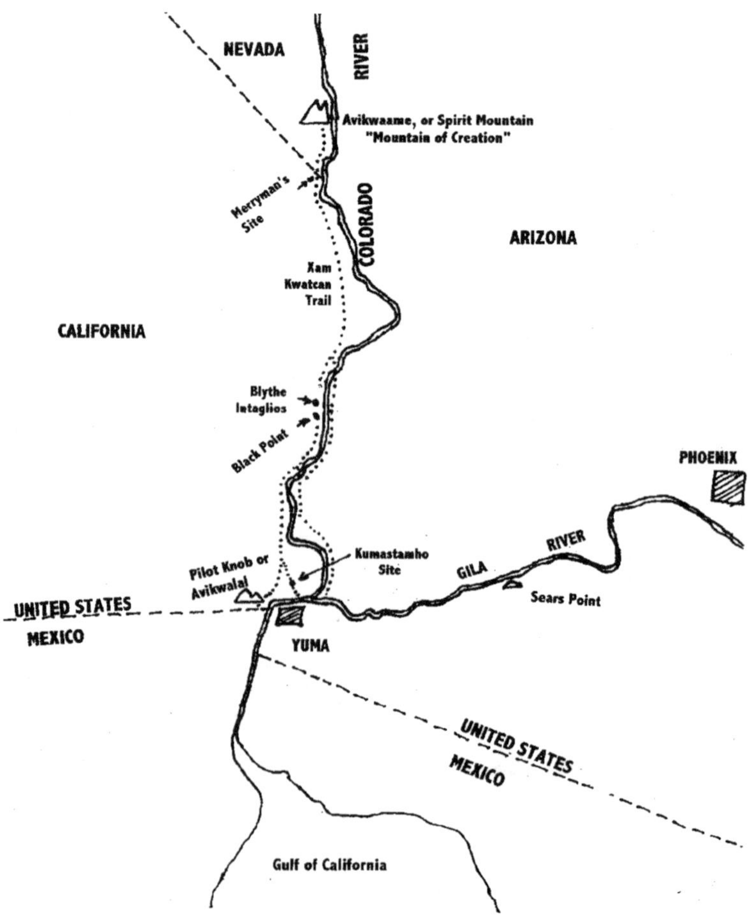

Figure 7. The Xam Kwatcan Trail.

The Trails, Center Places, and the Great Cross. For the historic Hopi and Zuni peoples and their ancestors, the Anasazi culture of the Southwest, the original Sipapuni or emergence place from the underworld is a small open-throated cinder cone located in Northern Arizona near the confluence of the Colorado and Little Colorado Rivers. This sacred location was selected by the Creator as the first "Center Place," from which two great lines extend NS and EW to form a Great Cross. The cinder cone at the intersection of the two Cross lines marked the center of the ancient Anasazi homeland. Along the NS line from Utah south to Mexico are an impressive series of places sacred to the Native Peoples yet today. At the Utah/Arizona border is Paria Canyon. Continuing to the south is a section of the Colorado River that runs NS. On the south side of the Grand Canyon are the Tusayan Ruins. At Flagstaff, Arizona, is located the most sacred of all mountains to the modern Pueblo peoples and their ancestors, San Francisco Peak. This sacred peak is where annually the Kachinas descend from the Sky World for the ceremonial season, followed by their return back to the Sky World. Continuing south on the same line we encounter Oak Creek Canyon and Sedona, then Montezuma Castle and Well, and Red Rock Canyon. Next to the south is the Verde Valley and its many pueblo ruins. Next we come to the Casa Grande ruins with the Hohokam Big House. Continuing toward Mexico we note three mountains on the same line still sacred to the Tohono People today: Picacho Peak, Kitt Peak, and Boboquvari Peak. The fact that this impressive series of sacred places is along the original NS Milky Way line of the Great Cross is likely not a coincidence. We can also note that outside this line are remarkably fewer sacred sites.

The same is true for the EW line from the Sipapuni. From the east to the west we find Chaco Canyon in New Mexico, and in Arizona we find Canyon de Chelly, the Hopi Mesas, the Moencopi/Tuba City area, the Hopi Salt Trail with its several rock art sites, the EW portion of the Grand Canyon and Colorado River, Havasupai, and the Lost City complex, the westernmost Anasazi Pueblo group located in the area where the Virgin River joins the Colorado in the Lake Mead area.

Two other regional NS Milky-Way-type lines are located in New Mexico. The easternmost is the great complex of ancient and modern Pueblos along the Rio Grande River Valley. The second is a recently recognized NS line extending from the Aztec Pueblo group in NW New Mexico south to the great Casas Grandes Pueblo in northern Mexico. In between these two places is the most significant cluster of pueblos in the greater Southwest at Chaco Canyon, with its master plan of aligned pueblos, kivas and NS road system. Farther to the south in SE New Mexico on the same line is the Mimbres Valley with its pueblos and famous pottery.

Southwest State Lines and the Great Cross. It has been said that during the late 1800's, a team of ill-informed state boundary surveyors created a new but false Great Cross when they tried to mimic the original Cross with the state lines for the Four Corners States. The problem was that the surveyors did not know the actual location of the original Cross intersection, so they arbitrarily picked a place today known as the Four Corners Monument.

Sacred Geography, Symbolism, and the Geoglyphs

Sacred geography is the use of the natural landscape as a metaphorical stage and backdrops for ceremonial reenactment of life's cycles. Added to natural landscape is the alteration of some element in nature to create a symbol, or an abstract or artificial use of some object to represent another. Most human communication relies on symbols of some type. Symbolism is the study of images, objects or other things and their intended meanings.

The concept of sacred geography, or spiritual landscape, is not only ancient in human thought, but also at the very core of the Native American worldview and cosmology. And yet only in recent years has archaeology as a discipline rediscovered this concept. We all remember well when archaeology concentrated almost wholly on analysis and explanation of physical things, mostly artifacts and features altered by humans. Today we are catching a new vision. Now many avenues of research involving much more than artifacts and features are woven together for a far more complete view of the human experience. As such, it is now OK to consider natural unaltered features on the landscape (stones, hills, mountains, rivers, etc.) as part of the archaeological record. However, far more significant to our understanding past human experience is the addition of mostly nonphysical dimensions of culture to the record. In this category are included verbal stories, legends, myths, belief patterns, and use of symbols and metaphors. Only by the study of Indian symbolism and metaphorical comparisons between the seen and unseen levels of their perceived reality can we hope to discover or comprehend the rest of Native American culture. Only by a backdrop of Pai/Yuman worldview and cosmology can we hope to comprehend their views about sacred places like Spirit Mountain, Pilot Knob, the Xam Kwatcan Trail, and the many geoglyphs, ceremonial and rock art sites along the Trail. Do you know, Jay, why it took so long to figure this out? It seems so clear and obvious now.

Sacred Mountains Along the Xam Kwatcan Trail. In Yuman stories, many mountains are part of their stories and mythology. Most important is Spirit Mountain, or Avi Kwaame in the Yuman language, the first center place for all Yuman/Pai Peoples. Not only did the Creator Kumastamho go upon this mountain and create all things in spirit form first, followed by a second creation of all things in physical form, but from here the Creator led the Yuman/Pai people to their traditional homelands. Also, Spirit Mountain, being at the north end of the Xam Kwatcan Trail, represents an earthly manifestation of North Star at the north end of the Milk Way. Near North Star is located the great portal where people entered this galaxy after leaving Seven Sisters. After traveling south along the

Milky Way, the people arrived at our Solar System and new place of learning on this earth. For the Quechan and Cocopah people in the Yuma area, traveling along the Xam Kwatcan Trail between Pilot Knob and Spirit Mountain is a physical reenactment of the past and future journeys of Spirit People along the Milky Way.

At the south end of the Xam Kwatcan Trail, Pilot Knob or Avi Kwalal, is considered a local Center Place and Big House by the Quechan and Cocopah. Also for most Yuman/Pai People, Pilot Knob is an earthly manifestation or mirroring of the portal near the "J" shape and Sagittarius at the south end of the Milky Way where people can leave this galaxy en route back to Seven Sisters. A number of years ago I located a large "J"-shaped geoglyph and nearby circular cleared area not far from Pilot Knob (south of the Kumastamho/Blind Old Man geoglyphs northwest of Winterhaven), which at the time didn't fit into any known Yuman stories. Since then We (John Fountain, retired University of Arizona astronomer turned archaeo-astronomer) and I have located about a dozen more "J"-shaped geoglyphs along the Colorado and Gila Rivers. Later we learned from the Native people how they consider the "J" and circular voids near Sagittarius to represent the sky hole for returning back to the universe. Since we now recognize what we are looking at, we now see them represented in rock art fairly often, usually in panels that reflect stories about the Great Journey. Amazing, is it not, that a large "J"-shaped geoglyph and associated cleared circle or portal symbol are located at the south end of the Xam Kwatcan Trail/Milky Way Pathway, exactly where they should be. Or is it? And oh, another correspondence! In many Indian stories, the Milky Way is referred to as the Great River and/or Serpent of the Night Sky.

The Milky Way, the Serpent, and Water. Since water (rain, rivers, springs, tanajas, wells, etc.), is a primary earth/blue sky lifegiving substance required by all living things, peoples all over the world have sought to explain its source and how to keep it coming. Most cultures in the world attribute rain from the sky as ultimately from the Creator. However, most peoples require a more detailed cause-and-effect explanation, a scenario to structure into all sorts of details to explain how the unseen world relates to the seen world and how it all works. Typically the serpent has been and still is a central symbolic figure representing the Creator of life. The body of a living serpent is rarely straight. Instead it curves and undulates, as do rivers and streams. The symbolic likeness of serpents and rivers compares the serpent guise of the Creator to His provision of water in rivers. Also the serpent/ Creator relationship is associated with symbolic connection between serpents as lighting, the cause of thunder, which in turn knocks rain out of the clouds. Finally, the stars of the night in the Milky Way form first, a "great river of the sky," the primary source of water that becomes clouds, and second, a "great serpent in the sky," the Creator Himself. In a series of logical steps it happen like this: (1) the Creator makes the Milky Way as an aspect of Himself as the "great serpent of the sky" and ultimate source of all life; (2) the serpent of the sky is also the river of the sky, source of water to form clouds; (3) the serpent/Creator sends small lighting/serpents from the clouds to earth to cause thunder; which causes rain which, in turn fills rivers, springs and water holes; and (4) finally in the last measure, the water allows life to form new life. As we know, Native American stories, ceremonials, and imagery are replete with the interplay of all the above aspects of how the Creator gave the gift of rain/life to all life forms. Failure to properly respect and reenact each stage of the process in ceremony could, as in the past, lead to drought and much deprivation to all life forms.

The Milky Way and the Geoglyphs. Perhaps the most complex and instructive ceremonial pathway geoglyph with a Milky Way component is on Black Point, a river terrace along the Xam Kwatcan Trail. Black Point is part of a large cluster of geoglyphs known as the Big Maria Mountains area group of geoglyphs. (See below a site description of Black Point with a summary of site elements or features, and what I have learned about some of their various ceremonial functions). A central feature at Black Point is the generally NS position of the Milky Way path within a circular pathway, the latter representing the greater universe (Figure 15).

In Native American cultural context, these pathways on the ground are mirror images of the real Milky Way and universe and ceremonially function to bridge the gap between earth and sky and between mortals and their ancestor spirits. At least six other geoglyph figures are recorded which illustrate combining of a NS path within a circular pathway (e g. Johnson 85: 47). Also in historic times several similar patterns have been recorded in ethnographic records as geoglyph/ground paintings used in puberty or Keruk ceremonies among Indian peoples in Southern California (Figure 8) (DuBois 1908, Waterman 1910 and Kroeber 1925).

At least two dozen more pathways, avenidas, or rock alignments without circular pathways are recorded with a general NS orientation. (e g. Whitman & Varner 1998). Some of these have limited ethnographic ties to the Milky Way, such as the 90-meter-long rock alignment near Lost Lake along the Xam Kwatcan Trail. A good likelihood exists that many of these images on the ground once held similar functions.

The Great Cross and Geoglyphs. Obviously, a cross consists of two intersecting lines at a 90 degree-angle. The classic circled Hopi cross with four prior world portals and a roundabout universe circle is an example (Figure 9). The use of a cross within a circle is common in rock art, but less common in geoglyphs. Examples of circled crosses in geoglyphs are illustrated in Figure 10, a–d, next page, and Figure 13, center. Yet it is becoming apparent that Native peoples often assumed a cross representation without displaying all lines or arms. The classic "T"-shape image is an example of a cross less one arm (Figure 11, a & -b). The odd arm, or "tail," opposite the missing arm, usually indicates the direction of the action inferred by an incomplete cross. In some geoglyphs, both lines of the cross are present, but more commonly one of the two lines is not present. Native persons have suggested that lack of one or more features within an image does not say that the artist/author was unaware that they were missing. Rather, all cultures allow in the concept of "artistic or poetic license" the choice of actual verses or metaphorical/symbolic verses for inherent display. Under this concept an informed viewer/reader is aware of what is clearly present, and what is missing, and what is actually inferred by a composition. In this light, not all elements of any composition need to be openly present. What is present is related directly to the primary subject being communicated, while secondary subjects are often assumed or inferred, yet are well understood in the greater context, but never ignored. In this context an intended or understood cross with only one arm is no real problem, only to those who fail to see the big picture. The reason for this discussion is that while we have come to understand NS lines as being associated with the Milky Way, at the same time we have not understood as well EW-oriented lines or alignments on the ground. What I have learned is that both NS and EW alignments are inferential of the cross and four directions regardless whether all lines or arms are present, or whether a circular Universe path is present. Many EW geoglyph patterns also exist with or without circle paths present. These EW patterns usually are associated with the transitory physical or mortal level of life we are now experiencing and not with the beyond world. Sunrise often equals birth and sunset equals death in this realm. (See Johnson 1985; 56, 106, 114, 118, 119 as examples of lines without circles, and many in von Werlhof 1987).

Summit Path Geoglyphs. Most known Summit Paths are found on the sides of large hills located primarily along the Gila River, which trends EW in Southern Arizona. So far only four are found along the Colorado River, all being in the Palo Verde and Cibola Valleys, and Big Marias areas not far south and north of Blythe, California. At least three are also reported in the northern Baja Peninsula. One may well be, at a half mile long, the largest known so far (Ewing & Patchen, 1991). Summit paths along the Gila River are mostly clustered in four areas: Sears Point, Painted Rock, Wintersburg Wash, and the Robbins/Powers Butte areas. Each area has six to twelve paths. Another area with at least as many paths is reported in the Tucson Basin along the Santa Cruz River; however, I am not yet acquainted with these sites.

Summit paths are typically wide (1 to 2 meters, some more) and mostly in straight lines extending from the base to the top of the hills on which they are located. About half of the paths appear to have been actively used as processional pathways where people would ascend to or descend from a ceremonial area at the top of the hill. Human alterations to the ceremonial areas on the hilltops consist of variously, cleared areas, dug-in pits, rock alignments, stacked rocks, rock cairns, short pathways to specific places, and rock art. However, the other half of these paths do not extend the full distance from the hilltop to the bottom, nor do they appear to have served as actual processional pathways. Furthermore, most of the latter types of paths have little or no evidence for a ceremonial area, perhaps meaning their function was more visual than ceremony-related.

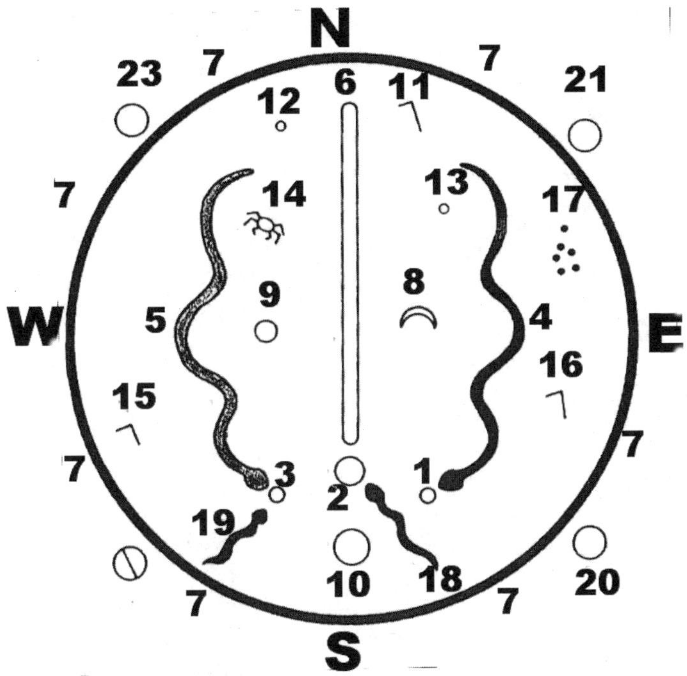

Figure 8. Earth painting map or diagram of the earth and night sky formed with colored earth and other materials by the Ipai of Southern California for use during the boys' initiation ceremony. Image identification: 1-3=datura, mortar and pestle; 4-5=twin rattlesnakes; 6=Milky Way; 7=rim or edge of the world and sky; 8=new moon; 9=full moon; 10=sun; 11=coyote; 12=buzzard star; 12=crow or raven; 14=black spider or spider woman; 15=wolf; 16=mountain sheep or Orion; 17=Plelades or Seven Sisters; 18=black snake; 19=gopher snake; 20-30=four sacred mountains at the corners of the traditional homeland. (After Waterman 1910).

Figure 9. The Hopi cross.

a) b) c) d)

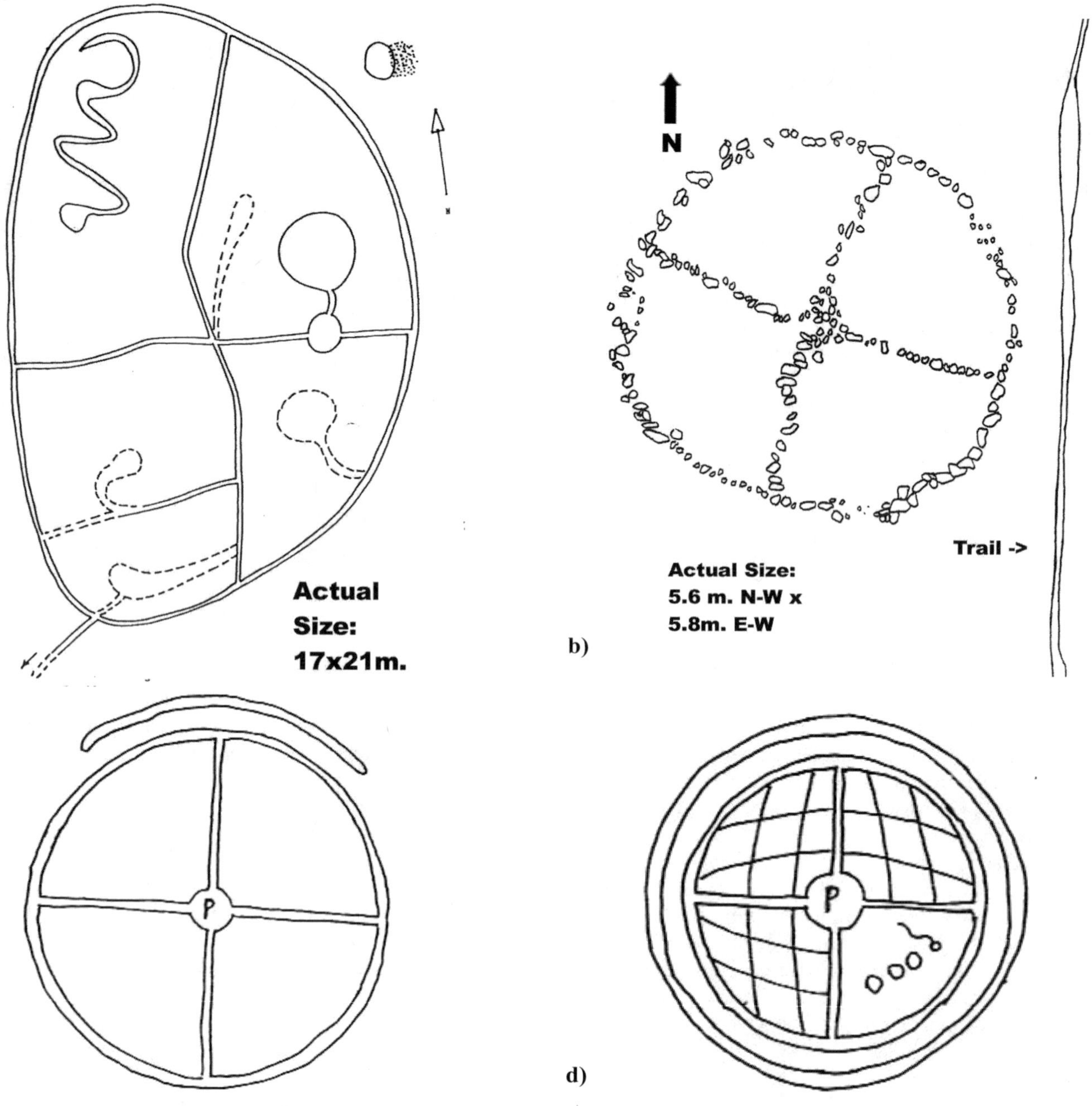

Figure 10. Circled cross geoglyphs; a) Dance pathways near the Xam Kwatcan Trail (Johnson 1985:96); b) Rock alignment pattern on the Goldwater Gunnery Range (BLM 3127); c-d) Ground paintings, Southern California Indians (After DuBois 1908:90,178).

With the exception of the Palo Verde Point Summit Path in the Palo Verde Peak area on the Xam Kwatcan Trail (Johnson, 2003), we have little specific information about the uses for the rest of these paths or areas. However, we can, at some risk, apply generally what we know about the Palo Verde Point Path to other summit paths, i.e., that summit paths functioned as formal hillside access trails to reach "high places" to conduct ceremonies not befitting more ordinary low places. This concept was first postulated by Pat Stein in the very early '70s, in an archaeological report for the land where the Palo Verde Nuclear Plant would be built. Pat recorded two hills within the plant area that had pathways going up the sides and that had alterations on the tops. Her comment was that the hills looked pyramidal, much like "temples" in Mesoamerica, and perhaps the pathways served a function similar to stairways on actual temples. I agreed with her then and I still do. Why build temples when you have natural hills to serve the same purpose? High Places for worship are almost as universal as mankind.

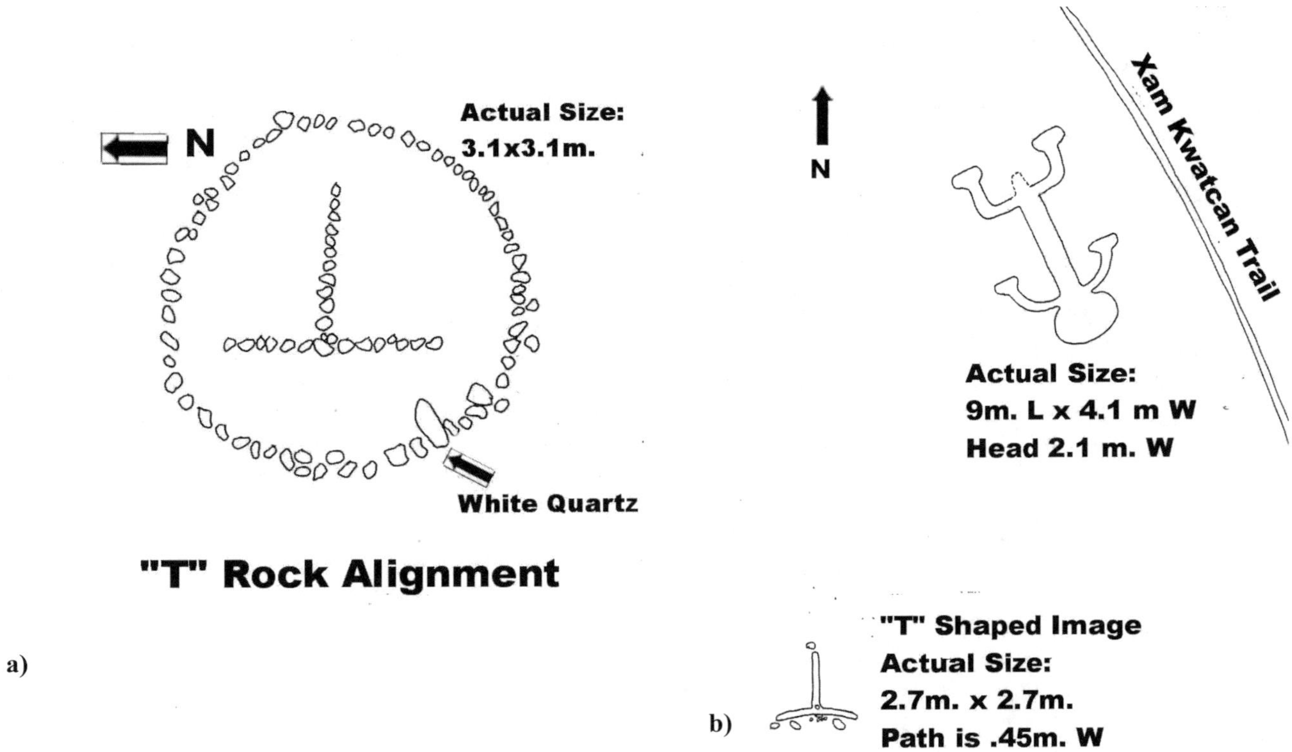

Figure 11a-b. Two geoglyphs with "T" shapes, north of Yuma, Arizona, near the Xam Kwatcan Trail (author's personal file).

There are two other observations that I can make about summit paths. First, I note that the general orientation of summit paths on hillsides align with either one or more of the four cardinal directions, or with the semi-cardinal directions, or with a prominent horizon marker. I suggest that summit paths were placed on these hills at certain orientations for much the same reason as the orientations of stairways on temples. Temple stairways are often oriented NS when associated with the Milky Way/Great Sky Serpent pathway for sky people travel, while EW stairways often associate with events of our present world.

The second observation I can make is about the locations of the Summit paths. I suspect that the earliest use of summit paths was not on the Gila River, rather to the west on the Colorado and/or in northern Baja. Next I suspect that the cultural use of these unique paths was taken east along the Gila and/or along the EW oriented Coco-Maricopa Trail. My theory rests on two related or overlapping observations. First, we know that in historic times a military coalition of Mojave and Quechan warriors drove their neighboring tribe, the Halchidomas, to the east away from their original Colorado River homeland to live with the Maricopa Tribe, who lived along the Gila River at the "Big Bend" area. Secondly, as we look at original homeland of the Halchidomas, we note that it was the same area on the Colorado where we have found the four summit paths. Also in this same area we find the western terminus of the Coco-Maricopa Trail. Next I note that as we follow the Coco-Maricopa Trail to its eastern terminus, we find not only the new Halchidomas homeland at the big bend of the Gila River among the Maricopa, but we also find in the same area the three major summit path concentrations in the Painted Rock, Wintersburg Wash, and Robbins/Powers Buttes areas. I think I can safely say that the Halchidomas had something to do with the building and use of Summit Paths in their two homelands; however, more significantly in their second adopted homeland.

Portals, Pits and Geoglyphs. Associated with many geoglyphs are features that look like a pit or basin-shaped hole dug into the earth. Some of these pits have a small mound of dirt or gravel very nearby, suggesting that it is the dirt dug from the pit. I shall never forget one time walking over the Black Point Ceremonial Site with a couple of members of the Colorado River Indians Tribes Reservation (C.R.I.T), Parker, Arizona, when I stated that it was a shame that this site was damaged by a bulldozer road to a power pole and two "pot hunting" holes within two obviously old circular pathways at the site. A return question was "Boma, what makes you think that these two holes are pot hunting holes?" Then I said, "They look like pot hunting holes, plus I don't think Indian people would typically dig holes in their own ceremonial sites." The comment I received in return was "Don't be so sure of that, Boma!" Further discussion then and at other times with other Indian people taught me that purposely dug holes or pits in the earth at this type of staging site are usually associated with a portal linking the previous life with our present life (see Figure 15).

Two classic models of Native American cosmology will illustrate their thinking on the subject: the Hopi Kiva, and the Kogi Temple of northern Columbia with their respective sacred architecture, symbolism, and mythology (Krupp 83: 231-241). Although most of the micro-scale details of these two worldviews differ, yet on a macro-scale level they are very similar. On a primary level, the Hopi/Kogi cosmos is divided into three realms: the Past or "underground" level, the Present or the here and now level, and the Future or beyond earth/sky level. On a secondary level these three realms are further subdivided into nine realms. The Human family is seen as being in the middle realm, or the Center Place, with four realms in the past prior to our physical birth, and four realms in the future beyond worlds after our leaving Earth. The architecture of the Hopi Kiva and Kogi Temple both directly reflect these nine levels or realms of life. And the rituals enacted within these sacred structures reflect the greater journey of life through the nine realms. Each realm of the journey is separated from the others by a "death/rebirth" experience, symbolized as a mythical ordeal, where persons must eventually pass through a series of nine portals or openings between levels.

In the Hopi Kiva the floor area symbolizes the division between the underworld and our present world. The sipapui or hole in the floor under the ladder symbolizes the four portals used in the past to reach Earth. The circular bench in the kiva symbolizes the world we now live in, and the area above the bench, called the kivaove, symbolizes the blue sky above Earth. The hole in the roof, where the ladder extends from the floor to the outside world, symbolizes the hole in the sky, a more or less generic representation the four sky portals set between Earth and Seven Sisters. The ladder, with its rungs and spaces between the rungs, symbolizes both the places of residence and the in-between "death/rebirth" portals connecting the journey from the underworld to the celestial world, and vice-versa. To descend from the outside down the ladder through the hole in the roof to the floor of the kiva was paramount to the first half of the great journey from Seven Sisters to Earth. To ascend the ladder to the outside was the second half of the journey, or a return to Seven Sisters. In some Hopi kiva coming-of-age ceremonies, actual hoops made of river birch branches were passed through four times by an initiate while coming down the ladder, and then again when returning up the ladder, all in an reenactment of our greater extended journey. The function of all this ritual was to teach initiates the complexity and sometimes difficulty of the trip and ways to reach our destiny without being caught in snares or on dead-end roads.

So how does all this ritual fit into the geoglyphs? It is fairly simple and straightforward. You do see where this is all going? A fair number of geoglyphs have holes or pits purposely dug into the earth surface to be used as part of the journey reenactment. Like the sipapuni and the hole in the roof of the Kiva, the pits in the ground are an essential part of the geoglyphs and also represent portals for emergence from one realm to another. (See Figures 12, 13, and 14, where "p" represents pit; see also Smith and Turner; 75:15). Pueblo kivas, temples in many cultures, including Pacal's Temple at Palenque, Mesoamerica, and many geoglyphs, are finite, partial or incomplete microcosmic representations of the infinite macrocosmic Universe. The design, layout, construction and usage of kivas, temples and geoglyphs, with their attendant symbolism, mythology and ritual activity, all have one major common denominator, a way to relate earth and earth people with ancestor spirit people and their dwelling places in the greater Universe. The highest spiritual value associated with ceremonial usage of a geoglyph complex, with all of its component parts (pits, pathways, alignments, images, mounds, and staging areas), is to provide earth people with a physical stage upon which to reach out to people and places beyond earth, and replicate in miniature the difficult journey from here to there. Also, central to the journey was the participation of key Deity personages (for the Pai, Kumat or Kukumat; the Father Creator, and His two twin sons Mastamho or Kumastamho, the elder, and Bakotahl or "Blind Old Man," the younger) who performed various acts of Creation or uncreation, depending on their role in the story. Staged reenactments of various Creation or Great Journey/Migration episodes at various geoglyph locations usually included acting out in ritualized drama, dance, song, prayer, and story telling the "hows," "the "whys," and the "wherefores" of these events as they once happened at specific locations on the sacred landscape. Like kivas and temples, many geoglyphs, but not all, served as bridges between the vast infinite orderly Universe and the somewhat insignificant finite and often chaotic earth. A primary order of business in most sacred ritual or ceremony is to move the human family from a chaotic profane, even mundane terrestrial realm to a greatly improved and orderly celestial realm. And Gods are needed for that.

The Black Point Ceremonial Pathway Site

Ancient Patayan/Yuman-speaking people of the Lower Colorado River region of Arizona/California, known today as the Cocopah, Quechan, Halchidoma, and Mojave, celebrated the multiple stages of life at a unique ceremonial geoglyph site today known as Black Point. Located about fifteen miles north of Blythe, California, off Highway 95, the Black Point site consists of a complex pattern of processional pathways and associated features or images laid out on a river terrace also known as Black Point (Figure 15). Through we do not know the details about the various ceremonial activities that took place at this site, we have learned many of the functions of the pathways and symbolic meanings of many of the images and features on site. In general, we recognize the layout of the paths as reflecting the Universe and the Milky Way. To better understand Black Point, we must look to a holistic overview of the cultural parts that are present or associated with the site. These parts are:

1. Place. The Black Point site is positioned on a broad flat terrace or mesa top that extends out into the Colorado River floodplain, affording an excellent view of the valley and mountains in all directions. Black Point is centrally located within perhaps the largest single cluster of geoglyphs on the River, among them the famous Blythe Intaglios being only a few miles to the north.

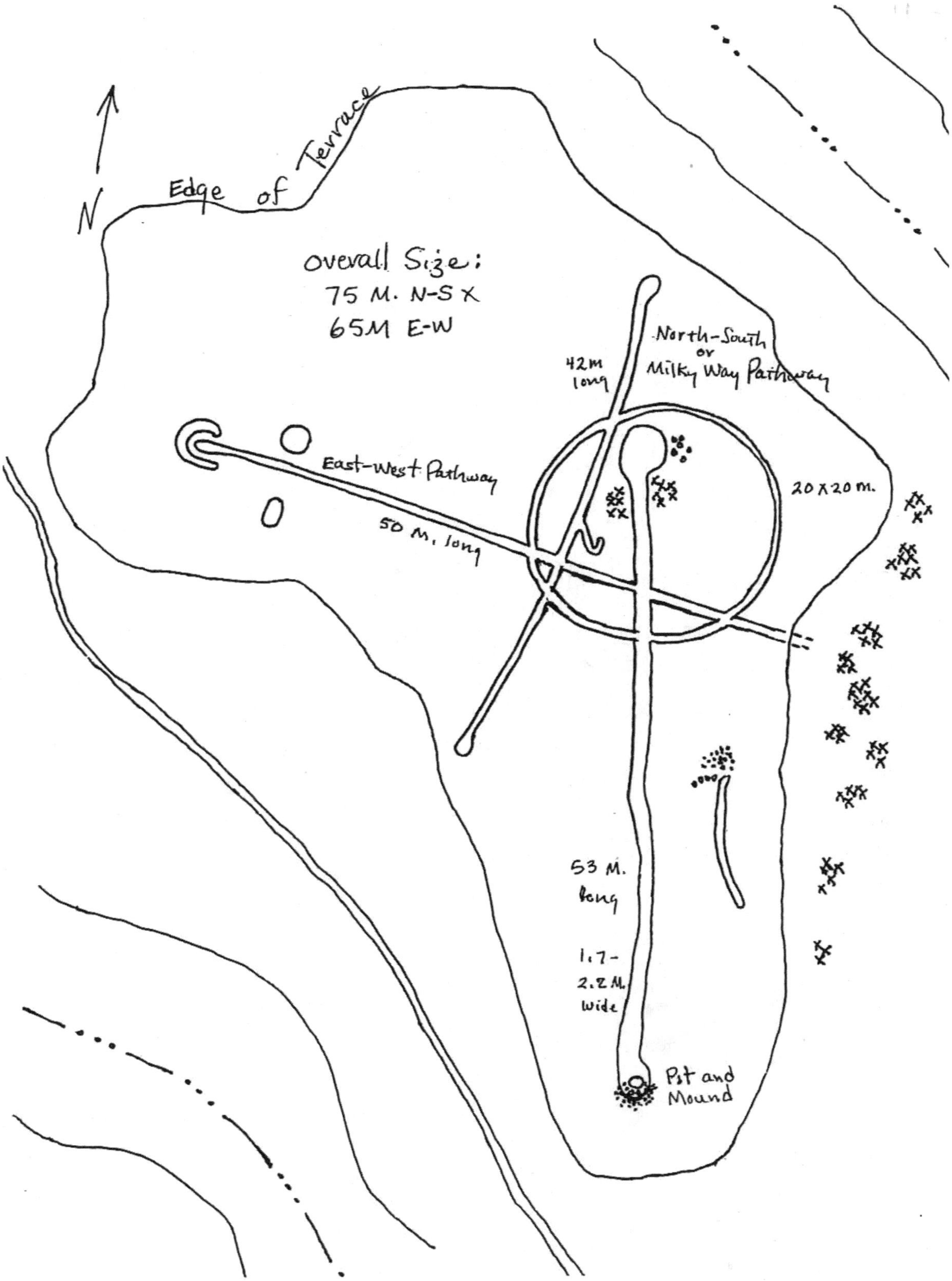

Figure 12. Multi-component ceremonial dance pathways (Johnson 1985:80, updated).

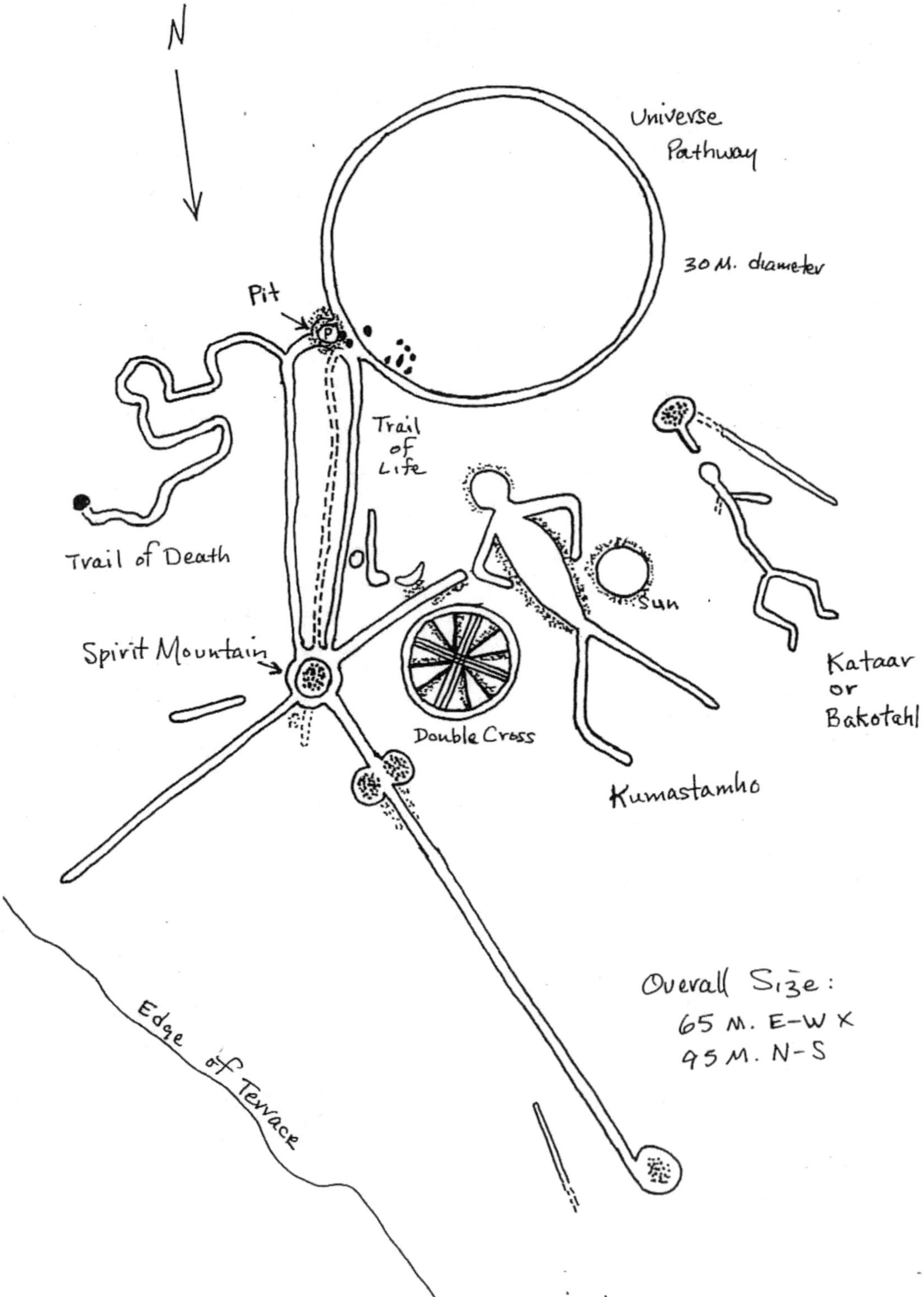

Figure 13. Feature B of the extensive Ripley Geoglyph Complex (Johnson 1985:69, updated).

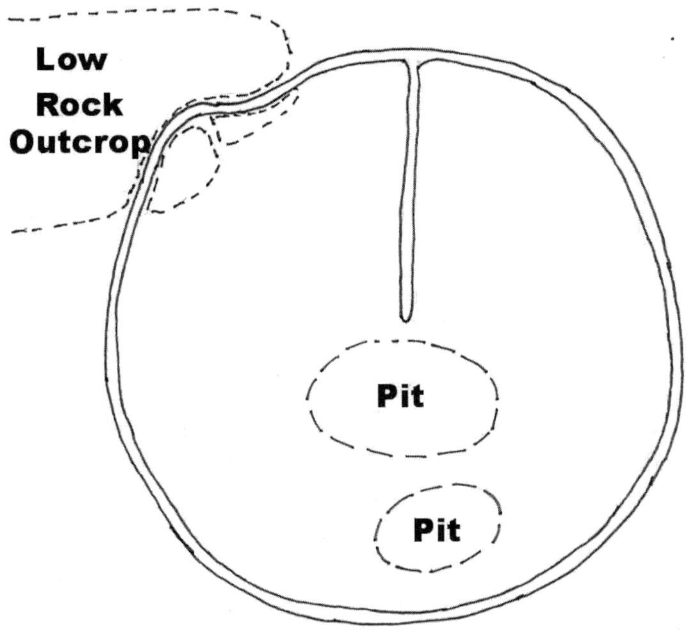

Figure 14. Davis Camp dance pathway. (BLM 1381 and author's personal files).

Collectively, this cluster of geoglyphs, known as the Big Maria area, is located on the Xam Kwatcan Trail at a spiritual halfway point between Spirit Mountain or Avi Kwaame and the Mojave villages to the North, and Pilot Knob or Avi Kwalal and the villages of the Cocopah/Quechan peoples to the South. Following the Creation, as Kumastamho led some of the First People south along the Xam Kwatcan Trail, one of the primary stopping and instruction places was at Black Point. In time the Black Point Site became second only to Spirit Mountain in sacredness.

2. Stories, Myth and Legend. According to Yuman traditions, the Creator Kumastamho under direction from the First Creator Kukumat came to a great void in space where He brought into being the Earth, the Sun, and the Moon, as part of our Solar System, and our Galaxy, the Milky Way. Next came all forms of life in their spirit form.

In the first cycle of Creation in the beginning, all life forms were in a spirit form first, including inanimate forms (earth, mountains, rocks), and eventually all animate forms, including First People. During the second cycle of Creation, the Creator ascended to the top of Spirit Mountain (at this point Spirit Mountain was still in spirit form, hence its name) where He added to all spirit forms a physical form, or in other words, physical matter as we call it, was added to spirit substance. At this second level of Creation all spirit substance gained a material framework and became matter, including mountains and us. In time First People grew into seven tribes living around the base of the Sacred Mountain of Creation. Kumastamho then invited two of these tribes, the Cocopah and Quechan, to follow Him south along a trail paralleling the River from Avi Kwaame to a new home area or center place in the region around Avi Kwalal. The Trail became known as the Xam Kwatcan, meaning in the Yuman language "we who came down from Avi Kwaame" or in short "coming down." Also, the Creator invited these two Tribal Peoples to make pilgrimages back to the Mountain of Creation to remember their creation and origins, and then to travel again south along the "Coming Down" trail to renew their memory of the first "Coming Down." Today the term "Quechan" is a Spanish corruption of Xam Kwatcan, meaning "Coming Down People."

3. Images and Features. The Black Point Site is a complex set of images and features, including various types of pathways, which were once used in ceremonial reenactment by people who visited Black Point Mesa for that purpose. The primary images and features at the site, with some of their functions, are listed below by alphabetic letter, and are located on a site map by a corresponding letter, (see Figure 15).

A. Several rock clusters or scatters that have been suggested as representing various nighttime star patterns; however, beyond that, they are unknown entities.

B. A circular pathway within the greater Universe circle that represents the past journey of spirit people from the great constellation Pleiades or Seven Sisters to our present resident Galaxy, the Milky Way, and also a return trip in the future. Within the circular pathway are two more features: a gravel mound, representing our former home at Seven Sisters, and a pit dug into the earth, representing the several portals passed through en route to our residence within the Milky Way system.

C. A processional NS-orientated pathway representing the Milky Way of the sky, or the pathway traveled by spirit people following their entry into this galaxy at the North Star portal before birth on Earth, and again in the future as people who have left Earth and travel south along the Milky Way to exit this galaxy at the portal near Sagittarius, then back to the greater Universe.

D. A great circular processional pathway representing both the greater Universe as a whole, and the path of our greater travels, real and metaphorical, within the Universe before and after life on Earth.

E. Several cairns of rock located along the greater Universe pathway which represent (1) actual or mythical places along the great journey, and (2) teaching stations where a teacher or shaman would instruct initiates about that place, what has or will happen, and what is the right thing to do.

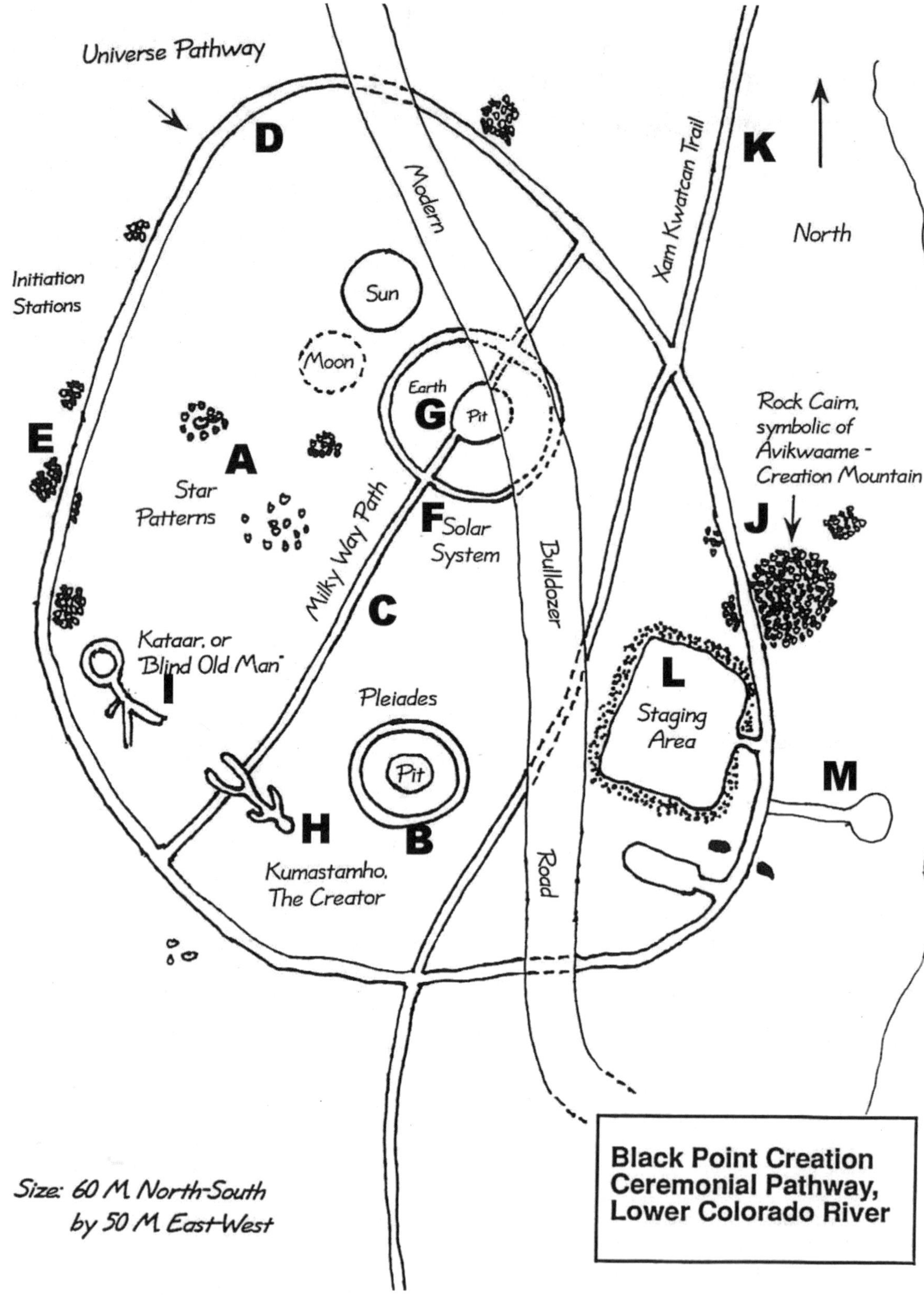

Figure 15. Black Point Ceremonial Creation Pathway, lower Colorado River.

F. A second larger circular pathway, also within the greater Universe circle, which represents our present solar system. This pathway is positioned along the NS Milky Way path.

G. A gravel mound and pit or depression in the center of the circular pathway representing our solar system. The mound represents our Earth, and the pit is equivalent to the sipapuni in the Kiva, or the symbolic portal where First People emerged in birth to life on this Earth.

H. The anthropomorphic form of Kumastamho, the Son of Kukumat. These two deities tend to share the work of Creation of Heaven and Earth and all in them. Note that the image of Kumastamho is positioned directly over the southern end of the Milky Way pathway just before reaching the Universe path. Also note that His two legs create a gap in the Milky Way path.

I. The semi-anthropomorphic form of Kataar or Bakotahl (Blind Old Man), the twin and evil brother to Kumastamho and enemy of Creation, choice, and all good qualities of people. True to the stories about Bakotahl, he is shown with a compromised head (in this case, a ring for a head) and one arm missing.

J. A large cairn of rocks, measuring four meters in diameter and almost two meters high, which represents the real Spirit Mountain at the north end of the Xam Kwatcan Trail.

K. A primary branch of the Xam Kwatcan Trail that crosses the entire Black Point Mesa and the site on a NS axis. Two other branches of the Xam Kwatcan Trail also cross Black Point Mesa, but purposely avoid the ceremonial site. One branch is on the west and the second is on the east, below the main mesa on a lower terrace. Apparently these alternate trails allowed travelers not involved in the ceremony to wisely avoid crossing through the more sacred areas of the mesa top site.

L. A large 10x10 meter cleared area with a heavy berm of gravel around the edges and two openings on the east edge leading directly to or from the Universe Pathway. This area was used for ceremonial preparation prior to entering the processional pathways.

M. A faint path extending east of the southeast "corner" of the Universe Pathway and leading to a smaller 2x3 meter cleared area at the very eastern edge of the Mesa. Though I am not clear about the function of this cleared area, I have learned that similar areas on the east of a site often have to do with Sunrise, even assisting the Sun to rise, as part of an opening of a ceremony.

Collectively these pathways, images, and features on Black Point Mesa form one of the largest, if not the largest, Native American ceremonial stages known in North America.

4. Ceremonial Activity. Yuman/Pai People who preformed rituals or ceremonial activities at Black Point were in effect reenacting on a microcosmic stage the drama of Creation, origins, and migrations for both on this terrestrial sphere and the celestial spheres beyond Earth. The original order and details of the ceremonial routine once performed on the stage at Black Point are uncertain; however, a general sequence of events is postulated from a context of the images, features, and pathways at the site, and from known Yuman/Pai statements, stories, myths, legends, and symbolism pertaining to the Creation and the human journey in time and space. People en route to Avi Kwaame from areas to the south would arrive at the Black Point Site via the Xam Kwatcan Trail after several days travel. There were yet several more days travel prior to arrival at the sacred Mountain of Creation. Black Point served as a mid-journey place of initiation and spiritual preparation prior to going for the real thing at Avi Kwaame.

At Black Point a series of ritual activities were preformed in the proposed following order: (a) At sunrise, as an opening for the days' activities, a song/prayer ritual was preformed at the east edge of Black Point Mesa; (b) The second order of ritual would be a gathering of teachers/shamans and students/initiates within the "staging area" for ceremony preparation. The preparation area in most ceremonial settings was for instructions to the initiates about what was going to happen and the serious nature of what was to be accomplished. Also, here was the making ready of costumes and other required ritual paraphernalia. According to general practice, such preparation areas were not to be seen by noninitiates, thus insuring privacy within, a wall of brush and other materials separated those within from those without; (c) Once participants were ready, they would exit the staging area and proceed onto the Pathway System and begin reenactment of past, present, and future aspects of the Great Journey. Perhaps among the first places to visit was the smaller of two circular paths within the Universe circle path. This path, on the south side, represented the pathway to and from Seven Sisters. Participants would do a "sideways shuffle" dance around the path to commemorate those stages of the Journey. Within this circle path is a pit representing the portal of the first birth of Spirit People at Seven Sisters, and successive portals experienced in the trip to Earth; (d) The next part of the journey reenactment probably involved proceeding north of the staging area on the Universe Path. In a short distance participants would come to the large cairn of rock representing Avi Kwaame. Here initiates would hear the story of the two Creation cycles and how the Creator led various tribes to their new homelands. Also it is most likely that certain songs were sung, and specially prepared prayer bundles could have been left at the cairn as a sacred offering; (e) Next, participants would proceed north along the Universe Pathway in a counter-clockwise direction, soon crossing over the Xam Kwatcan Trail and coming to the north end of the Milky Way Path. At this intersection, two options were possible: one, continue along the Universe Pathway, or two, go south along the Milky Way Path. For now I opt for continuing west, then south along the Universe Path. (f) Along the west side of the Universe Path are several smaller rock cairns positioned next to the path, each which served as a representation of a significant place and event that has happened or will happen along the real macrocosm in the realms beyond. Participants would stop at various rock cairns, or "teaching stations," to receive instruction and hear stories about these important places and events. Also on the west side, participants would be introduced to the "Un-Creator" or Blind Old Man, adversary of Creation and all that is good. (g) At some point, participants would enter the Milky Way Path at the north end and proceed south to the circular path representing our solar system. Here participants would learn of the creation of the Earth and how the pit in the center of the Solar System path served as a birth/death portal symbol for our emergence from the underworld at birth, then later followed by the reentry of the body into the underworld again at death. (h) Continuing south on the Milky Way Path, participants were introduced to the human figure of Kumastamho, the Creator, who stood over the Milky Way Path to assist humans in preparation for reentry back into the greater Universe via the symbolic Universe Path.

5. The Spirit of Place, Myth, Image, Ceremony and Black Point. Human populations in pre-modern cultures have traditionally invested great amounts of time, energy and resources to creating complex scenarios that would allow the survival of human intelligence, knowledge, accomplishment and life beyond the grave. Great cultures of the past (China, Egypt, Mesopotamia, Mesoamerica, etc.), and countless other often-unnamed cultures have to one degree or another sought to prepare their people for a better life beyond this one. So the really BIG question to many anthropological researchers is this: Why do we often deny the same rights to more rural small-scale societies?

We often carry on as if these lesser-known folks were too dumb or impoverished to think about an afterlife. And we get annoyed when a researcher suggests that a good percentage of their cultural baggage is all about ceremony, not just to insure rain and sustenance, but to also to insure an organized way in which departed ancestors can achieve an immortal status hereafter. As a way to honor status, we usually allow Pharaoh types, and sometimes their courts, the privilege to travel west to a blissful nether world, but heaven forbid if commoners or the masses ever think of doing the same. We like to act as if the archaeological record has everything to do with physical survival and nothing to do with spiritual survival or the survival of spirit people. We are sometimes obligated to allow Shaman types the right to do things in a non-physical realm beyond earth, but even then we are yet convinced that what really happens is only in the shaman's mind, or a trance-like experience that has little to do with anything except chemical alterations in the brain.

Generally speaking, Native peoples around the world do not agree with our need to account for all human experience under a cloak of scientific method. I hear them telling me that much of what humans can experience is neither chemical, nor can it be explained fully by our present formulas of science. We obviously do need to balance this equation a lot better or we will continue to overlook, misunderstand, or misinterpret much of the archaeological and ethnographic records.

Contrary to the opinions of some researchers, the true spirit of Native American belief patterns does very much involve close attention to the principles of cause and effect. People in all cultures in time and space have always sought to understand and develop "rules" about their observations of all types of actions and their results. Individuals or societies who do otherwise earn the Darwin Award for being "unfit" sooner or later exit the scene. The primary difference between so-called "modern" and "primitive" techniques of observations and resulting conclusions is not so much about being "scientifically correct" as it is about covering all basis of explanation. "Primitive" people regularly arrived at "correct" conclusions in their observations of cause and effect. Scientific methods did not begin with the coining of the phrase.

The flip side of this coin is that modern science cannot yet explain all human experience and observation, and in all likelihood may never be able to do so. It is a matter of degrees along a continuum from "soft" to "hard" observation and explanation. Proper science does well with the hard stuff since here we are dealing with quantifiable measurement. When we come to the soft stuff, true "hard science" must back off and give room for other kinds of explanation, sometimes called "soft science." But in much observation the term soft science is a misnomer because there is little to no quantifiable measurement.

So what do we do now? We do exactly what people of all ages have done; we also must have explanation at all cost. So we do the best we can with what we have; we revert to alternate types of explanation usually known to be less than scientifically accurate. Some like to call this research "pseudo science" or false science. But that is not a solution because the observations and conclusions are not always false. All cultures, even our own, still try to cope with the shortfall of not having all the correct answers by resorting to myth, as in mythology. All cultures, even our own, practice as much mythology as we do good science, and all because good science cannot explain it all.

Since observation demands explanation, and since cause and effect must have acceptable relationships, we have learned to accept myth as almost as "real" as good science. Myth is a strong form of great power in all cultures, an enabling force that takes over when the five senses and logic begin to fail us. Myth can deal with concepts like "sacred," "spirituality," "spirit people," "Gods," and "Eternal Life." Science waffles on these topics because they are "un-quantifiable," "un-testable," and "figments of our imagination." Well, many subjects and experiences in life are not testable or duplicable in a lab, but that does not make them false or second-class truth. The truth of the matter is that myth is (a) the mortar that holds the bricks of culture together; (b) the threads that give meaning to life and tie together the past, present, and future; (c) the fabric that provides social/ cultural identity and continuity for individuals, tribes, communities and nations; (d) a framework for explaining the unexplainable and comprehending the incomprehensible; and (e) the mechanism that links the mundane and sacred; material and immaterial, carnal and esoteric, visible and invisible, youth and age, time and space, life and death, chaos and order, and a long list of similar paradoxical conditions (Bierlein 1994). Myth is rarely an inexcusable misrepresentation of truth or outright fabrication. Rather it is more like a saying we hear: "If it isn't true, it should be." And that is the way it will remain until science can handle it all. Good luck, science.

In the Big Picture, the spirit of the Black Point Ceremonial Site and of the other ceremonial geoglyphs in the region is really much the same as any ceremonial site or place in any part of the world. Ordinary people ban together as participants in life to achieve their collective goals. One of the first realizations of the group is they become aware that they cannot reach many of these goals on their own, so they solicit help from compassionate higher powers ranging from their own spiritual teachers to chosen animal spirits to departed ancestor spirits to lesser gods and finally to the greater Gods. Before talking to the higher powers, a plan of action and desire results must be understood by all parties. The "plan" requires a special place or stage where the people and higher powers can assemble for communication. All parties know that most of the communication will be from the people to the higher powers, and that the communication must be acted out in reverent/sacred ceremony and ritualized drama. The plan for ceremonial drama must be well conceived and carried out, leaving no doubt about what the people expect as the desired result. Highly ritualized and dramatized activities, including song, prayer, dance, oral presentation, and symbolic costumes and backdrop, serve as the people's way to communicate with the higher powers. Any ceremonial shortcuts, omissions, lack of attention or mistakes can be perceived by the higher powers as less-than-sincere communication on the part of the people and a sure way to produce failure. The entire package of group-staged ceremonial drama, as acted out to achieve desired results, represents a people's shared sacred beliefs, stories, and events, or in short, these aspects of their culture becomes their MYTH. Together the interaction of myth, ceremony, imagery, place, and participants combine to create the SPIRIT of what people consider to be of greatest long-range value.

To the spirit and myth of Black Point, we must add that the whole complex pattern of geoglyphs, along with the Xam Kwatcan Trail and the sacred mountains, are each part of a larger pictured myth. Black Point is only one of many geoglyph-type outdoor "classrooms," each with variant themes and imagery. Together they form an even greater story for us to comprehend. Once we learn that myth is not a misrepresentation of true or wisdom in Native Culture, and is a desired consequence, we can begin to realize that mythological stories are acceptable. Therefore when some Indian peoples talk of Seven Sisters as our place of origin and perhaps our destiny, we need not feel sorry for what some researchers assume to be naïve, immature, and unprovable beliefs. After all, are not many of the concepts, precepts, and myths of most modern religions, including the concept of heaven, in the same way unprovable beliefs? At least the Indians picked a finite, real, and observable place for their heaven. Care to locate the whereabouts of the Judeo/Christian Heaven?

References

Bierlein, J. F.
1994 Parallel Myths: A Look At The Common Threads Woven Through The World's Greatest Myths, and The Roles They Have Played. Ballantine Books, New York

DuBois, Constance
1908 The Religion of the Luiseno and Diegueno Indians of Southern California. University of California Publications in Archaeology and Ethnology 8 (3) Berkeley

Ewing, Eve and Marvin Patchen
1991 A Summit Path Geoglyph from El Marmol in Northern Baja Mexico. Rock Art Papers 27 (8) 9-16 San Diego Museum of Man

Johnson, Boma
1985 Earth Figures of the Lower Colorado and Gila Rivers: A Functional Analysis. Arizona Archaeological Society 20. Phoenix
1992 A Unique Expression of the Venus Star Symbol Among Petroglyphs on the Lower Colorado River. Unpublished Report on file at the Bureau of Land Management Office, Yuma, Arizona, and with the author.
2003 "Geoglyphs Associated with the Xam Kwatcan Trail in the Palo Verde Point Area, South of Blythe, California". Chapter 11 In: A View Across the Cultural Landscape of the Lower Colorado Desert; Cultural Resource Investigations for the North Baja Pipeline Project. Cleland, James H. and Rebecca McCorkle Apple, Editors. EDAW Inc. San Diego, California.

Kohler, Kenneth O.
1989 Other Truths Not Guesses: The Story of the Sevier Fremont Indians. Classic Printing, St. George Utah

Kroeber, A. L.
1925 Handbook of the Indians of California. Bureau of American Ethnology, Bulletin 78: 663.

Krupp, E. C.
1983 Echoes of the Ancient Skies: The Astronomy of Lost Civilizations. Harper and Row, Publishers. New York

Smith, Gerald A., and Wilson G. Turner
1975 Indian Rock Art of Southern California. San Bernardino County Museum Association, Redlands, California.

Smith, Gerald A. and Steven M. Freers
1994 Fading Images: Indian Pictographs of Western Riverside County, California. Riverside Museum Press, Riverside, California.

von Werlhof, Jay
1987 *Spirits of the Earth*: A Study of Earthen Art in the North American Desert. Vol. 1, The North Desert. Imperial Valley College Museum, El Centro, California.
2004 That They May Remember: Spirits of the Earth, Vol. 2. Imperial Valley Desert Museum Society, El Centro, California.

Waterman, T.T.
1910 The Religious Practices of the Diegueno Indians. University of California Publications in American Archaeology and Ethnology, 8 (6) plate 25.
1920 Yurok Geography. University of California Publications in American Archaeology and Ethnology, 16 (5) 475.

Whitman, Omar and James Varner
1998 A Possible Milky Way Geoglyph at the Olsen Canyon Sites, Corona, California.

CULTURAL RESOURCES OF THE CHOCOLATE MOUNTAINS

Cheryl Bowden-Renna and Rebecca McCorkle Apple, M.A.
EDAW, Inc.
1420 Kettner Boulevard, Suite 620
San Diego, California 92101

Although archaeological investigations in the Chocolate Mountains started a little over 30 years ago, it is investigations in the late 1970s that first thoroughly document resources in the area. Initial reconnaissance work in 1973 was conducted by Charles McKinney for a Geothermal Leasing Program. Unfortunately, McKinney's original records have been lost, but the limits of a National Register District in the southern portion of the Chocolate Mountains were established based on his work.

The earliest defining work in the area was conducted by Jay and Sherilee von Werlhof in 1977 and is still referenced today. The von Werlofs employed a sampling strategy for approximately 67 square miles within the southern portion of the range. The mountain range was divided into eight geographic areas (Figure 1). Within these areas, intuitive and random sample surveys were conducted within the four distinct topographical settings: the upper levels of the mountain areas (above 1,300 ft), canyon mouths (pass areas), terraces, and wash areas within canyons and alluvial fans (low elevations). Table 1 provides a breakdown of the survey areas.

Table 1. Square Miles Examined (after von Werlhof and von Werlhof 1977:B-40)

Area	Total	Mountain	Canyon Mouth	Terrace	Alluvial Plain
Mammoth Wash	9.64	2.8	1.0	4.85	0.99
Salvation Pass	6.41	2.58	1.28	-	2.55
Acolita	7.78	-	1.25	2.75	3.8
Mary Lode	14.62	0.34	1.72	2.15	10.41
McKinney	2.84	-	-	1.38	1.46
Mt. Barrow	3.16	0.63	0.63	1.9	-
Little Mule Mountains	4.37	1.73	-	1.8	0.84
Surveyors Pass	18.68	2.06	2.86	2.63	11.13
Totals	67.5	10.14	8.74	17.46	31.18

Site distribution within each surveyed area was analyzed by elevation to help develop an overview of settlement and subsistence patterns within the Chocolate Mountains. In all, 146 sites were recorded.

The sites were categorized by cultural elements. These included rock circles, rock rings, cleared circles, trails, lithic scatters, ceramic scatters, petroglyphs, geoglyphs, and historics.

Cleared Circles and Rock Features

Rock rings, cleared circles, and rock alignments were found in all four topographical areas. Rock rings and cleared circles were the most frequently encountered features, and sites with these elements were considered occupation sites. On the lower terraces, from approximately 375 ft to 575 ft above mean sea level (AMSL), 30 sites were recorded, with 27 sites reported at elevations between 575 ft and 775 ft.

The majority of sites contained multiple rock features. The number of sites with these features decreased as the elevation increased. The average diameter of rock or cleared features ranged from 1 to 2 m in diameter to 6 to 8 m in diameter with no regard to elevation (Figure 2).

Trails

Twenty-two trails or trail segments were recorded (Figure 3). The majority of trails (n=15; 68%) were located on alluvial plains. Four others (18%) were recorded on terraces, and three (14%) were reported in the mountains.

Lithics

Lithics were generally sparse. Eight lithic scatters were recorded (another 10 sites contained lithics with other materials). Most of the lithic scatters consisted of little more than a core, scraper, or other tool with associated debitage. Half of the sites were recorded on terraces, between 575 ft and 700 ft AMSL. Another 30 percent were found near sea level. The remaining 20 percent were encountered above 1,000 ft. It is interesting to note that all the sites near sea level had associated rock features and/or ceramics, compared to the sites located on the terraces. One of the sites above 1,000 ft had approximately 20 associated flaking stations. Cryptocrystalline and quartzite dominated the observed assemblages.

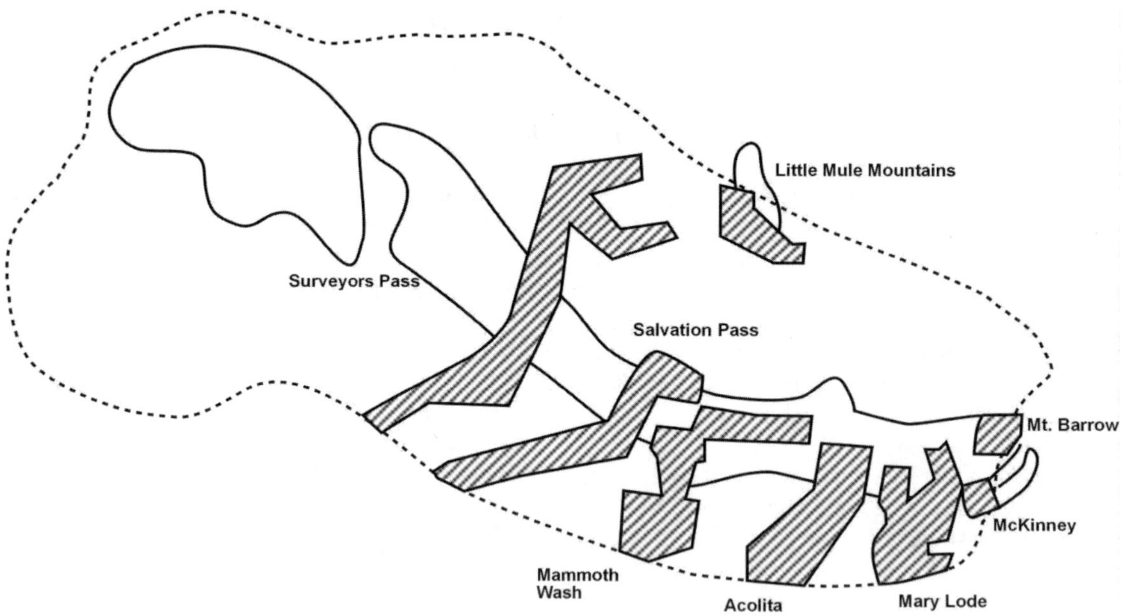

Figure 1. Geographic areas – Indicating eight geographic areas.

Figure 2. Rock ring.

Figure 3. Trail.

Ceramics

Eleven ceramic scatters were recorded, four (60%) located between 300 ft and 700 ft AMSL. Another five were found between 800 and 900 ft. One scatter was documented at 1,700 ft AMSL. Ceramics noted consisted of Tizon Brownware and Buffware with red dots.

Petroglyphs and Geoglyphs

Two geoglyphs were observed during survey investigations. A rock ring is associated with one of the geoglyphs. Both sites were located below 900 ft AMSL (370 ft and 840 ft).

Historics

Three historic sites were recorded during this effort. These consisted of a miner's camp, a military camp, and Tabaseco Tanks. Others, such as the Bradshaw Trail and Pegleg Well, were noted but not recorded.

Summary

The expanse of area covered by the von Werlhofs still comprises the majority of the work conducted in the Chocolate Mountains to date. The majority of the sites they identified were small settlements or campsites consisting of rock features or cleared features. These were primarily found on the pavement-covered terraces. Few sites were noted in the mountains and these were mostly trails. According to the sample survey results it was concluded that the mountains were primarily used as travel corridors. Based on the site densities encountered during the surveys (Table 2), the von Werlhofs concluded that topography rather than geographic areas was the more important factor in the settlement patterns and adaptive strategies of the prehistoric inhabitants of the Chocolate Mountains.

Table 2. Site Densities per Square Mile by Area

More Recent Investigations	
Mammoth Wash	4.77
Salvation Pass	2.49
Acolita	5.14
Mary Lode	2.66
McKinney	5.28
Mt. Barrow	0.63
Little Mule Mountains	0.93
Surveyors Pass	0.21

To date, several smaller, more intensive surveys have been conducted within the Chocolate Mountains (Figure 4). As an indication of the lasting legacy of their work, the results of these subsequent surveys tend to support the range of site types and the general pattern of site locations identified by the von Werlhofs.

References

von Werlhof, Jay and Sherilee von Werlhof

1977 Appendix B Archaeological Examination of Certain Portions of Chocolate Mountains. Prepared for U.S. Naval Weapons Center. Prepared by Imperial Valley College Museum, El Centro, California.

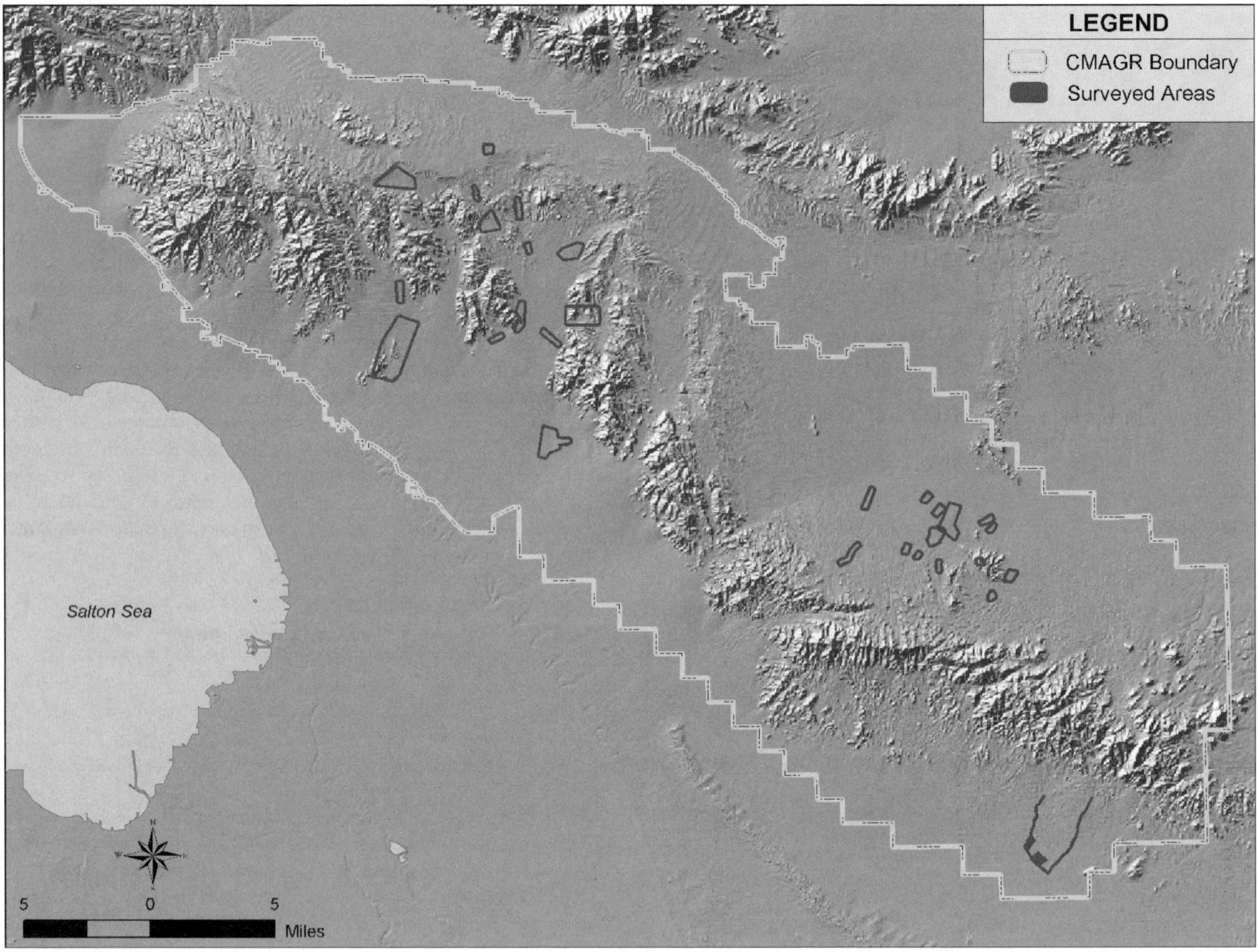

Figure 4. Recorded sites.

EARLY MAN AT SAN DIEGO
A GEOMORPHIC-ARCHAEOLOGICAL VIEW

George F. Carter, Ph.D.
Texas A&M University
Department of Geography
College Station, Texas 77843-3147

Abstract

The relatively stable, slowly uplifted San Diego costal belt has preserved ancient landforms to an exceptional degree. Much of the area can be read as modified by a regressing sea throughout most of the Pleistocene. The glacial and interglacial sea level changes left classic alluvial covers over low sea stand beaches and valley fills recording high sea stands. The latest costal terrace alluvial covers over high sea stand beaches contain an archeological record. The valleys have fills attributed to episodes of high sea stand, and an archeological record is found in these fills also. The record of human occupation runs from an interglacial time into the present.

Introduction

At San Diego a complex set of processes have preserved a very long record of man's presence and allow considerable insight into how much time is involved. There never was any glaciation at San Diego but the effect of the glaciers was felt in two ways: through climatic change and through sea level changes. The climatic changes are recorded in soils and sediments, and the sea level changes are recorded in marine terraces and by valley fills. The variety of interrelationships allows a crosschecking set of observation.

The human record has been preserved by geological processes that left ancient landforms relatively little disturbed, and man's presence at San Diego was assured by the meeting of land and sea. Where these realms meet, man has the better of two worlds. The bounty of the land is supplemented by the wealth of the seashore. Men with minimal equipment could gather roots and shoots and seeds on the land and the shellfish of the intertidal zone along the shore. The junction of land and sea was always of maximum attraction and it was this Man-Land-Sea-Shore interaction at San Diego that helped to create a remarkable record extending from the present back into Pleistocene time.

The Archaeological Record: Two Views

The generally accepted archeological sequence at San Diego for nearly 50 years has been as follows. The record begins about 4000 years ago, later modified to about 9000 B.C. (Rogers, 1966) and presently to 12, 000 years (Masters, 1983). These were the San Dieguito people, as originally defined. Rogers (1966) later used San Dieguito as a catchall term for everything from earliest to relatively late, using I, II, III for ill-defined phases. A very good review of Roger's shifting terminology has been written by Warren (1967) and anyone interested in detailed discussion should consult this source. I will restrict the term San Dieguito here to the classic phase described by Rogers (1929). The classic San Dieguito used fine-grained volcanics for their stonework, made many extremely finely made scrapers, some bifacially flaked knives and some odd crescentic stones. Whether they had the mano and metate is controversial. In my view, they did (Carter, 1977).

They were succeeded by the La Jollan people after 8000 years ago. These people rarely used the fine-grained volcanics but instead depended on the inferior cobbles of quartzite and porphyry so abundant in the region. Instead of fine scrapers, they made crude ones. Bifacial knives of the San Dieguito type disappear. The metate was of great importance in their economy that emphasized seed gathering, much use of acorns, much small game, and moderate amounts of deer-sized game. On the coast there was much use of shellfish. (This is a cultural decline. Or, perhaps a replacement of the earlier people by a culturally less advanced group.)

The historic people differed little from the preceding La Jollan people. They simply added pottery making and use of the bow and arrow. They took up the practice of cremation in place of burial. The original way of life simply was carried on with little change.

I have for some decades put forward a quite difference picture (Carter, 1950, 1951, 1952, 1957a, 1959, 1978, 1980). The Diegueno are the historic people, Yuman in speech and culturally enriched from the east, but otherwise little different from the La Jollan people who clearly preceded them in time. This, so far, is also the generally accepted view.

Differences appear in treatment of the San Dieguito problem. They seem to me to be a very weak impulse reaching this area from the east, a pale reflection of the Clovis-Folsom revolution with its production of beautifully made stone points and with some emphasis on big game hunting, though even for the Great Plans and The East, it is probable that the hunting has been overemphasized. Perhaps few people brought these advanced stonework techniques or perhaps it was only a weak flow of ideas. At any rate, it died out leaving little mark on the old life way.

It is my contention that the basic life way at San Diego for some tens of millennia was fundamentally that of the La Jollans. The San Dieguito phase was then a brief interlude that did not greatly change the La Jollan life way. That the La Jollan life way or cultural pattern is both earlier and later than the San Dieguito has been controversial among local archeologists. Evidence testing the idea of great time depth for the La Jolla pattern will appear below. No one questions its later phase or its virtual continuum into the historic period.

This is a minor difficulty compared to the next step. My contention is that not only did the La Jollan pattern extend far back, but also that it was preceded by still earlier people. No name has been given to these people, and their presence at San Diego is best known from sites in the lower valley of the San Diego River, the Mission Valley. The classic sites are the Texas Street site (Carter, 1957) and the adjacent Buchanan canyon site discovered by Herbert Minshall (1976). Reexamination of the evidence has led to some movement toward greater antiquity, indeed toward vast antiquity (Reeves, Pohl, and Smith, 1986). The conflict in estimates of the cultural sequences and of their age can be tested by considering the geomorphology of the San Diego region. This requires a brief survey of some of the pertinent features of the region.

The Stability of the San Diego Area

The Mesas:

San Diego is marked by broad, relatively flat landforms called mesas. They are festooned with long sand ridges that run roughly north-south and parallel to the modern coast which today is also marked by long sand ridges, our modern beaches. The sand ridges on the mesa surfaces are the beach ridges of an ancient past, when the mesa were formed by marine planation of this area. Their preservation is one evidence of remarkable stability, a record primarily of slow uplift.

The sand ridges are progressively older as elevation increases to the east. One of the marks of their age is the soil phenomena associated with them. The older sand ridges are deeply weathered to bright red, and the weathering increases in intensity and depth as one goes from west to east, from roughly 300 feet to 800 feet elevation. These deeply weathered ancient soils have well developed iron concretions: up to walnut size. Beneath the red oxidized zone lies a pallid zone penetrated by roots casts testifying to a past forest cover. These features indicate that a long period of weathering with accompanying climatic change followed the time of marine planation.

A measure of the age of these soils, which have to be younger than the landforms that they developed on, is the existence of an endemic flora and fauna that is found in the ephemeral ponds associated with the mima mound topography that characterizes the broad areas between the ancient beach ridges. Cemented hardpans are overlain by thin clay-based soils with thinner sandy topping. This sandy topping is piled up in mounds, the mima mounds, with shallow basins in between. The effect from the air is a quilt-like landscape.

In San Diego's brief rainy season the hollows fill with water and the impervious cemented floors of the hollows hold the water for weeks. Then there is an explosion of biota uniquely adapted to feast or famine and able to lie dormant for long periods, even years of little rainfall. Assemblages of endemic species adapted to such special circumstances require immense periods of time to develop. It is a million-year type of phenomenon and has to follow the development of old age soils.

When I escorted Dr. Robert L. Pendleton over these sand ridges about 1955 he exclaimed, "I would expect to see such soils only in some area such as Panama, a land of high temperature and a rainfall of around 50 inches per year." (San Diego's present annual rainfall hovers around 10 inches.) Pendleton earned a Ph.D. in soils science at UC Berkeley under the great Hilgard and was on the original soils survey of San Diego. Thereafter he spent his life in the humid tropics and was acknowledged to be one of the world's authorities on tropical soils and especially old tropical soils, laterites. His observations are not to be taken lightly. The implication is that these landforms, despite their youthful appearance in the Davisian cycle, are very old and their soils faithfully record past climates and vast expanse of time. But this can only be true if certain geological facts are also true.

The San Diego area must be relatively stable geologically. These ancient landforms with their beach ridges preserved could not exist if this area were like the Los Angeles area where the San Andreas and the Garlock fault systems have combined to turn sedimentary strata up at giddy angles. Even casual observation at San Diego shows that Eocene strata are not greatly deformed and are covered by less deformed Pliocene strata which are overlain by Pleistocene forms that show almost no deformation except along the Rose Canyon fault line.

At or near the end of Pliocene time, the region was uplifted to form the mesas. Then, or shortly thereafter, foundering of the blocks between Point Loma and the mesas occurred. The timing of this foundering and the uplift of Mt. Soledad have been little discussed and needs investigation.

The point of all of this is to establish that San Diego is remarkably stable, that the principal thing that has occurred through most of Pleistocene time has been the slow uplift of the region. If Pleistocene time is something like 1.6 million years, then man is surely not present until toward the end of this time, but just when requires attention to younger but equally interesting features in the landscape.

Younger Terraces, Valley Fills, and Alluvial Covers

The vast mesas have tended to hypnotize students of the geology of the area. This has led to neglect of lesser features that also are the work of the sea on a rising coastline. Mount Soledad is so festooned with terraces that it looks, to a geomorphologist, almost like a wedding cake. It has high terraces that correspond to high levels on the Linda Vista Mesa, and these high terraces have equally weathered soils. At lowered elevations there are other terraces and the weathering phenomena on them becomes progressively less as the modern sea level is approached. It is the younger terraces that need more work. The earliest observers tended to see only one terrace.

Geologists tend to explain most everything in terms of crustal movements. For some areas and for vast periods of time, this may be the best explanation. On a seacoast, however, a different situation prevails. Crustal movement, except on major fault lines, is relatively slow. Sea level on the other hand responds to glacial and interglacial changes with fluctuations around 300 to 400 feet from glacial lows to interglacial highs. On a coastline, multiple sea level changes can be assumed. Crustal movements have to be demonstrated.

Thinking about this has changed hugely since my early work at San Diego. It was then thought that there were but four glaciations with three interglacials, the present being a fourth interglacial marked by high sea stands. At that time I could not explain the multiplicity of marine terraces in the area. It is now clear from theoretical considerations and the sediment record cored on the deep ocean floor that the Milankovitch solution is closer to the truth and that there were frequent glaciations and interglacials. This calls for much more actively rising and falling sea levels, and many terraces are expectable. Frequent sea level changes are now established. The best date for a young terrace at San Diego is about 120,000 years (Ku and Kern, 1974). There are at least two younger terraces (Carter, 1957).

With a model of a constantly rising land and a sea level fluctuating with amplitudes set by glaciation and deglaciation, the expectation is for many episodes of high and low sea level stands with geomorphic consequences: wave attack of all headlands and drowning of all the river valley mouths during high sea stands. This must alternate with lowered sea level at which time rivers will excavate their valleys to adjust them to the increased gradient caused by the lowered sea level. At the same, time wave-cut platforms in front of the headlands will be exposed and thereafter covered with alluvium. This explanation is generally accepted and has been applied even in so active an area as Los Angeles (Davis, 1933; Birkeland, 1972).

The foregoing has an interesting crosscheck in soils and sediments. During glacial times large shifts of climatic zones occurred in California. For example, redwoods extended south into the Los Angeles area. From soil phenomena it can be calculated that annual rainfall at San Diego probably was around 40 inches during glacial times. San Diego then probably had a Torrey Pine-Oak forest cover. The Cuyamaca Mountains must have had high rainfall, and the San Diego River must have had a year-round flow with high volume and high velocity since its gradient is relatively steep. All of these factors are readable in the present landscape. The alluvium in the major river valley terraces alternates between heavy gravels with very coarse sand reflecting large volumes of water at high velocities (the glacial condition) and very fine silty sands (the present interglacial conditions).

We are now in a period of high sea stand due to most of the glaciers having melted, and the lower parts of all the streams reaching the sea are sediment filled. Only a few thousand years ago, the rapidly rising sea level of de-glacial time flooded these valleys creating lagoons that rapidly silted up and were during that time exceedingly rich sources of shellfish. Clams galore! The human record of this is found in the extensive accumulations of Indian occupation marked by heavy accumulations of clam shells mixed with stone implements: choppers, scrapers, manos, and metates for grinding wild seeds, and occasional sear or dart points, but no arrow points or pottery, for these items appeared at San Diego only belatedly. The interglacial relationship of land, sea, glaciers, and man is obvious here, and time begins to enter the human equation.

Man, Land, Time: The Clam-Rich Lagoons

The land, sea, glacial, sea level equation says that these shell middens (the accumulated debris of human occupation) should date to the time of rapid sea level rise at the end of the last glaciation. The sea level rose rapidly after 10,000 years ago to reach its present level about 4000 years ago. This rapid rise was far faster than the rivers could fill their valleys by sedimentation even in the restricted areas of their lower narrow valleys. This was particularly true of the lesser streams. This was also a time of decreasing rainfall and decreasing stream flow. These conditions should have occurred each time that the sea rose rapidly at the end of the glacial period, and if man were present at such times the record should be in the older valley fills. This point will be returned to later. When the rapid sea level rise after 10,000 years ago created an ecological wealth, man immediately took advantage of this as is recorded by the extensive shell middens around these lagoons. Middens should date to this period of rapid sea level rise, and at falling in the terminal period of the glacially wetter conditions at San Diego. This evidence is subject to cultural crosscheck. These people, the San Dieguito, lacked pottery and arrow points, made bifacially flaked heavy knives, and had small cresentic bifacially flaked objects of uncertain function. These crescents are important for they are also found in the Mojave Desert area where they overlap the appearance of fluted points. Fluted points mark the Clovis-Folsom period, which falls in minimal glacial time—roughly 9000 to 12,000 years ago.

We have an abundance of late sites, those of the Dieguено People. The sites are, as Rogers noted, located near modern water sources. They are black, loose and fluffy in texture. They have arrow points and pottery and evidence that the people practiced cremation. They are a clear record of recency both culturally and in terms of climate and of soil phenomena. For the La Jollans the evidence from their shell middens indicates considerable age, for some middens are colorless, have strongly carbonate-cemented lower strata. At times young Huerhuero-type soils have developed on their village sites. These soils are characteristics of the soils of the intermediate-aged younger terraces. This evidence would seem to put the early La Jollans at least into a time frame comparable to the San Dieguito.

If this seems to be beating the question of the time and earth science records and the associated cultural evidence almost to death, it has a purpose. These relationships are unchallenged, noncontroversial. Material to be presented now moves out into controversial realms, but the application of earth science observations will remain much the same. The times, however, will become lengthy and the involvement of man will enter controversial realms.

The Narrow Shelf and Submarine Archaeology

For man possibly living at San Diego as early as 12,000 years ago, the sea level would have been considerably lower than the present. It would be expectable that man would then have lived along the seashore just as he did in later time. A favorable circumstance has actually made it possible to find some of this record. At La Jolla submarine canyons extend almost into the outer breaker line. At these spots the shore line moves very little as the sea level changes, for a lowering of the sea level would only move the seashore to the lip of the canyon. Whether sea level was high or low, man would be attracted to this area, and the record should be clear, if obtainable.

For many decades, swimmers have found archeological material on the ocean floor in this area. Stone bowls, metates, and eventually even stone stools were found (Moriarty and Marshall, 1964, 1983). The deepest items were at a depth of 100 feet and virtually on the lip of the submarine canyon, and some were found within the canyon. Man could live no further out at this particular spot.

Alluvial Covers Over Marine Platforms

If man were living at San Diego while sea level was down due to the removal of water to feed the glaciers, where would one look for such evidence? The prologue briefly presented the case for a slowly rising, little faulted block at San Diego against which the rising and falling sea levels of the glacial period etched their record. What is the record of lowered sea level, and under what circumstances is that record made available?

When the sea level is down, the platform cut into headlands by the previously existing high sea stand is exposed. This a gently seaward-sloping platform littered with the debris of the old sea floor. Just such a platform lies off the west side of Point Loma and La Jolla today. A returning glacial epoch could lower the sea and expose this area. Erosion of the land would then begin to cover this flat zone. If man were living at San Diego during the time of lowered sea level then the record of his presence should be enclosed deep in the sedimentary accumulation on the old marine platform. To test this, one needs exposures through these alluvial covers for examination.

The present high and rising sea has truncated the alluvial covers that accumulated over the broad bench cut by the sea in the last glacial epoch and has exposed miles of cross sections. This is true not only at San Diego, but also all along the California coast and on the offshore islands. A classic example existed on the Scripps Institution of Oceanography campus just north of the pier and a lengthy exposure formerly was available for examination for a quarter of a mile to the south of the campus. Both areas are now hidden by sea walls and housing.

On the Scripps campus the alluvial cover had a shell midden at the surface that was dated at 2000 years on the surface, and 7000 years at the base. It was a typical La Jollan midden with a dearth of developed tools, but a quantity of percussively broken rock. Beneath this, hearths were exposed at varied depths in the alluvium. These were dated by C^{14} to the limit of the C^{14} method at that time. That they actually were hearths was accepted by such experts as Alex Krieger, Luther Cressman, and others. Luther Cressman has given a particularly useful description of his going to Santa Rosa Island as a skeptic and being converted by the evidence for man in the alluvial covers there. He then visited Carl Hubbs at the Scripps Institute of Oceanography and accepted the hearths there. Thence he went to Texas Street and examined a hearth, giving his reasons for considering it a hearth (Cressman, 1977).

Collapse of the cliff dumped out a Pleistocene horse skeleton, which wasrescued at risk of life and limb by Jeffrey Bada and his graduate students. This was dated by Bada using the protein racemization method to 50,000 years. Although Bada's method has been criticized, this date is in keeping with the C^{14} ages on the overlying hearths, and with the obvious geomorphic age of the formation.

Such finds are not unusual in the alluvial covers along the Southern California coast. I have reported repeated findings of manos and metates in similar geomorphic positions around La Jolla and on Point Loma. Similar finds of flaked stone near mammoth bone has been made near Oceanside.[1] Cooked elephant has been found in comparable alluvial covers on Santa Rosa Island (Orr and Berger, 1966; Carter, 1987; Berger, 1971), despite challenges (Cushing et. al., 1986; see Carter 1987 for a rebuttal). Berger reported finding a hearth with a mammoth leg in it that was accompanied by stone tools and abundant charcoal that was radiocarbon deficient (greater than 40,000 years). The C^{14} confirms the Santa Rosa Island work, and fits exactly into the geomorphic picture at San Diego.

---

[1]Photos and field notes sent me by Richard Cerruti of the Natural History Museum. The archaeologist from San Diego State University called in put in his notes. "I wish someone would explain this." The geomorphology is simple. It is an alluvial cover over an interglacial marine platform. The only question concerns which interglacial beach it covers.

The material from all of these sites fits the La Jollan in a general way. Local cobbles are used for the production of flakes that are used with little modification. The metate is the basic food-processing instrument. It is abundantly clear that men of La Jollan cultural type occupied Southern California during much of the last glacial epoch. But this implies that man was already south of the great continental ice sheet. While it has become evident that man could have followed a coastline pathway from Alaska even during glacial time (Fladmark, 1979), there is evidence that man was indeed present in America, and at San Diego, even during interglacial time.

Interglacial Man at San Diego

At San Diego there is evidence that man was present not only during the last glacial times, but in interglacial times also. For this evidence we turn to the river valleys. Preservation of a record usually requires sequential deposition. For observation, one needs exposures. In river valleys near the sea, deposition occurs at times of rising and high sea levels. If in early post-glacial time the river valleys were areas of attraction to man, and this is evident from the location of the late Indian villages in the river valleys, they surely would also have been attractive during earlier similar times.

The present is a time of aridity at San Diego, which actually has a steppe (semi-desert) climate, and water supplies are scarce and concentrated in the major river valleys. If man were present during a past episode of such conditions, it is in the major river valleys that one would have the best chance to find the record. The valleys at San Diego are or were (many have been removed by road and real estate developments) rich in valley-flanking terraces, the remnants of ancient valley fills. With a little bit of luck, one could find in one of the engineering exposures evidence of man in the ancient valley fills preserved in the terraces. The situation would be exactly comparable to the alluvial covers over the interglacial beaches.

The Texas Street Site and Buchanan Canyon

I have described at some length the accidental discovery of evidence for man in a valley terrace of the San Diego River (Carter, 1957, 1980). I entered the Bond Pit (a borrow pit used to obtain fill dirt for various works) to take advantage of an exposure of an interglacial valley fill. As a footnote to San Diego State University history, Alvena Suhl Storm marched her geography class of 1931 down Texas Street, then a little wilderness, and up over this obvious alluvial terrace. That was my freshman year, and I was in that class. Nearly 20 years later, in pursuit of soils, terraces, and time in relation to dating early man in America, I was to find here one of the most controversial archeological sites in America.

This valley fill would be most interesting even if there were no human involvement. At the base there are coarse sands and heavy gravels. This is clearly a record of a glacial-pluvial time. The San Diego River was then carrying a load totally out of keeping with present. The present bed load is fine silty sand. Above the coarse sediments just such fine silty sand appears, clearly marking a climatic shift toward aridity. The deposit is at the mouth of a small steep canyon and coarse gravel from that canyon is interbedded in fine silts. At some time the stream from the canyon cut a gully through the deposit. This would be an appropriate action to occur at a time when the San Diego River valley was deepened during a glacial period. But it could also have occurred due to a local cloudburst. Thereafter, a strong Huerohuero soil profile developed on the shoulder and flanks of the gully. It is not a young feature. The surface of the terrace, once freed from deposition, developed a weak mima mound topography never seen on the youngest terraces in the region.

There is a third riverine deposit that has never been adequately dealt with. This is a coarse red sand deposit. It probably represents a glacial pluvial time when it was exposed to lengthy surface weathering in order to be oxidized. If it represents a glacial-pluvial riverbed separate from the similar unoxidized and exposed at the base of the excavated zone and separated from these sands by fine silty deposits, then the age of the fine silty sediments with their enclosed human record would become even older than prior estimates. My failure to solve this problem is because of the site being but part of a larger study and of my having lived either a whole continent or half a continent away for the past nearly 50 years.

The borrow pit was benched to control erosion when abandoned. There are three artificial levels, and they are used as reference bases for the archeology. The original owners were the Bonds, and as boys they found manos and metates on the surface of the site. Amateur collectors since then have found dart points in the same area. No pottery or arrow points have been found. This indicates a late La Jollan type occupation of the type found around the lagoons, which can date to 8000 years ago. At a depth of 20 feet, hearths marked by circles of stone and with ash and charcoal-stained soil were exposed just east of the church on this site. In one of them a typical La Jollan split cobble tool with unifacial flaking was found. Herb Minshall tells me of his finding other such artifacts at this level. These are clearly early La Jollan of some considerable age.

Artifacts and Hearths from Beneath the 20-Foot Bench

From the face of the second bench and the upper part of the third bench (the lowest of the three), hearths and artifacts occur widely both laterally and in depth. They are evidence that man lived on the valley floor as it built up, an interglacial phenomenon. The attraction may have been the gully with its supply of cobbles for making tools. The artifacts have been widely attacked, but more often than not by critics that have never been on the site nor ever handled the objects in question. The artifacts were instantly accepted by John Witthoft (1955), a pioneer in lithic technology. More recently Barney Reeves of Canada, after three field sessions in San Diego, has accepted both the dating and the artifacts and thereby the evidence for early man (Reeves, 1977; Reeves, Pohl and Smith, 1986). The characterizing artifact is a cobble with either a natural or a prepared platform from which long parallel-sided flakers, technically blades, have been struck. There are other artifacts, notably cleaver-like heavy items, resembling a tool called a skreblo in Siberian archeology, a resemblance noted by Herb Minshall (1974, 1975, 1976, 1986). And there is much use of sharp-edged flakes and cores. There is an absence of manos and metates, bifacially flaked points, or any other tool typical of any of the later people.

Archaeologist critics have at times questioned the age of the site. No geologist, geomorphologist, or soils man who has been on the site with me has ever suggested that it was a recent deposit. Some have suggested that it could be older than my estimate of 100,000 years. It is a quite obvious valley fill that has been left as a valley-flanking terrace. Its history may be more complex than I thought, and if so, it is somewhat older.

The artifacts are attacked as the work of fire on cobbles, the action of tumbling in a violent stream flow, the work of faults on cobbles cemented in indurated sediments. I did experimental work on fire and its effects on quartzites, a favored material for tools here. The action of natural fires is also obvious in the present landscape and spalls are created that are superficially like man-struck blades. But there are clear diagnostic differences. Fire reddens rock. The spalls tend to follow the contour of the rock; there are no prepared platforms, no striking platform, no eraillures, force lines, or bulbs of percussion in fire-spalled rock. Human work is marked by the absence of the fire marks and the presence of the attributes of percussive flaking, eraillures, etc. The works of man and of fire are easily separated.

Tumbling in torrents is easily disposed of as a hypothesis. A survey of the geomorphic literature produces no support for such percussive breakage of rocks, but quite the contrary. The flat statement is found that a mountain torrent in full flood has one-tenth the energy needed to do such work (reviewed in Carter, 1980, p. 87-132). Herb Minshall put this to the test in nearby Buchanan Canyon. During a local cloudburst when the canyon was filled with the sound of roaring waters and the rattling of boulders in the torrent pouring down the steep gradient of the canyon, he stepped into the maelstrom and felt no impact more forceful than as if a fish had nuzzled his rubber boot.

Whence then the noise? Flume experiment (Carter, 1980, p 102, citing Schumm and Stevens, 1973) has shown that at a critical velocity cobbles become subject to lift. This is the same force that drives a sailboat. No one who has sailed has anything but great respect for the power generated by a curved piece of canvas in so light a medium as air. Immensely denser water flowing over the curved surface of cobbles lifts them. Since the water flow is turbulent, the velocity varies greatly. The stones are then set to dancing up and down. By actual measurement they move great distances vertically while being displaced laterally hardly at all. They wear with great rapidity but they flake, break, or shatter hardly at all. The reality is the opposite of the usual archaeological expectation. Instead of cobbles being broken to facsimiles of artifacts, the reverse is true. Artifacts entrained in such a mill are rapidly reduced to ground gravel. For an extreme test of this, note sometime what happens to glass in a beach. This ultrabrittle material is rapidly dulled, frosted, and turned into beach gravel.

That leaves the claim of fault action on cobbles cemented into indurated layers. It is an odd argument to advance in an area that in general lacks frequent evidence of faulting. The indurated areas so strong that cobbles shear where machine tools break such formations are limited to the mesa surfaces. They are startlingly devoid of fault fractures. At the Texas Street site, there are no known faults, and the deposits are unconsolidated. The claim has its amusing side. How easy for early man! He could haunt the minor faults in search of micro blades, and when in need of macro blades, he could visit the San Andreas Fault. It is a nonsense argument without any observational basis.

Other Sites

There are or were many ancient sites at San Diego. The mesas retained on the surfaces every hearth and artifact ever dropped there. Huge collections could have been made, and an occasional vocational archaeologist has done so. Richard Cerutti probably has the best collection. Herb Minshall had a small but very select collection.[2] The canyons were the easily accessible source of supply of cobbles for the early men as well as the later men. Buchanan Canyon, discovered and reported by Herb Minshall, is a classic example, but there are innumerable others. Many of the valley fills contained a record duplicating that at Texas Street. When Fashion Valley Road in Mission Valley was built and the terrace in that area bulldozed away, the resultant disturbed area was strewn with cobbles worked in a fashion identical to those at Texas Street. As B.O.K. Reeves (Reeves, Pohl, and Smith, 1986) has commented, a priceless record was lost through the studied ignoring of the evidence.

La Jolla Shores and Crown Point

No discussion of early man in San Diego can be complete without at least mention of two enigmatic sites, La Jolla Shores and Crown Point.

When Malcolm Rogers was first getting interested in archeology and was still an orange grower in the Escondido area, word reached him that real estate development of La Jolla Shores was under way and that in the process skeletons were turning up at some depth and

[2]The Minshall collection is now at the San Bernardino County Museum.

in an old landform. He visited the site as often as possible, and the engineer in charge gave him every assistance. I have reviewed this material courtesy of the San Diego Museum of Man staff, particularly Rose Tyson, but have not published on it. I have vacillated in my analysis of this feature but would tend to place it in the high sea stand between the two Wisconsin Peaks of glaciation (20,000 and 60,000) and hence about 40,000 years ago (Carter 1957, 1980, p. 169-170 for revised thinking). There are accelerator mass spectrometry dates for some of this material, but critical discussion of them is too complex for this essay.

Crown Point has another such record. First, Crown Point has not at all a simple formation. Landward it has moderately old soils, indeed sequences of them. The outer end has two shell deposits recording repeated bay bar building southward from Mt. Soledad into the Mission Bay area during high sea stands. The older shell deposit is cemented and highly eroded. The younger shell deposit is not. The present sand cover overlies the later shell deposit and represents a time when much of the Mission Bay area was an embayment being filled with silts from the San Diego River, just as the present Mission Bay was being filled with sediments in recent times, until dams on the San Diego River decreased the rivers flow and now dredging has reversed the filling. Mission Beach, of course, represents bay bar building at the present sea level.

I found a hearth (a circular arrangement of large fire-spalled boulders) at the base of the late sand deposit of Crown Point, and a mano cemented into the top of an eroded section of the cemented shell reef, obviously younger than the time of erosion of the reef but prior to the deposit of the upper sands of Crown Point. During installation of sewer lines, two metates were found right on top of one of the shell deposits and reported to the San Diego Museum of Man. Malcolm Farmer, then the director of the museum, told me of these finds. There is further archeology within the sands on Crown Point, and I located a burial site on the east side of the point and showed it to Dr. Clement Meighan of UCLA, but nothing was done. So, poorly studied, poorly reported, but very definite evidence of the presence of man is or was observable at Crown Point at a time when the sea level stood higher than now. As the continuum of archeology up through the upper sand indicates, man long continued to visit Crown Point. It was not an area of permanent occupation in later times for it had no water supply. It was handy to the rich clam flats of the Mission Bay, but the nearest water was at the mouth of Rose Canyon, and a huge archeological deposit of at least three ages is located there (Carter, 1957, p. 294). The tragedy is that few areas in the Americas had so rich a record of early man and that this one has been so sadly neglected.

On Early Man in General

Our knowledge of the origin of man and of man's spread over the face of the earth has been changing with great rapidity, and some of this reflects on man and America, and San Diego. Stone toolmaking now seems to have appeared as early as three million years ago. Fire was most probably in use by at least one million years ago. Most startling and significant for man in America is the Russian claim that man was living near the cold pole of the northern hemisphere in the northeast Siberia as early as one million years ago (Daniloff, 1986). At this date one has to be dealing with Homo erectus. The temperature at nearby Verkoyansk in January averages minus 58 degrees F. To survive, man would have to have fire; shelter; foot, hand, and head gear. The tools illustrated are simply pebble tools, cobbles with a flake or two struck off. The site is said to be beneath the Bruhnes magnetic polar reversal. I have seen only popular accounts of the Siberian evidence, but it is evidence that deserves watching.

We have never dreamed that Homo erectus had equipment such as I have noted above. But I am told that more such data is emerging in northwestern Europe on something like a 500,000-year time level. In Southern California we have the as-yet-untested claim of the killing and dismembering of a mammoth in the Borrego paleontological beds at the seemingly incredible date of 400,000 years (L'Honnediu, 1988). For America it should be noted that the Bering Strait would be like Miami Beach compared to Verkoyansk. The possibility of Homo erectus reaching America is very real, if these recent reports stand up. The absence of any Homo erectus skeletal finds is simply negative evidence, and given the American archaeologists' aversion to looking in early formations, weak evidence indeed.

There are indications that the winds of change are beginning to stir. There is some acceptance now of sites on the 30,000- and perhaps 50,000-year level (Dillehay, 1986, 1989; Bryan, A. manuscript, 1991; Guidon, 1987). But there are sites out there with suggestions of time on the 100,000-year scale, and the problem may become only that of determining how many 100,000 years.

I have for more than 25 years maintained an interest in the Calico site in the Mojave Desert. The artifacts there are so fine that it will be a classic study for someone to review their rejection over the past quarter century. Momentarily they are being accepted, but at times with the cry from some that "they are too good to be that old" (correspondence with Emily Haury). But this is demonstrably false, and how odd that objects that were only naturally broken rocks for decades have suddenly become too good to be true.

At the Calico site artifacts have been dated from the carbonate rinds on artifacts at 200,000 years (Bischoff et. al., 1981); this is a uranium series date. I have participated in recent years in three field trips into the Manix Lake beds in the Calico site vicinity. We found artifacts similar to the Calico site artifacts 20 feet beneath a volcanic layer in the lake sediments. This layer has been dated by K/AV at 185,000 years. A very thorough review of the Manix Lake area in relation to this site and a great deal more is to be found in Fred Budinger's unpublished Master of Arts thesis at California State University, San Bernardino, 1991.

In the Old World, Louis Leakey estimated the age of the Olduvai Gorge site as 200,000 years. A test of the lava at the base of the site yielded 1,7500,000 years. This was greeted with incredulity. Fission tracking, a system completely independent of the K/AV dating method, was then tried and it gave a confirming date. This turned Old World archaeology around. The dates at Calico present a similar case. They are by unrelated systems but come out with comparable dates. If these dates are sustained, and they are seemingly unchallenged (indeed Oberlander (1982) has used other data that tend to confirm these data), then the whole picture of man in America is overturned. Texas Street and Buchanan Canyon and the record in the alluvial covers over the last interglacial beaches are NOT unexpected.

For San Diego it suggests that energetic work on the uniquely preserved ancient landforms datable by their relationships to the changing sea levels and preserved on a little deformed but lowly rising coast line could lead to important discoveries of the type foreshadowed by the Texas Street site.

REFERENCES

Birkeland, P. W.

1972 Late Quaternary Eustatic Sea-level Changes Along the Malibu Coast, Los Angeles County, California, Journal of Geology, 80, 432-448. Marine terraces correlated with valley fills due to eustatic (glacial) high sea stands. A careful modern study.

Bischoff, J. L., R. J. Shelmon, T. L. Ku, R. D. Simpson, R. J. Rosenbaurer, and F. E. Budinger, Jr.

1981 Uranium-series and soil geomorphic dating of the Calico archaeological site. Geology, 9:576-682.

Bryan, Alan L.

N. D. The Fluted Point Tradition in the Americas—One of Several Adaptions to Late Pleistocene American Environments." M.S. in page proof courtesy of Alan Bryan. This is of great value not only for the date presented but also for its examination of the critics of pre-Clovis finds. Their methodology is shown to be non-scientific.

Budinger, Fred E.

1991 "The Geoarchaeological Potential of the Manix Type Section, Central Mohave Desert, California" M.A. Thesis, 1991. California State University, San Bernardino.

Carter, G. F.

1950 "Evidence for Pleistocene Geographical Review," Vol. XL, pp. 84-102. This early paper details the stratigraphy of the alluvial cover south of the Scripps Institution of Oceanography.

1951 "Man in America: A Criticism of Scientific Thought, *Scientific Monthly*, Vol. LXXIII, pp. 297-307.

1952 Interglacial Artifacts from the San Diego Area. Southwestern Journal of Anthropology, 8(4): 444-456. This reports the initial find of interglacial artifacts from the Sweetwater River gravels.

1957 "Pleistocene Man at San Diego," The John Hopkins University Press, Baltimore. A book length treatment of the geomorphology of the San Diego area and the place of the archeology in that system. Geology, soils as recorders of time and climatic change, terraces and sediments are discussed at length. This is the basic reference for statements on such data for this discussion.

1958a "Is There an American Lower Paleolithic?" Miscellanea Paul Rivet, Octogenario Dicata, Vol. I, XXXI Congress Internaceonal de Americanistas, pp. 227-243, Universidad Nacional Autonoma de Mexico.

1958b "Archeology in the Reno Area in Relation to the Age of Man and the Culture Sequence in America." Proceedings of the Am. Phil. Soc. 102:174-192. This expands the archeological sequence and relates it to Pleistocene Lake Lahoutan.

1959 "Man, Time, and Change in the Far Southwest." Annals Association of American Geographers. 49(e):4-33. A broad view of the picture accompanied by counter views from Alex Krieger.

1977 "The Metate: An Early Grain Grinding Implement in the New World," pp. 693-712, in, *Origins of Agriculture*, Charles Reed, Editor, Munton, The Hague.

1978 "The American Paleolithic," pp. 10-19 in, Bryan, A.L., Editor, Early Man in American From a Cirum Pacific Perspective. Occasional papers of the Department of Anthropology, No.1, University of Alberta, Edmonton, Canada.

Carter, G. F.

1980 "Earlier Than You Think," Texas A&M University Press, College Station, Texas. An update and restatement written just before the publication of the 200,000-year uranium date for the Calico.

1987a "Fire, Mammoth and Man on Santa Rosa Island." A discussion of Cushing et. al., Pacific Coast Archaeological Society Quarterly, 23(3):58-62.

1987b "Opposition to the Pre-Clovis." Pacific Coast Archaeological Society Quarterly, 23(3): 63-71.

Cressman, L. S.

1977 "Prehistory of the Far West." University of Utah Press, Salt Lake City, Utah, pp. 67-70

Cushing, John, A. W. Wenner, E. Noble, M. Daily

1986 A Groundwater Hypothesis for the Origin of "Fire Areas" on the Northern Channel Islands, California Quaternary Research, 26:207-217, Academic Press, New York.

Daniloff, Ruth
1986 "Digging Up New Theories of Early Man." U.S. News and World Report, Sept. 1, pp. 62-63. Reported as geologically dated at 1.8 million years, plural sites, 2500 stone tools that resemble those from Olduvai Gorge, Kenya, Africa. The tools (illustrated in color) are duplicated at Buchanan Canyon, Texas Street and in early sites in the Imperial Valley. The article contains useful discussion by several Russian authorities.

Davis, W. M.
1933 "Glacial Epochs of the Santa Monica Mountains, California." Bull Geology Society of America, 44:721-734. The title conceals the subject—marine terraces interpreted as due to glacial-interglacial; records on a rising coast. Also see Birkeland.

Dillehy, T. D.
1986 "The Cultural Relationships of Monte Verde: A Late Pleistocene Settlement Site in the Subantartic Forest of South Central Chile," in *New Evidence for the Pleistocene Peopling of the Americas*, edited by A. L. Bran, pp. 319-337. Center for the Study of Early Man, University of Maine, Orono.

Dillehy, T. D.
1989 "Monte Verde." *A Late Pleistocene Settlement in Chile*, Vol. 1, Paleo-environment and Site Context. Smithsonian Institution Press, Washington, D.C.

Fladmark, K. R..
1978 "The Feasibility of the Northwest Coast as a Migration Route for Early Man." *Early Man in America from a Circum-Pacific Perspective*. Alan Bryan, editor 119-128. Occasional Papers No. 1, Department of Anthropology, University of Alberta, Edmonton.
1979 "Routes: Alternative Migration Corridors for Early Man in North America, "American Antiquity 44:55-69.

Guidon, Niede
1987 "The First Americans." Natural History, 96(8):6-12.

Hubbs, Carl L., B. S., Bien, and H. E. Suess
1962 "La Jolla Natural Radiocarbon Measurements IV," in Radiocarbon 7:66-117.

Ku, T. L. and Kern, J. P.
1974 "Uranium-series Age of the Upper-Pleistocene Nestor Terrace, San Diego, California." Geological Society of American Bull, 85:1713-1716.

L'Hommedieu, Helena
1988 "Evidence Indicates Man Was in Area 500,000 Y. B. P. " *Borrego Sun*, 37(13)3-4.

Masters, Patricia M.
1983 "Detection and Assessment of Prehistoric Artifact Sites Off the Cast of Southern California," in Quatemary, coastline and Marine Archaeology, P. M. Masters and N. C., Fleming editors. Academic Press, London, New York. Thirty-four underwater archeological sites in Southern California are reported. Settlement on the shelf when sea level was lower 6000 to 4000 years ago is suggested.

Milankovitch, M.
1936 "Mathematische Klimalehre und Astronomische Theorie der Klimas Swankungen." In Handbuch der Kliamtologie, Vol. 1, Berlin. Verlag von Gebruder Bornfrager.

Minshall, H. L.
1974 "Early Man Sites at Texas Street and Buchanan Canyon in San Diego, California," Anthropology Journal of Canada, 12:18.
1975 "A Lower Paleolithic Bipolar Flaking Complex in the San Diego Region: Technological Implications of Recent Finds." Pacific Coast Archeological Society Quarterly, 11:45-56.
1976 "The Broken Stones," The Copley Press, San Diego California. This popular but well-informed work presents both data and the history and personalities involved. As with all the works cited here, there is an extensive useful bibliography.
1986 A comparative study of some apparently very early updated assemblages from western North American and similar Chinese early Paleolithic industries. Pacific Coast Archaeological Society Quarterly, 22(2):1-19.
1989 Buchanan Canyon: Ancient Human Presence in the Americas. Stawson Communications San Marcos, California. Of particular value for comparisons with the Chinese Paleolithic based on Minshall's visit to China.

Moriarity, J. R., and Marshall, N. F.
1964 Principles of Submarine Archeology, Pacific Discovery, 17(5):18-25.

Orr, Phil C. and Rainer Berger
1966 The fire areas on Santa Rosa Island, California, Proceeding of the National Academy of Science, Part 1, 56(5): 1409-1416; Part w 56(6):1678-1682, Washington, D.C.

Reeves, B. O. K.
1977 Mission Valley-Mt. Soledad Early Man: Geological-archaeological Sites and Sections—Descriptive Listings. Manuscript on file with the San Diego Museum of Man. Of particular interest for the early conclusion for very early evidence by a geologist-archeologist.

Reeves, B. O. K., J. D. Pohl, J. Smith

1986 The Mission Ridge Site and the Texas Street Question. In Pleistocene Peopling of the American Series, Alan Bryan, editor, Center for the Study of Early Man, University of Maine, Orono. The Texas Street artifacts are accepted and a cultural continuum from mid-Pleistocene to the historic period is claimed.

Rogers, M. J.

1929 "The Stone Art of the San Dieguito Plateau," American Anthropologist, 31(3):454-469.

1966 *Ancient Hunters of the Far West.* The Union-Tribune Publishing Company San Diego, California. This posthumous work by Rogers with contributions by Wormington, Davis, and Brott is invaluable for its record of thought at that time. The great difficulty in using it lies in the shifting terminology. San Dieguito becomes a term applicable from the most ancient to the appearance of late type dart points.

Schumm, S. A., and M. A. Stephens

1973 "Abrasion in Place: A Mechanism for rounding and size reduction of coarse sediments in rivers. Geology, 1:37-40.

Warren, Claude

1967 "The San Dieguito Complex: A Review and Hypothesis," American Antiquity, Vol. 32, pp. 168-185.

Withoft, John

1955 "Texas Street Artifacts," New World Antiquity, Part 1, 2(9) pt. 1:132-134, Part 2, w(12):179-184.

THE INTERRELATIONS OF ARCHAEOLOGY AND GEOLOGY IN THE YUHA DESERT

W. Morlin Childers
Curator of Paleoanthropology
Imperial Valley College Museum
El Centro, California 92243

ABSTRACT

Outwash from heavy precipitation in the Coastal Mountains exposed a cross section of Palm Spring Formation and stratified alluvial materials. Eighty specimens of flaked stone and flakes that are the technological work of man were collected, after being photographed and documented in place, near the base of the exposure. The earliest material was taken from 25 meters below the upper surface, a few cm above the Palm Spring Formation within the exposed strata of the Yuha Pinto Wash. The artifacts are core tools and flakes reminiscent of Middle Paleolithic materials found in the Old World (Brace 1971). In order to realistically estimate the age of the earliest archeological material exposed in the Yuha Desert, it is necessary to determine at least a minimum age for the geological deposit in which they are located.

GENERAL GEOLOGY

The general geology of western Imperial County, containing the Yuha Desert, has been well described as "A crystalline complex of pre-Cretaceous metamorphic rocks form the basement of which Miocene volcanic rocks and younger sediments lie unconformably. These sediments are marine sands and yellow clays (Imperial Formation), age Pliocene (Hanna 1926) silts with fine sands and red-brown clays of the Colorado Delta, (Palm Spring Formation), age Pleistocene (Woodring 1935) and fanglomerate of local origin (Canebrake Formation), age middle to late Pleistocene" (Dibblee 1935).

One of the main physiographic features is the flank of the southern California Peninsular Ranges, which rises in rugged relief above the west side of Imperial Valley. The margin of the valley, in this region termed the Yuha Desert, is low hills and intricately eroded alluvial badlands (Yuha Badlands). An erosional, nearly flat swale or valley 3 to 5 kilometers wide parallels the mountains on the western side of the Yuha Desert (Childers 1974). Historically the term Yuha Desert has referred to the southwestern extremity of the Salton Sink, east of the Peninsular Range, north of the International boundary and south of Coyote Wash, with the eastern boundary being the valley farmland; however, in this report the area is extended to include the adjacent area in Mexico to 8 kilometers south of Highway 2.

> The Yuha Alluvial Formation consists of coarse and fine-grained, well-rounded to angular boulders and cobbles, that vary in size from 1 inch to 14 inches in diameter. Grey to green angular prophyry is the most prevalent stone. The more rounded fist size and larger cobbles in the Yuha Desert are usually older fine-grained volcanics [Poway type referred to later in this paper]. Granites are poorly preserved and usually can be shattered easily with a rock hammer. Decomposed granite and reddish clay materials serve as a matrix for the Yuha Alluvial Formation. The upper 6 to 30 inches of the Yuha Alluvium is well cemented with a calcium carbonate caliche. Exposed cross-sections of the Yuha alluvium reveal the nature of this caliche deposit. The basic structure is horizontal lensing with vertical columns up to 4 inches in diameter bisecting and protruding a depth of as much as 30 inches into the alluvium. (Childers 1974.)
>
> The columnar caliche structure represents calichified root casts (Stanley 1966).

The Yuha Desert area represents the most difficult type of geologic problem, one that is detailed over a large area. Five independent but interrelated variables have to be studied simultaneously: alluvial fan formation, lakeshore and lake deposit formation, faulting, doming, and erosional processes. The following comments are at best a preliminary sketch of some of the interrelationships.

Elevations of the Yuha Alluvial Formation range from 150-foot a.s.l. in the basins to more than 1,2000-foot a.s.l. near the Coastal Range, which overlies the Palm Spring Formation and older Imperial Formations (Childers 1974).

The underlying Imperial and Palm Spring Formations have been moderately to intensely deformed by folding and faulting within the Yuha dome. Sometime following the deformation, there was a period of relative quiescence when streams draining northeastward from the mountains beveled a gently sloping plain (a pediment across the deformations.) Alluvial fan deposits (the Yuha Formation) accumulated on this pediment. At about the same time, the 150-foot a.s.l. Lake LeConte beach was established in the Salton Sink.

Subsequent changes, probably the withdrawal of the 150-foot lake, resulted in incision of the pediment so its remnants now stand as flat-topped ridges within the Yuha Wash and are referred to as the Yuha Terrace.

The Yuha and Coyote Wash Terraces existed as drainage basins that impounded drainage and created transient lakes or ponds behind the Lake LeConte beaches. Seasonally the level of the ponds fluctuated with the amount of regional precipitation and perhaps during times of flood, water from the Colorado River overran the Lake LeConte beaches.

With the withdrawal of Lake LeConte within the greater Salton Sink, erosional processes removed the beach bars and the greater portion of the Yuha and Coyote Wash Terraces. However, the Yuha Basin, except for the main drainage channels, did not receive heavy runoff during the erosional removal of the Yuha Terrace, so only the fine materials were removed from there. The heavier materials were lowered by much the same process that forms desert pavement (Rogers 1966). Stone materials including Malpais artifacts, fossil bone, and petrified wood have been lowered 40 feet to 100 feet, where they are now aligned at the base of the hardened slopes of tilted Palm Spring sandstone or scattered over the erosional lowered surfaces that stand a few inches or feet above the more active drainage channels.

The Yuha alluviation from Plaster City to 8 kilometers south of Mexico's Highway 2 has been intermittently subjected to ponding, erosional channel cutting, that partially resulted from fault controlled land forms, and fluctuating levels of lakes in the Pattie Basin and Salton Sink. Pedogenesis of alluvium at various localities across the Pinto-Yuha drainage also indicates intermittent local deposition relating to geological changes from area to area.

The 150-foot Lake LeConte beach deposits may also include sediments for more than one depositional episode. The Yuha Formation is locally truncated by wave erosion, with beach deposits at the base of the scarp; elsewhere it is apparently graded down to the 150-foot a.s.l. beach. The beach deposits locally contain an interbed of red-brown fine-grained sediment, which may represent a period of nonbeach deposition. A somewhat similar deposit locally underlies the Yuha Formation around the northern scarp of the Yuha Basin. Whether the earlier and later periods of the Yuha Formation and 150-foot a.s.l. Lake LeConte beach deposition were synchronous and whether the two fine-grained units are correlative are yet uncertain.

Sometime after the deposition of the Yuha Formation, gentle doming occurred, which upwarped it at the same time that the 150-foot a.s.l. lake withdrew and Yuha Pinto Wash was diverted to the south.

Following upwarp, the Yuha dome was breached by drainage, which rapidly eroded the weak Imperial and Palm Spring Formations and formed the Yuha Basin within a structural dome. During a period of base-level stability associated with middle to late stages, possibly the 100-foot level of Lake LeConte, a pediment surface veneered with alluvium (the Yuha Terrace) developed within the basin.

Annular drainage that occurs on the north side of the Yuha dome where less resistant beds have eroded to form linear valleys partially encircling the dome represent ancestral Pinto drainage. Their relative age is uncertain except that the valleys probably are younger than the earliest unwarping of the Yuha Formation and older than the geographical changes that diverted the Pinto Wash south near its present course. Physical evidence suggest that the most recent annular drainage valleys are probably associated in time with the Coyote Terrace, the Yuha Terrace, and the 100-foot a.s.l. Lake LeConte.

Faults

At least four major faults plus a number of minor faults have intermittently diverted the Pinto-Yuha drainage during the late Pleistocene time.

The Yuha Basin is a breached dome of anticline associated with the Elsinore Fault. Several authors show the fault passing through or along the west side of the dome (Biehler, 1971; Elders, et. al., 1972; Sharp). A southwest and a west-trending fault are shown within the dome (Sharp, 1972; Kilkenny, et. al., 1973); the latter (perhaps the Elsinore Fault) trends south out of the domed area through the low pass south of Sunrise Butte.

Late Pleistocene upwarp of the Yuha Formation possibly exceeds 50 feet near the Elsinore Fault; however, at the Lake LeConte beach berm 5 miles to the east, uplift is not apparent.

A Sierra Juarez associated fault is mapped just west of Pinto Mountain with vertical movement, the west side down (Gastil, et. al., 1975). A fault is exposed by channel cut in the north bank of the Yuha Pinto Wash north of Pinto Mountain. A 10-centimeter gouge zone shows drag, the east side down; strike is to the north 57 degrees west, and the fault slope is 67 degrees to the southwest. The total vertical offset of materials by the fault that cuts through cross bedded Palm Spring sediments and bouldery stream deposited alluvium is at least 5.7 meters, 3.2 meters of offset are exposed by the channel cut, and 2.5 meters were exposed below the floor of the wash by a backhoe operation. This hitherto unreported fault is referred to herein as the Pinto Fault; however, it may be related to the Elsinore Fault system.

A half mile east to the Pinto Fault the Laguna Salada Fault crosses Pinto Wash, north to south, the east side of the fault down (Bernard 1968).

Origin and Distribution

Riverine boulders and cobbles, "Poway type gravels," distributed from near the summit and on the east slope of the Jacumba Mountains and the western highlands of the Yuha Desert underlie the Jacumba volcanics that are pretertiary in age (Minch 1973). The source of these riverine deposits has been a controversial subject.

However, the pretertiary bedrock of the same geological age and lithological materials at Poway, California, has characteristics that are found at a location north of Caborca, Sonora, Mexico (Merriam et. al., 1968). "The distribution of the Poway type gravels of the Coastal Range predate the separation of Baja California and were distributed across the Jacumba area to the Pacific Coastal region on an essentially parallel surface just south of Jacumba near La Rumurosa. All materials representative of the early river deposits are extremely indurated meta-volcanics, meta sedimentary rocks and quartzite that are not found locally." (Minch 1973.)

The major portion of the "Poway type" material in the Yuha Desert lie on the slopes and near the base of the mountains; however, outwash has transported isolated cobbles out into Salton Sink to below the 40-foot Lake Cahuilla beach berm.

Interest in these old riverine materials is further stimulated by the fact that wherever early people found these cobbles, they were used for tool making. Although most of the material is not prime knapping quality, in the Yuha area it was the best material available. The local rock was mostly granite medi-sedimentary rock that is not suitable for flaking. The most used locally derived materials were vein quartz, basalt quartzite and prophyry from the Colonio Progress volcanics south of Signal Mountain.

The Yuha Buttes are uplifted marine sediments Imperial Formation correlative with the opening of the gulf trough and encroachment by the ocean to create the Baja Peninsula. Erosion of the overlying Palm Spring Formation within the Yuha anticline has exposed the fossiliferous Plio-Pleistocene Imperial clay and sandstone of the Yuha Buttes area.

A different erosional force appears to have been involved south of Pinto Wash northwest of Mount Signal. Here the Yuha Formation is absent and presumed to have been erosionally removed. The surface of the area is covered with sand and well rounded pebbles that overlie unconformably the deformed Imperial and Palm Spring formations. Fossils incorporated in some Pebbles strongly suggest a Colorado River deposition for these pebbles that vary in size from 3 centimeters to 30 centimeters and are generally smaller than "Poway Type" materials; both earlier and later people in the area usually found them too small to shape artifacts from.

The average size of the Colorado River materials grade larger to the southwest toward the Pinto Mountain and the head of the Laguna Salada.

Similar pebbles are found on the cuesta-like Palm Spring Formation erosional remnants northwest of Mount Signal with alluvial cappings (Barnard 1968). This evidence represents a geologically recent Colorado River flow from possibly a southerly direction that postdates the Yuha Formation and the Yuha Terrace, but prior to the most recent uplift northwest of Signal Mountain.

Assuming that a lake in the Pattie Basin drained south to north into the Salton Sink, it is extremely unlikely that cobbles and pebbles would have been transported across a lake to the area north of Mount Signal. Any hypothesis for the problem also has shortcomings. However two types of locally derived stone indicate a possible answer. Angular materials, gray-green prophyry from the Colonio Progress volcanics just to the south of Paso Inferior in the Sierra De Los Cucapas, and silicified sandstone from a site on the northwest end of Paso Inferior are distributed over the surface northwest of Mount Signal, lying side by side with Colorado River materials. Conceivably an ancestral Colorado River could have crossed the Cucapa Mountains through Paso Inferior on its way to the Salton Sink. However, any evidence to support this has been erosionally removed or is buried. Further investigation may reveal the route by which the riverine materials were brought into the area.

Older Lakes

For purposes of clarification, the 100-foot to 150-foot Pleistocene lakes in the Salton Sink are herein referred to as Lake LeConte levels; the term Lake Cahuilla is reserved for the younger lake that existed until approximately 500 years B.P. at approximately 40 feet a.s.l. (Macdougal 1914).

Lake terraces on the Yuha Formation north of Pinto Wash near the mountains stand at 420 feet and 440 feet a.s.l., and bisect the drainage. Similar lake terraces near the same levels exist on eastern slopes of Coyote Mountain and on the north end of Superstition Mountain with a series of lake terraces that range in altitude from 250 feet a.s.l. to nearly 500 feet a.s.l. on the north and west sides (Loeitz, et. al.). These terraces may represent single bodies of water.

Hayden (1976) indicated a younger age for the higher 420- to 440-foot a.s.l. lake levels than for the 150-feet a.s.l. Lake LeConte inundation. This appears to be an error both from a deomorphological standpoint and the fact that he erred in documenting the material dated (a caliche sample collected by Childers and dated by Geochrone Laboratories at 24,000 (plus 2000 or minus 1000) years B.P. (Childers 1974). Hayden erroneously referred to the material tested as lake tufa.

Geological evidence indicates that the Colorado River during late Pleistocene time flowed alternately into the Gulf of California and the south end of Sierra del Mayor maintaining a large deltaic system that sustained a lake in the Pattie Basin at last 60,000 years B.P. The age of the lake is based on uranium series ages of calcareous paleosols (Carter 1975).

A special phenomenon is the appearance of a seemingly fresh water tufa in the portion of the fan lying to the west of the Elsinore fault 2 miles east of Sites 4-Imp-905 and 906. The 37,000 and 50,000-year-old beach lines lie 15 kilometers to the east of the Elsinore fault. The fault zone is marked by a jumbled topography; a belt of irregular hills that the Yuha-Pinto Wash has cut through. During a late stage in the fan's building, moving along the Elsinore fault dammed the wash and created a lake that persisted long enough to deposit a tufa formation that today forms a minor cap rock. This lake deposit overlies the Palm Spring Formation and underlies alluvial materials, and extends from northwest of Sunrise Butte south across Pinto Wash for a mile into Mexico and contains volcanic tuff intermixed with diatoms and calcareous matter that has been infiltrated by lake or sea borings. Samples of volcanic tuff are being analyzed for age. The geographical extent of the lake associated with the deposit is not known. However, it may represent a northwestern extension of the 60,000-year B.P. lake in the Pattie Basin reported by Carter that was separated from the Salton Sink by the Yuha Anticline and associated uplift.

Further investigation may reveal whether the lake deposit and the 420- to 440-foot a.s.l beach berms are related. At this time, however, the evidence serves only to document a lake near the Coastal Range in the Yuha Desert that is younger than the Palm Spring Formation and geologically equivalent in age with the alluvial deposits that overlie archaeological materials at Sites 4-Imp-905 and 906.

At least four Lake LeConte beach berms range in elevation from 150-feet down to 100-feet a.s.l. just west of Plaster City. The 150-feet berm is the most prominent. A near-stable lake must have stood at the level for thousands of years. Carbon 14 dates on shell (Anadonta Californiansis) and lake tufa indicate an age from 37,000 to 50,000 years BP for the 150-foot a.s.l. Lake LeConte stand in the Salton Sink (Hubbs, et. al., 1965). The lower beach strands are smaller, less obvious and more dissected than the 150-foot a.s.l beach, suggesting shorter depositional periods that represent withdrawal stages of the vanishing lake. Carbon 14 dating of the 100-foot a.s.l. Lake LeConte is 26,800 plus 1570 or minus 1325 (Childers, in press).

Geomorphology Weathering and Time

The term geomorphology is used broadly to include a whole group of earth sciences from landforms to weathering phenomena, all attempting to provide an understanding of the structure and evolution of landscapes. These offer opportunities for estimating at least relative ages. While the whole matter of time determination in the Yuha Desert is in its infancy and all of the conclusions reached here must be considered tentative, we nevertheless can indicate some orders of probable magnitude. We have a few geochemical dates that offer our first beginnings towards an absolute chronology. The Yuha Desert is an area of great tectonic instability, so direct sea-level correlations cannot be assumed; however, we do have a series of beach lines each showing weathering phenomena peculiar to their respective elevations. As the elevation of the shorelines increase, time increases.

In the Yuha Desert the principal time markers are those derived from three beach lines dated at 1510± 180 years before present (YPB) (Hubbs, et. al., 1963) 26,800 plus 1570 or minus 1325 YBP (Childers in press), and 37,000-50,000 YBP, hereinafter referred to as 50,000 BP (Hubbs, et. al., 1965). In addition, we have successive younger Carbon 14 dates, obtained from caliche samples collected at varying levels between 1200 feet to 400 feet on the surface of the Yuha Formation. Carbon 14 date indicates the dryer times started prior to 30,000 years BP and continued to 20 thousand BP (Bishoff, et. al., 1976). The geomorphology of the Yuha Formation suggests that these are minimum ages and that the actual age may be somewhat older.

Rounding of numbers is used since even our fixed dates are far from being absolute, and our interest at the moment is in establishing the magnitude of time rather than absolute dates.

The 12.3 meters, 1510± 180-year-old loose sandy beach shows virtually no weathering phenomena. The beach forms are perfectly preserved; bay bars, storm beaches, spits, and cusps. There is no visible evidence of $CaCo_3$ movement and concentration—no yellowing or discoloration of subsurface sand. Artifacts on the surface show slight to zero sandblast or polish.

Relatively delicate objects such as shell beads, bone tools, fine arrow points. and shards litter the surface of these beach lines for miles all around the Salton Sink. Manos and metates of sandstone or granite show fresh surface with unobscured grinding and sharpening marks. No evidence of weathering; no yellowing, no etching out of the feldspars, no rotting of the mass of the stone; indeed, the surfaces retain polish and sharpening marks. The total weathering picture is of slight magnitude and is undetectable to the naked eye.

The 30.48-meters beach with an age of 27,000 years shows some changes, but startlingly little. Faint evidence of development of soil horizons appears. A few inches of loose, often moving sand, overlie a few inches of very weakly, and barely discernible, packed sand. $CaCo_3$ movement is apparent. A layer of weakly lime-cemented sand is intermittently present. Strong lime cementation is occasionally present, especially as root replacement. Color differentiation is weak. The sands are gray with no clear shift towards yellows and nothing reddish at any point. Biotite mica flakes are sharp and clear-edged. One would describe the soil profile as incipient rather than established.

Desert pavement is unsorted. Rock debris is scattered and has undergone enormous sand scouring, especially in the Plaster City area. Under such conditions artifacts and ventifacts are hard to distinguish. In other examples, only the outline of a core or a blade remains, and the shiny rounded form reminds one of an artifact made of wax which had been heated until the flake scars were almost blurred to extinction. Due to this sandblast attack, no desert varnish or ground stain has formed. Artifacts are being extinguished, and it is clear that in a sandy environment as at Plaster City, artifacts will be reduced to unrecognizable ventifacts before a desert pavement is formed.

In areas richer in gravels, pavements may form more rapidly and artifacts may then be preserved despite various stages of sandblasting. For good preservation of artifacts in these sandy desert environments, burial of placement on an already-formed desert pavement is required.

The 45.72-meter beach with radiometric ages of 37,000 to 50,000 YBP shows a further progression in soil development. Consolidation is now strongly evident in the subsurface and $CaCo_3$ migration, and concentration is marked. Where these beach ridges have been used as gravel quarries, steep faces are left standing in the gravel pits and a resistant rim of moderately cemented materials forms a resistant ledge around their upper margins. $CaCo_3$ cementation is at times strong, especially along root channels where columns and lenses of cemented sands may be formed. Color changes are still very weak. Light yellows in the subsurface are the rule. There is nothing brown, and the coloration only faintly reddish. There is little or no evidence of clay mineral formation or migration. Mica flakes show a very slight fraying on the edges under 10X magnification.

Sandblasting of surface at the 150-foot beach makes artifact identification conjectural. The degrees of weathering on these three beach lines are consistent with their relative ages in that there is a steady progression from the youngest to the oldest. The degree of weathering, except for sandblasting, is startlingly small. The two older surfaces probably experienced glacial-pluvial climates, which must here have been marked by some increases in precipitation. The 50,000-year-old beach seems to have recorded more rainfall as well as more time.

The Pinto Mountain—Davies Valley Fan (Yuha Desert)

Interest in this feature stems from the presence of erosionally exposed artifacts at Sites 4-Imp-905 and 906 near the very base of the fan. Davis Valley lies west of a northwest-southeast trending splinter of igneous rock, hereinafter Splinter Mountain, on the west side of the Yuha Desert. On the Mexican side of the border, Pinto Mountain is a continuation of Splinter Mountain, and angles southwesterly parallel to the Peninsular Range. Yuha-Pinto Wash drains the sizeable area behind Pinto Mountain and Splinter Mountain, including the very steep and rocky eastern face of the Peninsular Range. All of this drainage passes through a gap between the two mountains.

A deep gap similar to the present one existed here in the past. Due to a combination of climatic and tectonic controls, fan building began, and an immense alluvial fan filled much of Davies Valley, nearly buried Pinto Mountain and the south end of the Splinter Mountain, and built easterly into Imperial Valley. The fanglomerates in the space between the two ranges are estimated to have exceeded three hundred feet in thickness. Then this process reversed and down cutting began. The higher part of the fan on the flanks of the Peninsular Range and most of the fill of Davies Valley have been removed. Pieces of the old fan on the south ends of Splinter Mountain and the north end of Pinto Mountain are preserved as broad sheets east of these ranges. Reservation of the old fan sheet is limited by erosion of the Yuha Dome to the north, by uplift and erosion along the Elsinore Fault to the east, and by erosion toward the Laguna Salada to the south. Beyond the Elsinore faultline, the beach lines previously described cut across the foot of the fan surface, supplying a minimum age for the formation of the fan. During the down cutting of the fan, a series of broad terraces was formed, discussed below.

The fanglomerates exposed in the banks of the Yuha Pinto Wash rest on the Palm Spring formation. Two members are well exposed here: red-pink clay, and the overlying fine gray sands; both tilted approximately 45 degrees west. Over this is the artifact-ladened, rounded boulder bed clearly deposited by torrential flow.

Vertical movement of the Pinto Fault occurred shortly after the deposition of the boulder-cobble strata with which the artifacts are associated. The evidence here is in alluvial deposits 30 centimeters above the artifact-laden stratum, which has not been offset by Pinto Fault movement. The upper deposits are sands and gravels with occasional boulder strata. There are no marked weathering horizons to suggest long still sands within the exposed stratum. The sands are coarse and clean and show considerable sorting indicating strong currents. At numerous points near the base where exposures are good, lenses of stained sands mark periods of manganese accumulation. This too indicates considerable moisture. The granite boulders, even very large ones of 40 to 50 centimeters in diameter, are thoroughly decomposed. This is indicative of pre-Wisconsin weathering. The total picture indicates a pluvial period for deposition and a pre-Wisconsin age (70,000 years).

The fanglomerates have been extensively faulted. On a foot survey of the section exposed in the narrows, a minor fault displacement in the underlying Palm Spring formation and the overlying fanglomerates was noted approximately every 20 meters.

Since the artifact-bearing stratum is the basal boulder bed, its age is of primary importance. The simplest assumption is that it is indeed the basal member of the fan and measures the fan's maximum age. There is, however, the possibility that the terraces may be constructional in-fillings of the fan recording alternate cut-and-fill episodes during the fan's dissection rather than benches cut into the fan. The following observations bear on this subject.

The numerous faults in the body of the fan with displacements of 2 to 6 meters do not appear on the terrace surfaces. Since they run at right angels to these surfaces, they should appear as a series of steps, but no such phenomena are evident. The faulting seems then to be older than the terrace-cutting episodes, and since the fanglomerates bearing the artifacts are faulted, their age seems better assigned to pre-terrace cutting and to the original period of fan building. Our conclusions are that the two highest and oldest terraces represent a period when the Yuha-Pinto drainage flowed northerly. The third terrace is the top steam-cut bank above Sites 4-Imp-905 and 906; Malpais Site 4-Imp-907 is situated on this terrace. The fourth and fifth terraces are within the banks of the modern Yuha-Pinto Wash and document a change in drainage to the present more easterly course. The five terraces suggest topographical changes related to variations in drainage base levels that equate with different points in time and geological events. All of the events occurred subsequent to the accumulation of the highest alluvial deposits that stratagraphically overlie the Yuha-Pinto artifacts.

There is also important evidence of this on the fan and terrace surfaces. None of the terraces or any part of the fan surfaces has maximum development of desert pavement and desert varnish. In general, the strongest pavement and varnish are found on the highest remnant of the fan where they are preserved on the southern end of Splinter Range. Moderate desert pavement has formed here. Desert varnish is light, but this may be due to the considerable amount of sand movement. On the higher surfaces of Pinto Mountain, there are areas of maximum desert pavement formation with maximum desert varnish accumulation. This is in a quarry area where there is virtually a man-made pavement. The lower terraces show progressively weaker pavement and varnish.

Soils observations indicate aging on the upper three terraces far beyond the 50,000-year-old beach line. Pits excavated into the surface of the terrace immediately overlying the discovery site exposed very compact soils showing light red color in the surface horizon. Beneath a foot-thick reddish, very hard zone, a strong $CaCo_3$ accumulation was evident for at least another foot. Very clearly, however, weathering had gone very much further in chemistry and translocation here than at the 50,000-year-old beach. The parent materials are the same at both places, and the weathering differences are therefore significant. The depth of translocation of carbonates and weathering seems out of step with the modern desert climate. Rather, a pluvial climate with marked increase in precipitation is indicated. Carbon 14 tests on the upper terraces have resulted in infinitesimal dates (Childers 1974).

We have, then, a number of factors indicating a wetter climate. The sands in the fan are clean and sorted. Manganese has been taken into solution and moved in some quantity. The large granite boulders are thoroughly decomposed. If the 50,000-year-old beachline is far younger than this fan, then it seems unlikely that the last pluvial Wisconsin II, Evans Cycle 1C (Evans, P., 1971) of 40,000 years ago would seem to be required to account for the surface weathering phenomena. If the fan accumulated during a preceding wet cycle, it would seemingly fall into Evans Cycle 3D, and this would date about 130,000 years ago. All such dates must be viewed as tentative but it is felt that they are more likely to be too low than too high.

ARCHAEOLOGY

Early reports on the archaeology of the Yuha Desert and surrounding areas were made by Malcolm Rogers in 1929. However, the archaeological potential of the Yuha Desert was brought into focus as the result of a cairn burial excavation that revealed human bones of an early period. Carbon 14 dates on caliche that encrusted the human bone indicated a minimum age for the burial at 21,500 plus 2000, minus 1000 and 22,000± 400 years BP (Childers 1974). Thorium 230 dates on the same material indicated 19,000 30,000 years BP (Bischoff, et. al., 1976). Amino acid dating is in progress and may soon be reported. There is insufficient collagen in the bone to permit Carbon 14 dating.

Other archaeological reports on the Yuha area are Weide et. al., 1974; Minshall 1976, Davis 1965; Treganza 1954; Childers 1976; D'Amari 1974; and Rogers 1966 and Hayden 1975.

A preliminary reconnaissance of the Yuha Desert has revealed that the archaeology of the area presents equally difficult problems, as does the geology. The earliest archaeological material is buried at the base of the alluvium and lies a few inches above the Palm Spring Formation exposed in the banks of Pinto Wash. The artifacts found there are flaked cores and flakes reminiscent of Paleolithic cultural materials found in the old world and may have resulted in more questions than answers regarding the time span of man's occupancy in the Western Hemisphere.

Outwash that resulted from Tropical Storm Kathleen in September 1976 exposed the earliest materials in place near the base line of a 25-meter cross-section of alluvial, aeolian, and riverine materials.

The exposed cross-section is in the north bank of the Yuha Pinto Wash, one kilometer north of the U.S.-Mexican border in the Coyote Wells Quad., SE Quarter of Section 30, T 17 S, R 10 E, Imperial County, California.

The worked stones are mainly amorphous green or gray prophyry and basaltic materials, which are vertical or shallow-edge flaked on the ends or sides. Experiments suggest that such stone forms can be obtained by a boulder-busting process—throwing one heavy boulder against another—to net a number of broken pieces. The most favored piece or pieces can then be additionally shaped for a specific purpose. The most diagnostic feature of these artifacts is that they are not diagnostic.

Eighty specimens, flakes and flaked stone, were collected under the direction of Morlin Childers and Jay von Werlhof under a special permit issued by the Department of the Interior, Washington D.C., November 27, 1976. The locations of all flaked stones in the exposed cross-section were documented by drawings and photographs prior to the removal and collection of materials. Materials noted and collected were restricted to an area not more that 3 meters above the present wash floor or over 21 meters below the top of the exposed strata. Hazardous conditions prevented extensive investigation of the upper portion of the exposed cross-section; however, a cursory examination from ground level did not reveal suspect artifacts in the higher strata. Most of the worked stone came from a boulder-gravel stratum exposed at two site locations separated by 100 meters along the wash.

Two sites, 4-Imp-905 and 906, were separated geologically by down faulting of the Pinto Fault (new name) that concealed the stream-deposited strata in which most of the artifacts were located. (The Pinto Fault is located near the east end of the 4-Imp-905 site).

Site 4-Imp-906 begins downstream at the point where the bouldery gravel stratum re-emerges from the wash floor and continues for approximately a hundred meters where the stratum is concealed because of overburden.

Site 4-Imp-905 lies west of the Pinto Fault, though one artifact was found 2 meters east of the fault in a sandy stratum. The base of the boulder-cobble stratum lies conconformably on tilted Colorado River-deposited Palm Spring Formation (Woodring 1931), of Middle Pleistocene age for the upper section. Up to 16 inches of the lowest exposure of the alluvial deposit at Site 4-Imp-905 has been hardened by a cement-like calcrete. Geologists have interpreted the calcrete as evidence of a spring related to an earlier period.

One artifact was removed from the calcrete deposit, several were located a few centimeters above the calcareous materials, and a bone fragment was found 60 centimeters above the indurated section. The bone may be a fragment of a human long bone; however, positive identification has not yet been possible. Two human phalangal bones were found at Site 906. However, they were not in place, and an extensive surface examination of both sites has not revealed the place from which they eroded.

Related Materials

Technologically similar artifacts coated with lake tufa are at depth, 3 meters below the surface of the relic beach deposit related to the 43.2-meter a.s.l Lake LeConte. A great number of crude flakes and cores have been removed by a commercial sand-gravel screening operation at a location just north of the point where Yuha Pinto Wash bisects Highway 98 north of Mont Signal. (This beach that has been Carbon 14-dated up to 50,000 years BP on shell and tufa has been discussed in the geomorphology section of this paper.)

Twenty-five meters above the floor of the wash, near the edge of the Yuha Pinto embankment, two groups of three circular clearings are separated by 30 meters and located midway laterally between Sites 4-Imp-905 and 906.

This site has been designated 4-Imp-907. Based on the archaeological materials associated within and near the area, the circles are assigned to Malcolm Rogers, "Malpais Culture" (Rogers 1929). One teshoe flake removed from a core and a crude chopper, 12 x 10 x 5 centimeters shaped from basalt, were noted but not collected. Three circles were cleared in the surface gravels and were 120, 140 and 160 centimeters in diameter. Located across a shallow modern drainage channel to the east along a secondary drainage way, stones up to 15 centimeters in diameter outlined the three additional circles.

Referring to the Malpais lithic materials, Malcolm Rogers stated, "If some precursor of homo sapiens were involved, the problem would be simplified; but that any type of modern man could exist with such a primitive working equipment is difficult to believe." He further stated, "In concluding the writer wishes to apologize for the indecisive and generally indefinite nature of the treatment of various Malpais manifestations. His excuse is that such seems to be their nature; and despite this condition, cognizance and attempt at explanation is demanded." (Rogers 1929.)

Circular 15-foot to 30-foot artificially disturbed areas on sloping Yuha Alluvial Formation present both geographical and archaeological problems. "Where these circular areas are located on steep slopes or drainage ways a terracing effect makes for easy identification. On more gently sloping terrain circular artificially disturbed areas are more difficult to identify, natural sand-filled sumps can easily be mistaken for archaeological features, however, when rock alignment appear as related features identification becomes obvious. Rock alignments often alternately bisect or circumvent a series of individual circular areas giving a serpentine effect. No specific pattern is discernible for these rock alignments that vary in length to over 100 meters" (Childers, in press). The custom of aligning rocks in association with circular disturbed areas may have been more extensive than present evidence indicate. Short sections appear to be remnants of longer alignments that have been erosionally breached by outwash.

Calcareous paleosols that formed along known fault zones strongly suggest that following the deposition of the Yuha Formation, possibly due to greater precipitation in the Coastal Mountains, a high water table drained underground to the Yuha Desert where it surfaced as the result of being impounded by movement along faults.

Earth that has shifted along fault zones will disrupt or offset water channels to such a degree that water is unable to flow in its usual route to other ground. Even channels not completely blocked often have crushed rock within the fault zone that may turn to clay (gouge) and form an impenetrable barrier. Under either of these conditions, water backs up on the uphill side of faults.

It is possible that early people used aligned rocks in conjunction with fine materials (the fine material now erosionally removed as are sections of the rock alignments) for religious reasons. The results of such effort would have, in some instances, served to control both precipitated and ground water that surfaced along faults to the benefit of plants that man sought for seeds and tubers. Ground so affected could have resulted in the circular disturbed area.

Technological differences of stone artifacts plus a great difference in mechanical and chemical weathering suggest that man inhabited locations associated with circular disturbed areas at least intermittently for thousands of years, however, ridgeback materials are the type most consistently found associated with the disturbed areas and rock alignments (Childers 1976).

Artifacts found on the Yuha Terrace and flaked stone which erosional processes has lowered from the terrace to the present floor of the Yuha Basin strongly suggest that Yuha terrace was living surface for Malpais and San Dieguito I people (Rogers 1929).

Malpais and San Dieguito I artifacts are scattered down the eastern slope of the Yuha area to at least present sea level, and Malpais materials have been noted at obsidian buttes at the south end of the Salton Sea. This evidence suggests that the early people followed a fluctuating lake over a great period of time. Surface Malpais and San Dieguito artifacts below the 40-foot level are generally incrusted with lake tufa. There can be little doubt that both the early people and recent Indian activities were somewhat affected by the size of flakes in the Salton Sink and Pattie Basin. San Dieguito circular clearings (not to be confused with the larger circular artificially disturbed areas) are found on drainage slopes down to the 40-foot beach level in the Salton Sink.

The greater portion of stone and cleared circles noted in the area are along ancestral drainage channels or on sloping terrain now pedimented by outwash from the mountains. This could suggest that people in the area were dependent, at least seasonally, on surface drainage from the mountains or groundwater that surfaced because of faults. Circular clearings noted below sea level are related to later cultures and must be younger than the 500-year-old 40-foot Lake Cahuilla.

Stone cairns that range in size from 18 inches to more than 12 feet in diameter have been constructed over a wide area of the Yuha Desert. They appear to relate to a variety of intent, most of which is not clear. The alignment of rock cairns 18 inches to 24 inches in diameter has interesting aspects (Begole 1974). Without exception those noted in the area, designated Yuha Desert, are oriented basically parallel to older drainage channels.

A number of recent Indian trials crisscross the Yuha Desert. Two of these trails converge northwest of the Yuha Well and lead to the 40-foot beach at the mouth of the Yuha Wash. Both younger and older trails lead westerly into the coastal mountains. Some are terminated where outwash has cut older alluvial deposits.

There is evidence of ponding that occurred behind the 40-foot a.s.l. Lake Cahuilla beach berm similar to the ponding that earlier inundated the 100-foot a.s.l. Yuha Terrace. Recent artifacts, metates, fresh-flaked stone, and pottery are found along the recent trials, the 40-foot Lake Cahuilla beach berm, and associated drainage ponds. A special feature, the Yuha Wash gravel effigy, at Site Imp-83 is also assigned to recent people and has been well described by Davis (1965).

Conclusions

The Yuha alluvial deposit is a post-Mid Pleistocene phenomenon. It was built under relatively humid conditions. Comparisons with the dated beach lines recommend 85,000 years for this weathering. The fan-guilding episode is probably best placed on a cycle back of this, or about 130,000 years ago.

Stones displaying the technological work of man are associated with the earliest alluvium and stream deposits that directly overlie the Palm Spring formation in the Yuha-Pinto drainage. Malcolm Rogers' Malpais surface material cannot be differentiated technologically from the buried Yuha-Pinto Wash materials. This indicates that man survived in the Salton Sink for many tens of thousands of years while the alluvial fans were building, with no apparent improvement in his stoneworking ability.

Dating of carbonates from strata associated with the artifacts is now being attempted and hopefully will supply at least minimum ages for the Yuha alluvial deposits and the Yuha-Pinto cultural materials.

References

Brace, C. Loring, et. al.
1971 Atlas of Fossil Man, Library of Congress Catalogue Number: 74-155873, Copyright 1971 by Hold, Rinehast and Winston, Inc. SBN 03-08548-2, p. 72.

Childers, W. M.
1974 Preliminary report on the Yuha Burial, California. Anthropological Journal of Canada, Vol. 12, No. 1, pp. 2-8.

Hubbs, Carol L., Biek, George S., and Suess, Hans E.
1965 La Jolla Natural Radiocarbon Measurements, University of California, IV, p. 90. Test LJ 954.

Rogers, Malcolm J.
1929 Early Lithic Industries of the Lower Basin of the Colorado River and Adjacent Desert Areas. San Diego Museum Papers No. 1, p. 22, San Diego, California.

Woodring, W. P.
1931 Distribution and age of the Tertiary Deposits of the Colorado Desert: Carnegie Institution, Pub. 418, p. 1-25.

MORLIN CHILDERS' THOUGHTS ON THE MALPAIS

Courtesy of Imperial Valley Desert Museum 1978

Karen Collins on behalf of the Imperial Valley Desert Museum generously provided the following brief statement about the Malpais "Culture." It is not complete but is a series of random thoughts about what Childers thought the Malpais to be. It is presented here for the purpose of having Childers' thoughts concerning the subject be available in a logical and accessible reference document. The editor acknowledges the generosity of the museum and its staff for providing such an important, albeit rough, document for the reader.

In the Imperial Valley we have few dates with which to assign specific cultural traditions; usually it is necessary to rely primarily upon tool morphology for comparative age assignments and grouping of artifacts. The problem is complicated with the apparent use of the same land surfaces at several different times (spatial stratigraphy). Based on sandblasting ground patina, desert varnish and changes of lithic technology, the habitation periods and intervals between the cultural sequences, at least in some instances, were widely separated and long intervals may have existed when man was not present at a Malpais site.

We do have some minimum dates for the earliest technological sequence (Phase 1) vertical or shallow edge. The earliest evidence for this type of flaking is exposed in the Pinto Wash of the Yuha Desert, 25 meters below the surface and a few inches above the lacustrine-riverine Palm Spring clays and sands, of middle to late Pleistocene age. W. Morlin Childers reported on this site in April 1976 at the South Western Anthropological Association meeting in San Diego.

Artifacts that are technologically the same are found on the surface 25 meters above the Pinto Wash exposure and also are noted 10 miles to the east at a point where the wash bisects the 43.72-meter-above-sea-level (awl). Lake LeConte beach. Vertical or shallow-edge worked artifacts are found as an intricate part of this 43.72-meter a.s.l. beach and often have man-flaked surfaces coated with lake tufa to indicate that the stone was flaked prior to the withdrawal of the lake. Carl Hubbs, et. al., have radio carbon dates on shell (Anadonta Californianis) and lake tufa to indicate the age of the 43.72-meter beach to be from 37,000 to 50,000 years B.P. Geologists who have examined the site that is 25 meters below the surface of the Pinto Wash site suggest a pre-Wisconsin age for the formation in which the artifacts are located; 70,000 to 125,000 years have been suggested.

Radio-carbon dates on caliche collected from the surface above the Yuha-Pinto Wash sites give infinitesimal ages of greater than 32,000 years (Childers 1974). This evidence indicates that man lived in what is now Imperial Valley for thousands of years while the alluvial fans were building and that during all that period he made no apparent changes in his stone flaking techniques.

The only other pre-point date that may be meaningful in the Imperial Valley is the 21,500-year-B.P. date postulated for the Yuha skeleton, which was excavated from under a man-constructed rock cairn. Three artifacts that have been assigned to the Ridgeback Tradition were found under the burial cairn during the excavation. No other lithic artifacts or other cultural materials were noted during the excavation; however, it is not possible to conclusively assign the Ridgeback Tradition to Yuha Man. We cannot positively preclude the accidental interment of the older stone flaked materials with the Yuha burial; however, again we are faced with a possible minimum age for the Ridgeback artifacts.

Documenting of the various pre-projectile lithic traditions of the Southwestern United States and Northern Baja California, Mexico, has been in limbo since the first field studies and identification of stone tools began.

Malcolm Rogers used various terms to describe stone knapping complexes and lithic traditions. He advanced the term Malpais only to withdraw it in favor of the term San Dieguito I. He included under this term all of the earliest stone tools. He classified stone tools as San Dieguito I primarily by location of tool, crudity of technology, and weathering by desert varnish.

Julian Hayden, a close associate of Rogers, states that Rogers had second thoughts after making the name change and was considering the reinstatement of the term Malpais at the time of his death.

Robert Begole in the 1970s revived the term Malpais. He assigned certain natural forms, which he found in association with cleared circles on the slopes of volcanic hills, to the Malpais Tradition. His reasoning was that the stones noted had similarities in form and that since no man-flaked stones were found in association with the stone clearing, man had chosen the peculiarly shaped stones as functional implements.

Working separately, Julian Hayden also reinstated the Malpais term. A careful review of Malcolm Rogers' early work, published in 1929, indicates that Hayden's use of the term Malpais holds quite close to Rogers' initial use of the Malpais term, which relied heavily upon site locations (older terraces) and liquid varnish to identify Malpais materials.

Julian Hayden's reporting has been primarily on the Pinicate, Sonora Mexico area where he indicates desert varnish or the degree of it can be accurately used to identify the age and cultural affiliations of flaked stone. Unfortunately, desert varnish or the degree of it cannot be used accurately for age determination throughout much of the Southwestern United States and Northern Mexico. This is particularly true on the western portion of the Imperial Valley, California, and the Laguna Salada area of Baja California.

Because of sandblasted surface materials, the burial by outwash of artifacts near the coastal mountains, and a number of variable conditions that exist in the Imperial Valley and the Laguna Salada region, weathering phenomena are unreliable for comparative age dating or documentation of cultural affiliations. Undoubtedly positive identification of lithic cultural traditions, using desert varnish as the sole identifying factor has shortcomings at most locations in the Southwest.

It seems that archaeology would be well served if descriptive means of identifying the early lithic work could be agreed upon by the various archaeologists, particularly for the pre-point traditions that are so nebulous with characteristics that vary from area to area within the Southwestern United States and Northern Mexico.

Malcolm Rogers stated in 1929 that "It has been obvious for a long time that a New World system of classification is needed to meet the pyramiding problems in American archaeology, yet it is doubtful that any group sitting in a conference could outline a system of any merit for even an area the size of Southern California, let alone the Southwest."

Confusion has been compounded by interpretative reporting of people from various schools and field experience. Widely separated sites that often exhibit divergent aspects of a knapping tradition have further complicated the ability of archaeologists to freely communicate without a lengthy discussion, each explaining what he means when he uses a term for cultural designation.

The following outline is suggested as at least a partial solution to the present dilemma.

Using a "phase" with descriptive terms for pre-projectile cultural phases could eliminate much of the confusion. Archaeologists discussing knapping techniques and degrees of chemical and mechanical weathering can have a meeting of the minds without a lengthy discussion as to ambiguous terminology such as San Dieguito, Malpais, and Amaragosan, etc.

Using a phase descriptive system would allow additions, deletions or variations at various localities that would permit communications between each archaeologist working separately to vividly understand what the other is saying. Each individual is privileged to include or exclude specific artifact types for a designated phase and document any additional characteristics if they are needed to identify it properly. As an example, if evidence available were prevalent and for some items it is, a vertical or shallow-edge flaked tool could be assigned to Phase V or VI, based on site location or other diagnostic evidence. Ridgeback might be relegated to Phase I if supporting evidence warranted, it being a product of an individual experiment or innovation. Under the proposed phase documentation, terms such as La Jollan, Amargosan, Pinto Gypsum, and Ventana Cave, etc.., could be used as indications of special or local cultural practices in the traditional manner.

The following list of terms is an example of descriptive terminology for pre-point lithic artifacts, although it should be stressed that the chronological sequence of stone tool types does not follow the same order from area to area and that the suggested list is for the Imperial Valley and immediate area south of the Mexican border.

At the top of the list is the earliest work recognized in the Southwest followed by the proposed sequence of knapping style and tool development. This is not to say that once a particular technique had been learned, the style could have been repeated during any later stage if the need arose.

> Proposed Phase—Descriptive Terminology for Pre-Projectile Lithic Artifacts as Proposed for the Imperial Valley and Northern Baja California, for more clarification and understanding.

Phase I

- Vertical or shallow edge retouched (to a free form)
  - Exact duplication of forms are rare or nonexistent

Phase II

- Ridgeback (prepared core techniques) Childers 1977
- End struck vertical cleavage (blade and core-anvil technique)
- Flake and core (little or no retouch)

Phase III
- Ovate biface
- Elongate biface

Phase IV
- Early milling culture
- Blade and core
- Uniface flattop scraper

Phase V
- Pony hoof scraper
- Bisquit scraper
- Finely made biface knife
- Mano metate
- Dart point

Morlin Childers' Thoughts on the Malpais

PRESERVATION OF QUECHAN CULTURAL SITES

Jamie Cleland, Ph.D.
EDAW, Inc.
1420 Kettner Boulevard, Suite 620
San Diego, California 92101

Jay von Werlhof has long been a champion for the protection of Native American values and cultural sites in the Colorado Desert. He was instrumental in helping to call attention to Quechan concerns for the Indian Pass vicinity when the Imperial mine project was proposed for this area (Bureau of Land Management 1997). When the Bureau of Land Management (BLM) retained my firm[1] in 1997 to conduct a cultural resources survey of the project (Pigniolo, et al. 1997), we knew immediately that it would be beneficial to seek Jay's input in the identification and evaluation of the cultural resources there. Although I had known him for many years, the Imperial Mine project was my first opportunity personally to work closely with Jay. Subsequently, I have had several occasions to call on his encyclopedic knowledge of cultural resources along the Lower Colorado River. By way of paying tribute to Jay, I would like to use this occasion to review three of the important cultural areas (Figure 1) where I have had the honor of following Jay after he had had a hand in identifying and recording them and most importantly in helping to preserve them for posterity.

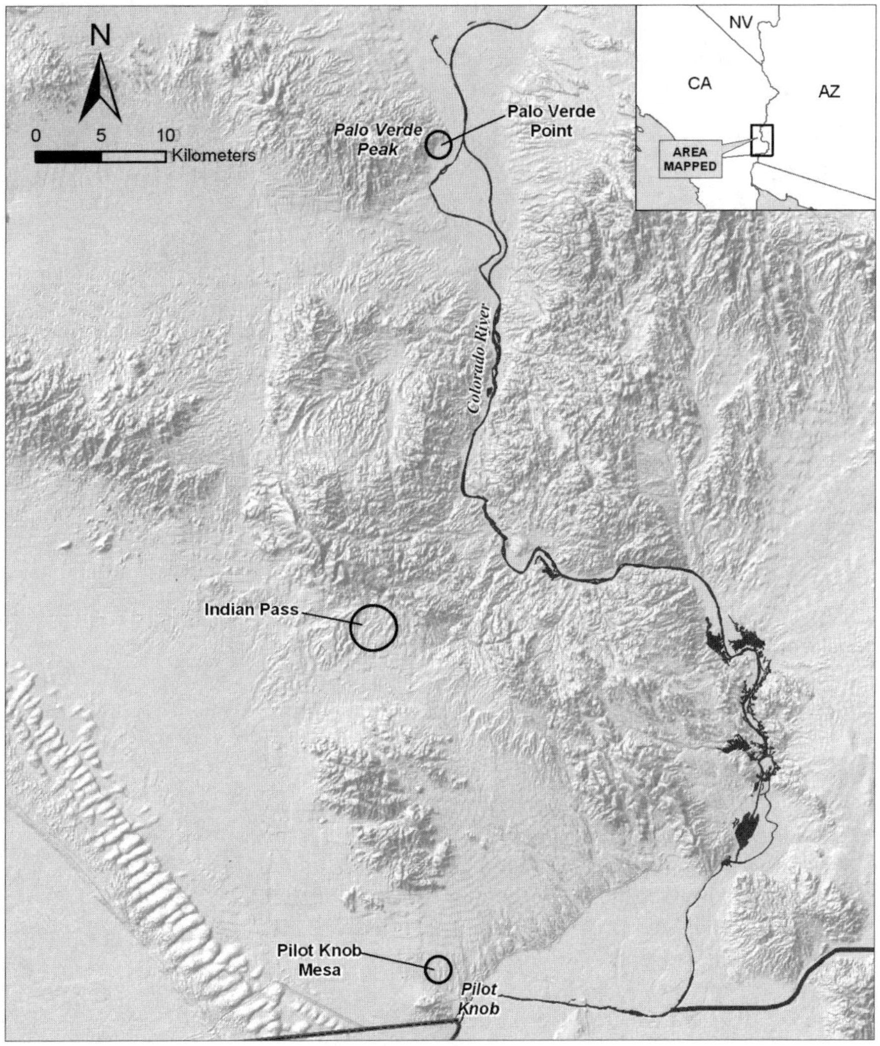

Figure 1. Cultural sites discussed.

Indian Pass (Figure 2) has long been known to be archaeologically rich as well as a favorite place for rockhounds (Weight 1949). In 1925 Malcolm Rogers excavated a trail shrine there (Rogers 1925; Waters 1982b) and found that it covered an offering pit full of Lower Colorado buffware that is now ascribed to Patayan I. Rogers' collections from the Indian Pass were critical in his seriation of Lower Colorado buffwares (Waters 1982a). In the 1920s whole and reconstructable pots could be found on the surface at trail shrines in the area. Photographic plates published in Waters' study of Patayan ceramics (1982a) show the quality of the materials that were out there in those days. Today there is still a very impressive concentration of ceramics at Indian Pass; the Imperial Mine survey identified 141 distinct loci containing ceramics in the proposed mining area (Pigniolo, et al. 1997), ranging from single sherds to large scatters.

When the proposal to develop the Imperial Mine was made public, members of the Quechan Tribe became quite alarmed by the massive impact that would have occurred with one of their most important cultural areas. They made their concerns known at public scoping meetings (BLM 1997) and enlisted Jay's support. Jay along with attorney Courtney Coyle helped to convince BLM that more intensive studies were necessary.

Our surveys (Pigniolo, et al. 1997; Underwood and Cleland 1998) provided dramatic archaeological support for the Quechan position that this area is indeed unique and should be preserved. Others have described the myriad issues associated with traditional cultural practices (Baksh 1997). My goal in this paper is to describe some of the rich material culture that is present in the area. I should note, too, that Jay worked several days a week with our crews at Indian

[1]Under a subcontract with Environmental Management Associates, Inc. (EMA) of Brea, California. The support of EMA is gratefully acknowledged

Figure 2. Archaeological survey at Indian Pass, 1997.

Pass— in July— helping to record sites. His stamina was truly amazing; people were wilting in the sun, but Jay kept right on going.

One of the qualities that makes the Indian Pass area so special is the well-preserved trail system (Underwood and Cleland 1998). Major transregional trails crossed there. Figure 3 provides a sketch map of a major trail crossing at the famous Running Man geoglyph. The Running Man was not on Malcolm Rogers' 1939 sketch map or his field notes on this site, despite the fact that he noted other smaller features in the immediate vicinity. In all probability the Running Man feature is post-1939; this shows that there is remarkable cultural continuity at the Indian Pass vicinity. Traditional cultural practices have never really ceased in the area. One can readily see from the quality of many of the constructions in the area that recent use, as well as ancient use, is evident.

Lorey Cachora of the Quechan Tribe has been eloquent in bringing Native American concerns for the cultural landscape to the attention of cultural resource managers (e.g. Cachora 1994). At Indian Pass, constructed landscape features, small rock circles, are abundant (e.g. Figure 4). Small circles, each about 1 meter in diameter, are interpreted as having been used in rituals conducted by single practitioners or small learning groups. Larger constructions are also common. Figure 5 shows one, again possibly relatively recent, that is made largely of white quartz. According Quechan cultural authorities, white quartz has special symbolic importance. The breaking of this rock was done ceremonially. Quartz flaking stations are abundant at Indian Pass. Small geoglyphs, particularly circular ones, are also present, as are cleared circles.

Near the pass itself scratched petroglyphs are abundant (Figure 6). Rogers found that scratched cobbles were used to seal the offering pit at the trail shrine he excavated (Waters 1982b). Although the concentration of these artifacts is at the pass itself, they are also found up to several miles to the south, including the Running Man locality.

The Imperial Mine saga is continuing. Under the Clinton administration, the Interior Department declined to permit the project. I understand that this was the first time that a project proposed under the 1872 Mining Act had ever been denied on environmental grounds, and cultural resources issues were huge in the decision. BLM moved quickly to withdraw the land to protect it from mineral entry. However, the Bush administration has moved to rescind the denial of the project. The National Trust for Historic

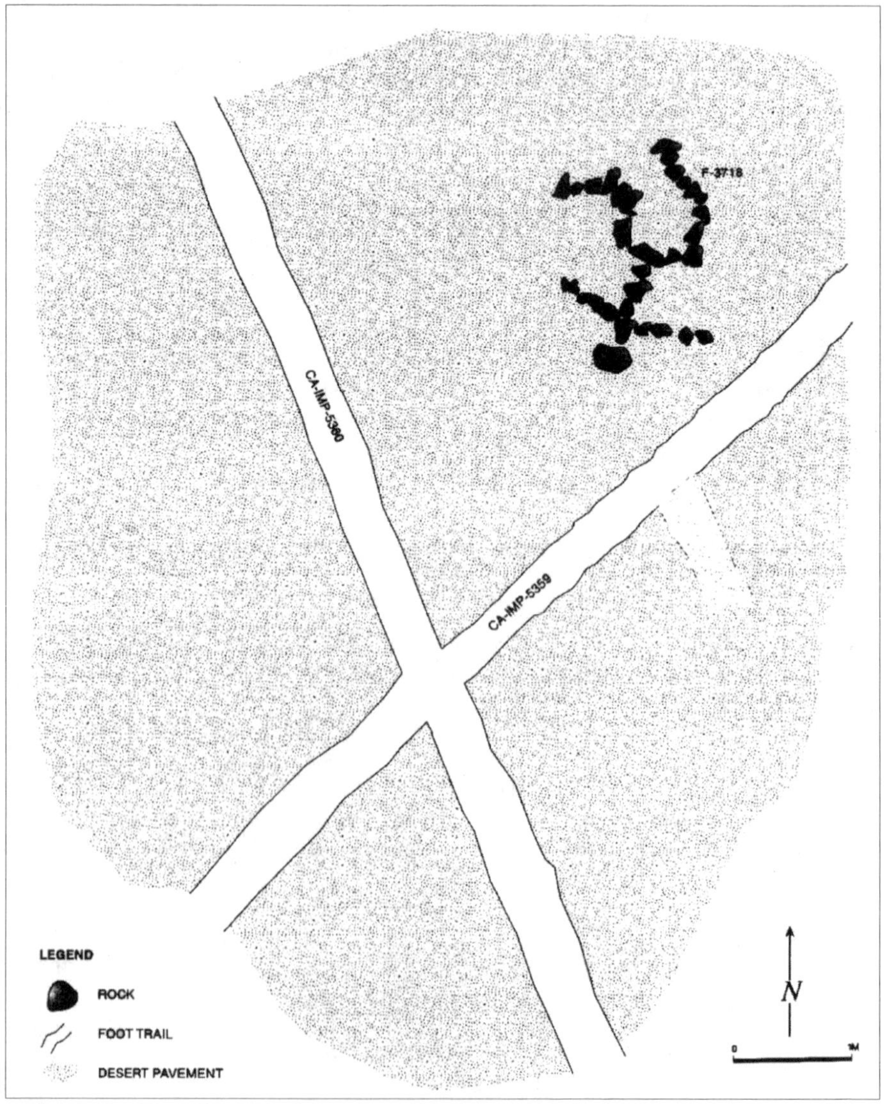

Figure 3. Trail intersection at Running Man geoglyph (CA-IMP-2727).

Figure 4. Examples of small rock rings at Indian Pass.

Preservation has placed Indian Pass on its national Most Endangered Places List. It's pretty rare to lose a site that makes this list, but the area is still threatened.

One of Jay's lasting contributions will no doubt be how he has inspired untold numbers of people in Imperial County to care about cultural resources. One individual who Jay worked with over the years was Harry Casey. Harry took high-quality aerial photographs of many of the geoglyphs and prehistoric trails in the California deserts (von Werlhof 1987, 2004). These photos serve as a wonderful archive of these important resources. My firm was able to make use of some these recently as we conducted our research around the Palo Verde Point area in conjunction with the North Baja Pipeline Project. Figure 7 is a cross-shaped geoglyph located north of the famous Palo Verde petroglyph site. We made a detailed GPS map of the site (Cleland and Apple 2003), but this photo remains a critical element of the site's record. We were not able to revisit another geoglyph site on the southeast slopes of Palo Verde Mountain (Figure 8), but because of the work that Jay and Harry and Boma Johnson (1985) have done, we can study it from high-quality aerial images.

Ever vigilant, Jay heard that the North Baja pipeline was going through the Palo Verde Point era. We got a call right away, and Jay was gracious enough to come and look for himself in April 2002. Among other sites, we visited CA-IMP-268/6905, probably the largest rock site in the Colorado Desert (Figure 9). Jay was concerned about construction activities in this vicinity. During our visit we observed a brush clearing operation that was being closely monitored by a cultural resource specialist. When a cultural monitor is on hand, the construction folks usually do what they are supposed to in terms of avoidance. Fortunately for us, we were able to convince Jay that the pipeline route had been moved far enough away from the sensitive resources to avoid any impact and that the monitoring effort was sufficient. I would like to say as a consultant who often works with private parties, it is always helpful to have concerned and knowledgeable public input in projects. This gives otherwise weak historic preservation mandates more teeth. It is a lot easier to convince project proponents to be sensitive when there are outside interested parties.

Most recently I have been serving as an expert witness for the Quechan Tribe in their claim against the Federal government for damage to cultural sites on the reservation. In this case, a preconstruction cultural resources survey (Moreno et al. 1995) identified ten National Register eligible sites on Pilot Knob Mesa. The archaeologists suggested that the sites might be part of a cultural district and indicated that traditional cultural values are present as well as archaeological information. The government agreed to provide monitoring of construction to avoid impacts. However, the agency then neglected to implement this mitigation measure, and several sites were damaged where they should have been preserved. The Quechan Tribe called Jay in, and of course Jay was there to help (von Werlhof 1999).

Figure 5. White quartz structure at Indian Pass.

The tribe has filed a potentially precedent-setting suit in federal court. The area in question is on the Reservation and was crossed by a major trail from Pilot Knob. The sites contain geoglyphs (e.g. Figure 10), cleared circles, quartz flaking stations, pebble mounds, and other features. Damage outside of the right of way is extensive in places and would probably not have happened if monitoring had been used as was originally planned.

In all these projects, and many others, I have benefitted from Jay's prior work. Archaeologists present and future owe him a debt of gratitude for his dedication to recording and preserving the fascinating cultural heritage of the Colorado Desert.

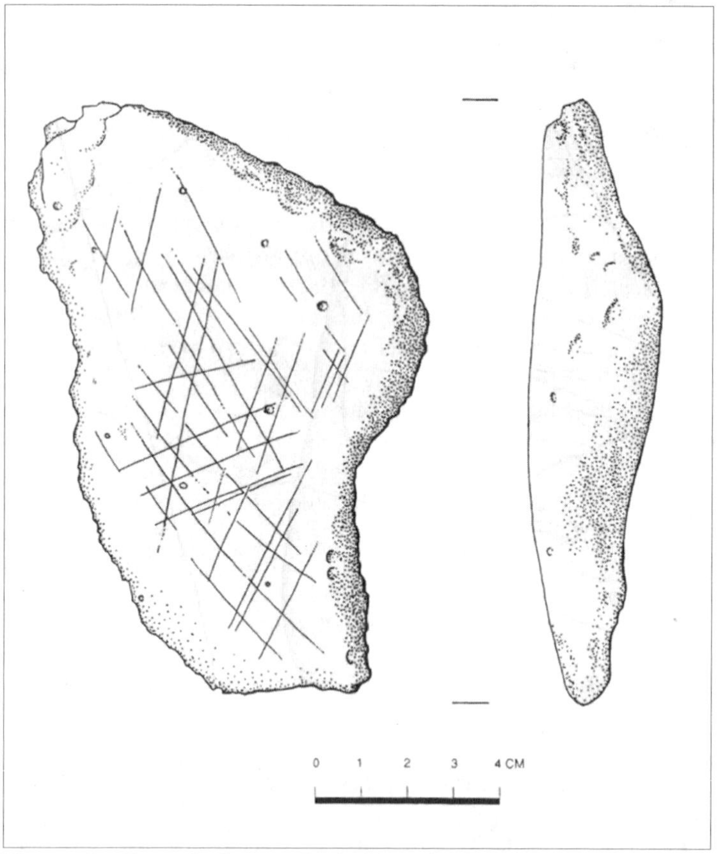

Figure 6. Scratched petroglyph from site CA-IMP-5067.

References

Baksh, Michael
1997 *Native American Consultation for the Chemgold Imperial Project*. Document on file with Bureau of Land Management, El Centro, and Tierra Environmental Services, San Diego, California.

Bureau of Land Management (BLM)
1997 Public comments in the matter of the draft environmental impact statement/environmental impact report for the proposed Imperial Project open pit, heap leach gold mine in Imperial County. Taken February 6, 1997 at Holtville, California and February 13, 1997 at La Mesa, California. Atwood, Jennings & Associates, Certified Shorthand Reporters, El Centro.

Cachora, Lorey
1994 The Spirit Life of Yuman-Speaking Peoples: Lower Colorado River Between Arizona and California. In *Recent Research Along the Lower Colorado River*, edited by Joseph A. Ezzo, pp. 13-14. Statistical Research Technical Series No. 51, Tucson, Arizona. Report prepared for Western Area Power Administration. Western Cultural Resource Management, Inc. Boulder Colorado.

Cleland, James H., and Rebecca McCorkle Apple
2003 *A View Across the Cultural Landscape of the Lower Colorado Desert, Cultural Resource Investigations for the North Baja Project*. Prepared for Tetra Tech FW, Inc. and North Baja Pipeline LLC. Prepared by EDAW, Inc. December.

Johnson, Boma
1985 *Earth Figures of the Lower Colorado and Gila River Deserts: A Functional Analysis*. Arizona Archaeological Society, Phoenix.

Moreno, Jerryll L., Dawn S. Snell, and Renee Kolvet
1995 Intensive Cultural Resource Inventory for the Western Area Power Administration Gila-Knob 161-kV Transmission Line, Imperial County, California and Yuma County, Arizona. Report prepared for Western Area Power Administration. Western Cultural Resource Management, Inc., Boulder, Colorado.

Pigniolo, Andrew R., Jackson Underwood, and James H. Cleland
1997 *Where Trails Cross: Cultural Resources Inventory and Evaluation for the Imperial Project, Imperial County, California*. Document on file with Environmental Management Associates, Brea, California, KEA Environmental, San Diego, California, and BLM El Centro, California.

Rogers, Malcolm
1925 Field notes from site SDM C-1 Black Mesa, Imperial County, California. On file San Diego Museum of Man.

Waters, Michael R.
1982a The Lowland Patayan Ceramic Typology. Appendix G in *Hohokam and Patayan: Prehistory of Southwestern Arizona*, edited by Randall H. McGuire and Michael B. Schiffer. Academic Press, New York.
1982b Trail Shrines at Site SDM C-1. Appendix F in *Hohokam and Patayan: Prehistory of Southwestern Arizona*, edited by Randall H. McGuire and Michael B. Schiffer. Academic Press, New York.

Underwood, Jackson and James H. Cleland
1998 Trails of the Indian Pass Area, Imperial County, California. Report prepared for Environmental Management Associates. KEA Environmental, Inc., San Diego.

von Werlhof, Jay
1987 *Spirits of the Earth, A Study of Earthern Art in the North American Deserts, Volume I: The North Desert.* Imperial Valley College Museum, Ocotillo, California.
1999 Archaeological Investigations of Impactions on Quechan Tribal Reservation. Report prepared for the Quechan Cultural Committee.
2004 *That They May Know and Remember. Vol. 2: Spirits of the Earth.* Imperial Valley College Desert Museum Society, Ocotillo, California.

Weight, Harold O.
1949 Rockhound Trail to Indian Pass. *Desert Magazine*, February, 1949.

Figure 7. Cross-shaped geoglyph and mail system, CA-IMP-4388/5237, at Palo Verde Point. Aerial photograph by Harry Casey.

Figure 8. Geoglyphs and trail southeast of Palo Verde Peak. Aerial photograph by Harry Casey.

Figure 9. Jay von Werlhof at CA-IMP-268-6905, April 2005

Figure 10. Spiral geoglyph at CA-IMP-7141, Pilot Knob mesa.

SO THAT WE MAY KNOW AND REMEMBER

Courtney Coyle, Esquire
Courtney Coyle and Associates
1609 Soledad Avenue
La Jolla, California 92037

My thinking on how to express my personal reflections about Jay was formed by a comment that ran in the last Society for California Archaeology Newsletter. Maybe some of you have seen it.

In any case, that comment helped me to explore how different people may approach their chosen profession of anthropology and what it means to inspire people and leave a legacy. I readily see at least four ways that Jay's approach differs from that of many of his peers.

I first met Jay, not too long ago compared with some of you, about seven years ago in the Colorado Desert—naturally. There he was—you guessed it—in a bluejean shirt, jeans, silver pendant necklace, pocket protector with pencils lined up, and a black and white student's composition notebook in hand. A desert apparition to be sure (Figure 1).

It was one of my first forays into desert archaeology and observing a real archaeologist in the field. I had heard some of the stories and understood some facts underlying the historical distrust between archaeologists and Indians. Nevertheless, I was impressed by Jay's ease among the Indians who were present that day and upon whose ancestral lands we walked.

The Indians also appeared comfortable around this mature white man, each sharing knowledge and ideas in a respectful way and later even including him in the circle of reflection after a Spirit Run by tribal youth and leaders through a sacred landscape.

I know at lease some of you have had similar observations and experiences. One is even referenced in his new publication, "That They May Know and Remember."

This brings me to difference Number One: Jay appears to genuinely work *with* the Indians and *respect* their beliefs and worldviews. While they might not always agree on an issue, interpretation, or approach (Figure 2), there is an overall climate of respect, equality.

Jay did not have to be there that hot, dusty day seven years ago. He was not consulting on a project, he was not being paid, but he was there and present in the moment, giving of himself because, I believe, Jay truly has a passion for the desert, the people who have lived there, and those who continue to live there.

Difference Number 2: Unlike how much of modern cultural resource management is practiced, the project-by-project basis which with we are only too familiar, Jay is also drawn to see the big board, the big picture, of how all the pieces may fit together. What do the geoglyphs mean in relation to the stories, songs, words—and in relation to one another? Jay has taken the time and effort to pull thoughts together in informal and formal writings and published volumes, with the new publication I encourage each of you to read and enjoy.

This synthesis of knowledge and pulling of pieces into a whole are very important things and something, I for one, hope we see more of from CRM professionals in the future.

Difference Number 3: Jay is very generous with his time and talents. To paraphrase from "The Lord of the Rings": We all have about the same amount of time here, the difference is what you chose to do with it. Jay *reaches out* to other professionals and disciplines to inform and check his thinking.

Figure 1. Jay having a quiet moment in front of the future home of the Imperial Valley Desert Museum, March 4, 2001. Photograph by C.A. Coyle.

Figure 2. Jay von Werlhof with Quechan President Mike Jackson; Tribal Singer Preston Arroweed; Tribal Consultant Lorey Cachorn; and runners, after Spirit Run at Indian Pass, November 28, 1998. Photograph by C.A. Coyle.

For a lawyer in a field—and a room—populated almost exclusively with highly educated anthropologists, Jay's ability and desire to *bring you in* and to believe that you have something intelligent and worthwhile to add to the discussion is welcome indeed. And I know that I deeply appreciate and personally have benefitted from that collegial environment.

Difference Number 4: I have seen Jay *fight* to protect historical sites and sacred places, sometimes, perhaps, to his fiscal detriment as a hired consultant. By this I don't mean doing the report to document the destruction of the features before they are gone, although there is a role for that,

Rather, I mean going out of his way, *using his professional credentials*, putting that substantial evidence in the record, writing or meeting with the Planning Commission, the Board of Supervisors, a Congressional Representative, a United States Senator or even the President of the U.S.—*doing* something to try and protect that ancient irreplaceable fish trap, dance circle, or rock alignment. Action, not just words and intentions, is something I associate with Jay.

Jay may not be paid in dollars for such activities, but his heart certainly must be blessed for standing up and being counted. We need so much more of this—now more than ever.

I think Jay realizes that to *do preservation*, is not just for us right now, but for the future generations who have yet to see this place and out of respect for the old ones who are now gone, so that they may know and remember. Jay knows there are times when we each must roll up our sleeves, get vocal, and get political. And, as was referenced in the banquet ceremony last night, we must build and show our constituency for preservation. We must be bold, as Jay has so often been.

If we can do this, we can all share in the wonder of enhancing the quality of our lives and spirits.

For these four reasons, and a few more I cannot share in polite company, I am delighted to help honor Jay von Werlhof in this way and call him

"Friend."

Figure 3. Jay von Werlhof and Carmen Lucas, Kwaayini Fish Creek Mountains, January 24, 2001. Photograph by C.A. Coyle.

AN OCCASION TO COMMENT ON THE MANY CONTRIBUTIONS OF JAY VON WERLHOF

William T. Eckhardt
Brian F. Mooney and Associates
9903-B Businesspark Avenue
San Diego, California 92131

Greetings. This is an occasion to comment on the many contributions attributable to Jay and his generous career in California archaeology. Jay, it is with gratitude that I have had this period of several months to reflect on this matter and the time to consider how many ways your practice has influenced me—and many of the rest of those in this room today. Thank you.

As I said when we spoke by telephone earlier in the week, I am looking for information on pigment mineral locations in the Imperial Valley. Late last year, I was reminded of some mineral samples I recovered many years ago from Painted Gorge in the Coyote Mountains, northeast of Ocotillo. Soon thereafter, I located the samples in an old cigar box and an accordion folder that had been left in a bottom file drawer cabinet marked "SNAKES."

Reflecting back to mid 1979, Jay invited my participation with the Museum to conduct an examination for cultural resources in areas proposed for development by an outfit named Texas Industries. At the time, the client was preparing to file for a license to quarry limestone, and the proposal included patented and unpatented mining claims demarcating the mountainside quarry (1,460 acres), a cement processing plant (240 acres) out at Plaster City, and a seven-mile route between the two for transporting ore (85 acres). Well, those were the client's intentions, anyway. To date the facility has never been built.

The project included roughly 1,785 acres for the proposed quarry, processing plant, and conveyance route. I remember my exuberance as you Huck Finned me into accepting responsibility for marshalling a field team across the highlands, while you set up headquarters down by the salt cedar windbreaks at Plaster City. We worked in tandem over many weekends, me, with my team and the Museum's three-quarter-ton ammo carrier and rucksacks up on the hillsides and you, Jay, with your travel trailer and lounge chair out on flats analyzing data and generating information. I know now what I did not know then. This is as it should be.

To this day, I am not sure that either you or I ever reviewed each other's sites in the field. My field notebook that includes that season was not with the copies of my teammember's notebook in the accordion file; maybe your field notes have an entry concerning this question.

This area is precious terrain; you know it. Rising up out of the Coyote Mountains, Carrizo Mountain stretches to an elevation of more than 2,400 feet above mean sea level. It is a rugged block of limestone sediments with a stark and steep eastern escarpment, falling more than 1300 feet in the short two-mile run to the valley floor. This, in contrast, to the western flank where access is moderately easy, to either the southwest over the Coyote Mountains or northwest through the badlands to Carrizo. (Crew member notes reflect that we had a beautiful clear view from the peak: the Santa Rosas to the north, Table Mountain to the west, and far, far out to the south we could see Laguna Salada in Baja Norte and we could see so far east that we probably looked into Arizona.)

At first glance, the property proposed for the quarry looked barren. Jay and Harry Casey flew over the site in Harry's 1936 Piper Cub and returned with film for 7.5-minute stereo pairs from the twin 20mm wing-mounted cameras. Ouch! Viewing that 10-strip photo mosaic through the stereo glass stand, reading the rugged topography, we learned that access was going to be a difficult and in some instances dangerous proposition. It is a good thing we used the aerial reconnaissance before finalizing the work plan and scope of work with Jeff Bingham and Dwight Dutschke at the California Office Office of Historic Preservation. We never would have been able to state that we had performed a 100% systematic survey of the quarry area.

The terrain was rugged and remote but not barren. Our survey recorded that the hillsides presented myriad locations for micro-environmental variation. Every steep pitch led either to a dangerous precipice, a small fan, or a level flat. Agaves, barrel cactus, and cottontops were throughout the property anchored by roots into the crevasses. Creosote, ocotillo, brittlebush, and more agaves spread across the landscape. The faunal lists include bobcat, ground squirrel, jackrabbit, sidewinder, and chuckawalla. We also viewed raptors. For the summer season, this abundance felt remarkable, and I cannot help but wonder what springtime on the mountain might reveal.

As I mentioned previously, the proposed quarry area comprised patented and unpatented lands. On the patented 640 acres, we recorded three archaeological sites, two isolated artifacts, and two recent water caches attributable to immigrants. These have been reported in the document along with the results of surveys for the plant site and conveyance route.

Field notes show that we returned to survey the 820 unpatented acres after we had drafted our report. From the notes and one hand-drafted table, the complete/combined inventory consists of three recent water caches, six shelter sites, 21 roasting pits, five trail fragments, five lithic scatters, one isolated milling surface, and three isolated flakes. Fragmented ceramics were observed in all of the shelters and also found with a few of the roasting pits.

These findings compared well to the sites recorded previously in the Coyote Mountains by D. L. True (1960), who recognized a number of rock shelters and temporary camps to the south of our project area. The distinctions between our record and True's are likely factors in the methods employed for survey. However, our findings extended the knowledge about the prehistoric land use patterns to higher elevations within the range. With shelters scattered everywhere fortune presented adequate opportunity, and a most-grand collection of nine roasting pits blanketed the peak of Carrizo Mountain.

As best as I have been able to detect, our fieldwork ended before Christmas 1979. There were no provisions in our contract for any testing or evaluation of the breadth, depth, and extent of these cultural deposits. In the intervening years, BLM has established the Coyote Mountain Wilderness Area–closing access from all but the eastern flank. I can only wonder what the passage up from the throat of Painted Gorge might be like today.

Anyway, about the file cabinet drawer marked "SNAKES." My file has only an incomplete copy of our report. The Table of Contents suggests we wrote 95 pages, plus references cited, and one appendix. I have the document through page 77, appearing to be the conclusion of our findings. Missing from my copy are our sections evaluating Native American values and National Register eligibility. Our recommendations for the mitigation of potential effects are also missing.

My recent inquiries at the Southeastern Information Center have been unsuccessful in locating any copy of the report—complete or otherwise. Making matters worse, our document text dealt with temporary numbers, not trinomials. I fear that there may be a boatload of site records in need of production, but that won't be certain until I compare the quad sheets in my accordion file with the archived maps at the Information Center.

My notebook for that period is accessible; I'll have to drag it out of deep storage. With that and the accordion file, I will access the record archives and make the comparisons. How frightful! The bright side may be that the Coyote Mountains and Carrizo Peak will require a revisit in order to set the record straight.

If you come across anything at your home that pertains to this project, I'd sure love to hear of it. Any photos, records, text, recollections...etc. You know, my ears are ringing in memory of Paul Ezell's persistent comment from the early 1970s "...an archaeologist may not always have a job, but they'll never be out of work."

What about the mineral pigment samples, you might ask? I'll treat that topic in my next correspondence. Until next time, Jay, thank you.

References

True, Delbert L.

1960 Site Record Forms CA-Imp-48 through CA-Imp-54. Primary site records on file with the California Historical Resources Information System, Southeastern Information Center, Imperial Valley Desert Museum, Ocotillo, California.

von Werlhof, Jay C., and William T. Eckhardt

1979 draft Archaeological Investigations of the Creole Corporation Portland Cement Facility near Plaster City, Imperial County, California. Report prepared by Texas Industries, November 30, 1979. Imperial Valley College Museum, El Centro. Partial text and graphics (79 pages).

A LIFELONG FRIEND

Gail Egolf
3620 E. 700 N.
Churubusco, Indianna 46723

It was in the fall of 1978 when I first met Jay and his wife Sherilee. This occurred following my transfer from the Klamath National Forest to the Bureau of Land Management in El Centro. The Area Manager, David Mari, and the District Archaeologist at the time, Russell Kaldenberg, suggested that a meeting with Jay could help jump-start my new endeavor. An invitation to dinner at Jay's and Sherilee's was forthcoming. Russ, State Archaeologist for BLM, Bill Olsen, and I were their dinner guests. Little did I know that it was the beginning of a longlasting friendship. The initial objectives were to brush up on my knowledge of southern California archaeology and acquire information about local and regional resources.

We arrived about 6:00 on Friday night. After a couple of drinks the conversation turned to a discussion of family and personal histories. The breadth of Jay's experiences was amazing and his interest in my family ancestry and me was sincere. Even though my tenure with the BLM was short lived, we've managed to stay in touch these last 27 years, and he's always asking about my family and their well-being. Jay places much importance on family and ancestors and their role in our lives, and he shows a genuine interest in the Egolf family.

As the evening progressed, Jay began reciting his vast amount of experience he compiled in researching the prehistory and history of California. His insatiable desire to learn as much as possible about what makes us who we are was obvious. It was a natural progression then for Jay to display his talents in the classroom. He talks often of his days at Imperial Valley College and the numerous field trips and surveys conducted by his classes. I'm sure these are experiences his students will never forget. In recent years Jay has put his heart, soul, time, and money in a project designed to inform the traveling public about the cultural history of the El Centro area. Almost single-handedly, Jay has raised the thousands of dollars needed for the project.

Back to the dinner party—the last few hours of the party were a blur. The drinks kept coming, and sometime around midnight a delicious dinner was served. I do remember a few things, the most important of which was that the only resource needed to get started in my new job was Jay von Werlhof. Another lesson learned was to eat before you go to a dinner party at Jay and Sherilee's home. Many more memorable latenight dinners followed that night. Thanks, Jay and Sherilee, for all of your help and the lessons you taught me. Thanks, also, for being my friends.

THE SAN DIEGUITO COMPLEX IN ANZA BORREGO DESERT STATE PARK

G. Timothy Gross, Ph.D
Affinis
847 Jamacha Road
El Cajon, California 92019

Claude N. Warren, Ph.D.
Emeritus
Department of Anthropology
University of Nevada, Las Vegas
Harry Reid Center
4505 Maryland Parkway, Nevada 89154

Joan S. Schneider, Ph.D.
California State Parks
Colorado Desert District
200 Palm Canyon Drive
Borrego Springs, California 92004
and
Department of Anthropology
University of California, Riverside
Riverside, California 92521

INTRODUCTION

In interpreting the archaeology of parks like Anza Borrego Desert State Park, one question comes up over and over: "What are the oldest remains found in the area?" For Anza Borrego and surrounding Southern California, most archaeologists would reply that the San Dieguito Complex probably represents the earliest archaeological material in the area. Dated from about 10,000 to 8,400 B.P. (years before present), the San Dieguito Complex is poorly understood and is controversial. Some archaeological material is suggested to be earlier than the San Dieguito in Southern California, although there is not universal agreement among archaeologists on these remains.

Malcolm J. Rogers defined the San Dieguito Complex based on field work in the first half of the twentieth century and recorded the first San Dieguito sites in Anza Borrego Desert State Park. Ranging widely over Southern California, as well as adjacent Baja California, Arizona, and Nevada, Rogers searched for evidence that would allow him to understand the prehistory of the area. Within what is now Anza Borrego Desert State Park, he recorded five sites which yielded artifacts that he recognized as San Dieguito, as well as numerous sites associated with the later inhabitants of the area.

Since Rogers' time, many more sites have been recorded as having San Dieguito remains, and artifacts from some of these sites have been collected and are currently curated at the park. Most of this work was done by Robert S. Begole, a tireless avocational archaeologist who has spent an incredible amount of time in hiking the hills, mesas, and mountains of Anza Borrego Desert State Park and its general vicinity. His work has also taken him into Imperial County, closer to the areas where Jay von Werlhof's museum and Imperial Valley College are. Although we present some conclusions that differ from Begole's, we want to stress here the respect we have him and the effort that went into locating and recording all of the sites that he has. Despite differences in interpretation, we find this effort of great value to the archaeology of the desert and of southern California, in general.

During an initial examination of the artifacts from the more recently recorded "San Dieguito" sites, Gross concluded they were not particularly similar to San Dieguito artifacts he had studied from coastal San Diego County, including those from the C. W. Harris site, the site on the San Dieguito River that has been called the San Dieguito "type site." Although the Harris site is assigned by Rogers to the third and last phase of his San Dieguito Complex, by Rogers' accounts the differences between San Dieguito III and the preceding San Dieguito II are primarily matters of additions to the pattern. The lack of similarity was, then, perplexing and led ultimately to the current project.

The project reported here explores the nature of the San Dieguito occupation of the Anza Borrego Desert State Park. The research includes evaluation of the more recently recorded San Dieguito sites and the material collected from them in light of Rogers' conception of the San Dieguito, as expressed in his publications, his field notes, and the collections from the sites he recorded. The project was funded by the Department of Parks and Recreation through a cooperative agreement with Scripps Institution of Oceanography, University of California, San Diego[1].

THE PROBLEM

The material curated at Anza Borrego Desert State Park appears to differ from San Dieguito material from the coast of Southern California. The goal of this project is to increase our understanding of the San Dieguito Complex in the park and to address the question of how closely sites recorded in last half of the twentieth century fit the San Dieguito pattern as defined by Malcolm Rogers. This approach does not assume that Rogers was correct in all of his interpretations of Southern California's past, but only uses his original work as a starting place.

---

[1]The project was funded in April 2000 and in Spring of 2001 California's budget problems resulted in money allocated for this project being reduced by half. A subsequent contract with the Center for Archaeological Research, California State University, Bakersfield, allowed for completion of the research in the context of a large project examining cultural landscapes in the Lower Borrego Valley.

What do the more recently recorded San Dieguito sites represent? Are they really expressions of the San Dieguito Complex? Are they older entities? Are they remains of more recent cultures? Do they represent coherent archeological assemblages or might they be amalgamations of materials from different periods? This last question is particularly important because the artifacts in question are surface finds.

To address these questions five tasks were outlined. These are discussed in the next section.

METHODS

The five tasks developed to achieve the project goals were 1) examination of the existing records and collections for San Dieguito sites from the Anza Borrego Desert State Park housed at the San Diego Museum of Man, 2) examination of archaeological sites in the field, 3) examination of artifacts from Park sites housed at the Park, 4) detailed analysis of a sample of artifacts from the recently recorded sites and from the Museum of Man collections, and 5) evaluation of the results, including comparisons of the results of the detailed analyses and synthesis of the data.

Existing Records and Collections

Prior to examining collections at the San Diego Museum of Man, the relevant publications of Rogers' work were studied. These include "The Stone Art of the San Dieguito Plateau" (1929), *Early Lithic Industries of the Lower Basin of the Colorado River* (1939), and *Ancient Hunters of the Far West* (1966). These provided insight into the evolution of Rogers' thought on the San Dieguito and also provided information on the sites he recorded in the Anza Borrego Desert State Park.

The initial visit to the Museum of Man for this project was made in May 2000, although Gross had spent considerable time in past years familiarizing himself with the San Dieguito collections at the museum, and Warren had worked extensively with the collections, as well, particularly those from the Harris site (Warren 1966). State Park Archaeologist Rae Schwaderer accompanied Gross on the initial visit. Subsequent visits were made in June of 2000, December 2000, and January 2001. On these trips research was conducted in the Museum's archaeology lab. There the site files were reviewed, as were copies of Rogers' narrative notes on his work in the Colorado Desert. The individual site drawers (Rogers' collection is housed in drawers in wooden cabinets in the lab) for the sites found in the park were examined, as well. The Rogers type collection, material selected by Rogers as representative of the various archaeological entities he defined, was also studied. This collection was organized by region as well as by archaeological pattern. Special attention was paid to the material attributed to the San Dieguito from the Central Aspect (essentially the Colorado Desert), but the entire type collection for San Dieguito was reviewed. In February 2001, Warren visited the collection with Gross and we examined and discussed the San Dieguito type collection.

Site Visits

The site visits were planned for the last stages of the project so that they would be conducted after familiarity with the artifact assemblages and the existing site records had been gained. With the reduction in funding, only one site visit was undertaken in the initial phase of the project. In July 2001, Rae Schwaderer and Gross attempted to relocate Rogers' Site C-127 at Dos Cabazes. On the same trip one of the recently recorded San Dieguito sites was also visited.

On the more recent contract, additional site visits were made. In December 2003, all three authors met in the field and were accompanied by Bob Begole to inspect Sites CA-SDi-785, -786, and -1211 in the southern part of the park. Sites CA-SDi-4160 and -4161 in the Lower Borrego Valley, both part of what Rogers called C-122, were examined.

Examination of Park Collections

The initial examination of materials purported to be San Dieguito occurred in 1999 and formed part of the impetus for this project. The collections were examined again in February 2000 on a trip to Borrego Springs that also included a meeting with Bob Begole. In February 2001, we examined the collections again in detail, discussing pieces and drawing some.

Detailed Lithic Analysis

The initial plan was to conduct a detailed lithic analysis of a sample of material from the Museum of Man collections and from the park collections to facilitate comparison. An analysis form I designed for flaked lithic artifacts from the C. W. Harris site was employed for this project. Many of the artifacts in the Rogers type collection that were attributed to the San Dieguito and came from the Colorado Desert were analyzed. Some artifacts from the San Dieguito in the Mojave Desert, as well as small number of artifacts from Arizona and Nevada were analyzed, as well. Between February and December 2004, Bonnie Bruce of the Anza Borrego Desert State Park staff analyzed 96 items using the forms. Data from the forms are being entered into a computerized data base at the time this is being written.

The data form includes identification variables, measurements, and coded variables. The identification variables are Catalog Number, Site, Repository, Date (of the analysis of the particular item), Complex (i.e., San Dieguito II), and Analyst. Measurements are Length, Width, Thickness (all in mm), Weight (in g), and Edge Angle for up to four edges. Coded variables are Material, Production Base (indeterminate, flake, core, cobble, or tabular piece), Faciality (unimarginal, bimarginal, bifacial, etc.), Plan Shape, Profile Shape, Shaping Intensity, Flaking Quality (the number of flake scars in 5 cm of worked edge), Dorsal Surface Characteristics, Ventral Surface Characteristics, Edge Shape for up to four edges, Battering (presence or absence) for up to four edges, and Platform Location for flake-based tools that retain evidence of the original platform.

Evaluation and Synthesis

The evaluation and synthesis of the results of this project are in progress. What follows is a summary of our results to date. These include insights we have gained both through study of the collections and documentation and through considerable discussion amongst ourselves.

RESULTS

The results of this projects will be presented by task as outlined in the Methods section, above.

Existing Records and Collections

One reason there is so much confusion about the San Dieguito Complex is that Rogers called this material by different names at different times. When he first described it on the San Diego coast he referred to it as Scraper Maker (Rogers 1929:458), and this name occurs regularly in the site records at the Museum of Man and in Rogers' notes. In discussing remains from the Colorado and Mojave Deserts in 1939, he described the Malpais Industry as the oldest material, followed by the Playa Industry (Rogers 1939). He indicated his conclusion that the Playa materials were related to the what he was by then calling San Dieguito on the coast by referring to the San Dieguito-Playa Complex. The Playa Industry was divided into two phases, Playa I and Playa II.

By 1950 Rogers was calling the Malpais and Playa material San Dieguito. This is presented in Haury's report on the excavations at Ventana Cave in Arizona, where Rogers saw resemblances between material from the cave and materials from Southern California. Haury (1950:192) says: "A re-examination of his own material further persuaded him [Rogers] to make the revisions which he has permitted me to include herewith." Haury then presents a table that depicts Rogers' revisions to his terminology. A portion of that table is reproduced here as Table 1. The equation of Playa I and Playa II with San Dieguito II and San Dieguito III, respectively, muddied the waters a bit.

Table 1. Malcolm Rogers Revision to His Terminology as Presented by Haury

Old Sequence	New Sequence
Playa II	San Dieguito III
Playa I	San Dieguito II
Malpais	San Dieguito I

Source: Haury 1950:Table 12

Hayden (1967) contends that Rogers did not intend to eliminate the Malpais as a separate industry, but had intended to keep it as an earlier manifestation. A review of the notes at the Museum of Man appears to contradict Hayden's contention, however, as in some of the notes the word Malpais has been crossed out and replaced with SD I, Rogers' abbreviation for San Dieguito I (see discussion of C-122, below). Rogers may have made these corrections and changed his position at a later date, but this is not reflected in the notes. For this project we have used the terms as presented in Haury and as described in *Ancient Hunters of the Far West* (Rogers 1966).

Review of Rogers' work also indicates that he recognized the fact that for at least the San Dieguito I, it was difficult to distinguish the artifacts from later post-San Dieguito material on typological grounds alone. This is illustrated by several quotes. In comparing the San Dieguito I material to the coastal La Jolla pattern, Rogers notes:

> A similarity to the San Dieguito I lithic pattern, with all its crudities and even a lesser variety of artifacts, is met in the San Dieguito confines of the Pacific Coast, where it is found in the La Jolla culture. (Rogers 1966:38)

He continues:

> If the entire assemblage from the La Jolla shell middens were to be scattered over an equal area of the San Dieguito I terrain, it is doubtful that more than a fraction of it would be found, or recognized as tools if found. The artifacts of the first La Jolla period are mostly one nature, namely primary flakes. Not until the second phase, La Jolla II, were flaked implements of the nature and technical quality of San Dieguito I artifacts produced. In fact, a chief difference between the contrasted patterns lies in the absence of the metate in the San Dieguito I pattern and its presence in the La Jolla pattern. (Rogers 1966:38-39).

He later compares San Dieguito I to the Late Prehistoric occupations of the desert, which he called Yuman. Speaking specifically of an area north of the Superstition Mountains, he says:

> The archaeological problem here is a complicated one, for during the Yuman II period the plus thirty-foot shore line was first occupied and later became inundated for a long period, after the water rose to the plus forty-five-foot level. During this period of submergence, the lime coating was precipitated on both rocks and artifacts. It is a simple matter to eliminate such typical Yuman artifacts as metates, manos, and pestles from the problem. But the flaked lithic patterns of San Dieguito I and Yuman II phases are almost identical, even to the crudity of workmanship. Both utilize the same local green prophyritic cobbles which are strewn over the gravel spurs. (Rogers 1966:56).

Rogers' criteria for distinguishing San Dieguito I artifacts used not only form, but degree of surface alteration by desert varnish and patination, as well as association with types of land forms and surface features such as "sleeping circles."

Rogers described the San Dieguito I as being a crude industry sparsely scattered over desert surfaces (Rogers 1966:37). The types associated with this phase include teshoa flakes, beveled flakes, hand-choppers, scraper-planes, and ovate bifaces (Rogers 1939:17-18).

San Dieguito II is seen by Rogers (1966:60) as being more refined in workmanship and occurring in concentrations uncharacteristic of the San Dieguito I. The types characteristic of San Dieguito II include choppers, scraper-planes, keeled scrapers, flake scrapers, plano-convex knives, and double-convex knives (Rogers 1939:28-31).

Even better workmanship on stone tools is seen as characteristic of the San Dieguito III. In this regard Rogers (1966:61) says:

> It is also noticeable that some ancient forms such as bifaces survived from the San Dieguito I period, but with considerable refinement in form and flaking technique. Other basic implements, such as flake scrapers, end scrapers, and concave scrapers, attained perfection as well during this closing phase.

Types listed for this phase include side-scrapers, tabular scrapers, rectangular scrapers, beaked scrapers, spokeshaves, end scrapers, ovate scrapers, borers, reamers, small knives, stemmed blades, and crescentics (amulets in Rogers' terminology) (Rogers 1939:32-36). Rogers (1966:60) also notes a change in raw-material use, with the San Dieguito III knappers making use for the first time of cryptocrystaline silicates in addition to the volcanic rock of the area.

Examination of the information on sites in *Ancient Hunters of the Far West* (Brott 1966:150-152) indicated there were five sites recorded by Rogers in what is now Anza Borrego Desert State Park that he attributed at least in part to the San Dieguito. These are sites C-94, C-122, C-127, C-166, and C-185 (Figure 1). The site forms and notes for these sites were reviewed at the San Diego Museum of Man. Rogers provided the following descriptions of these sites in his notes.

C-94 Seventeen Palms

> Very little evidence of occupation here–consists only of a few sherds of both Kamia and Desert Cahuilla. On the gravel mesa back of the spring a San Dieguto plane (?) was found by Charles Sch____ [illegible in notes] in Feb. 1939. (Rogers N.D.:138)

C-122

> This number covers 9 sq. miles of territory in the Granite Wash area all of which is Eolithic Culture [in what appears to be Rogers handwriting the heading for this site bore the label Eolithic, which was crossed out and replaced by Malpais, which was also crossed out and replaced by SD I. GTG]. A strong Indian trail from the Narrows to Harpers Well region crosses this site. This may account for the occasional sherds that are found in the area. This area includes at least 800 house sites. Counted 300 in a 2 hour walk. Some are grouped in units, occasionally with walls merged. This area offers the greatest diversification in house form seen: besides the usual circular and oval types are many rectangular houses and some with small cache circles attached to the outer walls and even a few to the inner walls. A few concentric circles occur.
>
> The [sic] is a great quartzite chipping industry about the mesa house sites which is probably in part Yuman. They used it for a quarry. This area also includes the mesas which flank the Barrel Springs Wash. On both the low terrace and the mesa back of the wash are very ancient hearths scattered over considerable territory. Composed of groups of boulders burnt red–No archaeological material of any nature associated with them. Only one was in the form of a circle 3 ½ feet in diameter. (Rogers N.D.:172)

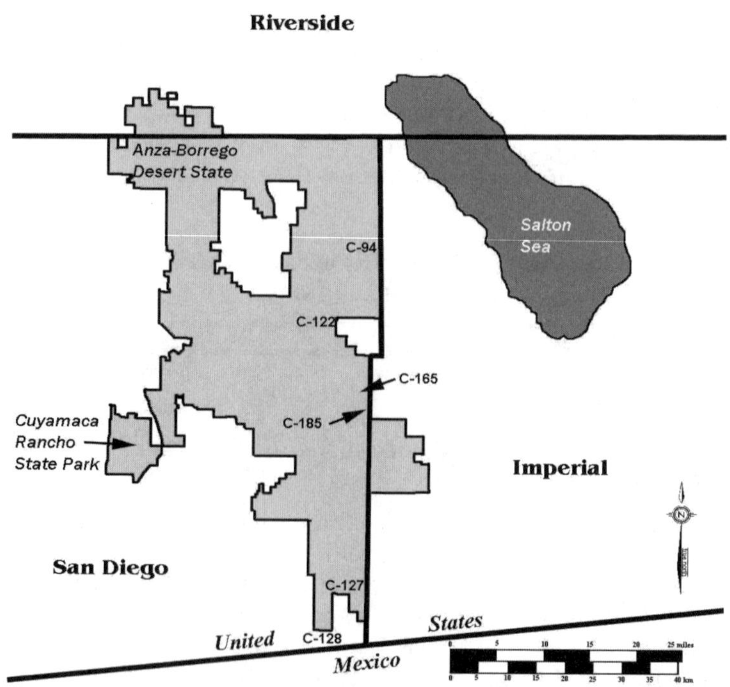

Figure 1. Location of San Dieguito sites recorded by Malcolm Rogers in Anza Borrego Desert State Park.

C-127 Dos Cabezas Valley

Only a small camp at the spring–main village in the central part of the basin on the north bank of the wash–scattered occupation for a mile or more to the east along wash. Indians lived under great granite boulders northwest of village also, and cached much pottery in the surrounding boulder piles. One large boulder has red ? [sic] pictographs on it. Bedrock mortars sparingly present at the spring camp but common on the north side of the valley. Also a few bedrock metates. On the north side of the valley in a cave with the back wall facing east are ten black pictographs–only eight can be distinguished readily–they are late work as a horse is shown. The black paint rubs off on the wetted finger and is probably graphite or wad.

The maximum depth of the occupation site at the spring is 24".

This region has a high % of desert ware–perhaps 40% but daubed ware is rare.

Following a brief description of the pictographs, the notes continue:

C-127-A Porphry flaking area on gravel mesas about the Dos Cabazes Road. This site number is given mainly to a San Dieguito area on a ridge of red loam located 2 miles east of Dos Cabazes. It lies between the Dos Cabazes Wash and the next big drainage to the southeast. It has a dune formation over it. Has some Diegueño porphry work and a few sherds on it at the east end. (Rogers N.D.:179-180).

C-166–Split Mountain Village

Located on east side of Fish Creek on a mesa at upper end of Split Mt. Mesa is 50' above the wash and very steep sided. Only one trail up the steep sides and this is flanked on both sides at the top by two defensive walls of boulders–badly eroded–probably never more than 3' high. About six house sites docuable [? word indistinct] in 1923. Outlined with boulders, elliptical shaped. 6 to 8 feet long. Quite a little obsidian flaking here in house floors–arrow points present. Main trail from Vallecito Creek drainage through to Salton Basin, passed by here. The mesa has small stacks of boulders scattered about in no apparent order. They are similar to those in the Sierra de las Tinajas, B.C., and may be of Eolithic period. Some patinated eoliths have been found on the mesa S of here and close by, also in ceremonial piles. (Rogers N.D.:228)

C-185–Coquina Mesa

North end of Split Mountain Ridge. Yuman II & III mescal camp. Stone house circles present. Approached trail from upper end of Split Mt. which passes by and above Long Tank. Very steep trail from this site goes to the east [word indistinct], with Fish Creek mesquite dunes villages. Small Tank one quarter way down trail. Another trail leads from here along the west flank of the Granite Mts. on its way south to Rincon Village.

C-185-A--Los Perditos Aite

San Dieguito I (with some II) site on gravel terrace east branch of Fish Creek Wash. This site is 2 ½ miles north of upper mouth of Split Mt. and almost due west of C-185 and several hundred feet below it. Between the terrace and a granite outcrop to the north are several long tanks in a chasm. These are almost duplicates of Long Tank. On a granite boulder here are a few petroglyphs. By the winding wash it is 4 miles from Split Mt. (Rogers N.D.:252).

These notes have been quoted at length to give the flavor of Rogers notetaking style and to show that all of the sites have material later than the San Dieguito occupations.

At the Museum of Man, artifacts are stored in drawers by site, for sites at which collections were made. The site drawers for each of these sites were examined. No artifacts were found in site drawers for C-94 or C-166. The three other sites had collections in the drawers, but it was not possible to determine which of the flaked-stone artifacts in the collections Rogers had classified as San Dieguito.

Rogers' San Dieguito type collection was then examined[2]. The type collection was organized by geographic area, or "aspect" in Rogers' terminology, and by phase. The central aspect is the area that contains Anza Borrego Desert State Park, and there were three drawers for San Dieguito from this area. One had San Dieguito I material, one had San Dieguito II and III material (undifferentiated within the drawer), and one had crescentics.

The San Dieguito I drawer (Drawer 520) has materials from sites in the Colorado Desert, but not items from the Anza Borrego sites. The drawer contained 21 items. The material was generally large flake

[2]There are actually two San Dieguito type collections at the Museum. As we understand it, one was assembled by Rogers as he was doing his research and the other was put together after Rogers death by Museum staff. It was the former collection that was used in this research because it was Rogers definition of San Dieguito that was important here, not others interpretation of his definition.

tools, cores, and cobble tools, as well as a teshoa flake. Most were heavily desert varnished. None of the items in this drawer looks much like material from the San Dieguito in western San Diego County, in our opinion.

The San Dieguito II drawer (Drawer 521) has 32 items, only two of which came from sites in the Park. A thin flake scraper made on a long flake with the distal end retouched came from C-127. A small domed scraper worked over about 85 percent of its dorsal surface was recovered from C-94, and may be the scraper mentioned in the notes. The collection was primarily unimarginal tools (scrapers), but there was at least one biface that was reminiscent of what Warren and True (1961) called Type II knives at the C. W. Harris site in western San Diego County. In general, many of the tools looked something like coastal San Dieguito material.

The crescents drawer (Drawer 522) contained 27 items, most of which came from the Mojave Desert. Specimens from Nevada and Baja California were also included, but none of the items was collected from the Colorado Desert. The crescents in the drawer differed from those found in coastal contexts in that the desert crescents tended to be simple crescents. Crescentics from the coast are usually eccentric crescentics with projections. This can be seen in a photograph from *Ancient Hunters of the Far West* (Rogers 1966:62) in which the Mojave Desert crescentics are illustrated on the right side of the photo and those from San Diego County sites are in the center and left hand columns. An eccentric crescentic like the western San Diego County specimens has been reported by Kaldenberg (1981). This came from the Pork Jerky site in the Kane Springs area of Imperial Valley.

The last visit to the Museum of Man for this project was with Warren and Gross in February 2001. We examined the notes together and looked at the site drawers and type collection from the central aspect. We also inspected the material from the western aspect (western San Diego County) which included material from the C. W. Harris site. This was done for comparison.

Generally, the material in the San Dieguito I drawer did not appear to us to be at all similar to San Dieguito as we know it from the coastal areas of San Diego (Figures 2 and 3), and especially from the buried deposits at the C. W. Harris site. In addition, our impression was that the materials did not form a coherent assemblage–one that made sense technologically or functionally. The material seemed to be a mix of expedient tools and early-stage quarry material (thick bifaces and unifacial cores). The San Dieguito II and III material, on the other hand, has some similarities to San Dieguito as we recognize it on the coast.

Site Visits

In July 2000, Rae Schwaderer and Gross attempted to relocate C-127 at Dos Cabazes. Although we spent some time hiking the area, we could not find the San Dieguito locus of this site. We did encounter some Late Prehistoric material, including sherds and flakes.

The later site visits to Begole Sites CA-SDi-785, -786, and -1211 and the sites that make up parts of Rogers C-122 (CA-SDi-4160 and -4161) were instructive. All these sites had stone circles, and there was considerable variation in the details of the circles. Some were just cleared patches where the desert pavement had been removed, exposing a sandy sediment. Others were ringed with rocks and were quite distinct.

The area around the circles at all the sites was examined for artifacts, as one of the goals of the site visits was to examine potentially San Dieguito artifacts in context. We found artifacts associated with all the circle sites, scattered between the circles. Artifacts were almost never found within circles. The artifacts were mostly large flakes, although some unimarginal tools (unifaces) were noted. Milling stones were found at many, including large rocks with slicks and basins, as well as manos. Sherds were found on some of the sites, as well, but never in great numbers. Entire reduction areas, places where a nodule had been flaked to produce a core or tool, were not uncommon, although most of these were relatively free of weathering and appeared to be relatively recent. These were marked by concentrations of flakes of the same materials concentrated in relatively small areas.

What was not seen on these sites were items that looked like San Dieguito as seen on the coast. While there was some superficial resemblance between some of the unimaginal items and San Dieguito scrapers, the flake scars were generally much larger and the unimarginal items from the Park lack the finished appearance of the coastal material. To the extent that the material on the sites we examined resembled anything from the material labeled San Dieguito at the Museum of Man, it was the material from the drawer marked San Dieguito I. As we have already indicated, we believe that this material is not part of a coherent assemblage and that it is most likely quarry debris of varying ages. The same explanation fits most of the flaked stone seen on the sites visited.

Figure 2. Unifaces from the Harris site.

Figure 3. Bifaces from the Airfield site, Carlsbad, California.

Examination of Park Collections

In February, 2001, the park collections were examined by Warren and Gross. Gross had examined them previously in a cursory fashion; however, this was a more detailed look at what the Park had in its collections. We pulled all the boxes with artifacts and went through each quickly to determine the general nature of the material. We then went through boxes again, pulling out each item, examining it and discussing it. Notes were made about many of the items.

The examination of the Anza Borrego artifacts came immediately after our last visit to the Museum of Man where we had reviewed Rogers' collections. We developed several impressions. First, some of the items were not necessarily artifactual. Items illustrated by Begole (1973) as Malpais appeared to be naturally shaped rocks. Begole (1973) says may of his pieces were, indeed, naturally shaped but that they showed signs of modification or use. We did not find definite evidence of such modification or use on these items.

Second, we encountered only one item that looked remotely San Dieguito. This is Catalog Number 270 SDi-1197B. It is a small uniface of green metavolcanic material. It is a flake-based tool with the entire dorsal surface flaked. We should note that Professor Warren is not as comfortable as Gross is with the resemblance between this item and San Dieguito scrapers.

There were a number of bifaces and unifaces in the collection (Figure 4), but these did not look much like the material in the Museum of Man San Dieguito Type Collection. Most of these looked more to us like the kind of material that would result from exploitation of lithic nodules on the desert pavement surfaces so common in the park or from cobble deposits. They are simple items with large flake scars such as would be made by people test-knapping pieces of raw material or removing flakes from nodules in the field for return to a base camp. In short, much of the material looks like quarry waste that may span the entire occupation of the area rather than the remains of a particular technology.

Detailed Lithic Analysis

The detailed lithic analysis is currently under way. Twenty-five items from the Museum of Man type collection have been coded on the analysis forms. Ninety-six items from the Park collections have been analyzed. The data from this step are being entered into a computer database for statistical analysis.

Originally it was thought that assemblages from sites could be compared to examine patterning in the occurrence of classes of artifacts. The nature of the collections makes this approach impossible. The Museum of Man collections are primarily items Rogers collected, and there is no indication what his collection criteria were. He appears to have picked up items he felt were characteristic of particular phases. In addition, except where items were placed in the Type Collection, it is sometimes not possible to tell which he felt were the San Dieguito items in collections from sites with multiple occupations.

The Park's collections also seem to consist of items judged by Mr. Begole to be important. It is not certain how these relate to the totality of the artifacts present on any of the sites.

The statistical analysis will be done at the item level. It will assess the similarity between individual items in the Museum of Man Type Collection and individual items collected from Anza Borrego Desert State Park sites.

Evaluation and Synthesis

Because some tasks are still in progress, only preliminary conclusions can be offered. These have been presented above, but will be summarized in the following section.

Summary and Conclusions

Several tentative conclusions have been offered above. These have to do both with Rogers' definition of the San Dieguito and with the nature of the sites purported to be San Dieguito in the Anza Borrego Desert State Park.

First, San Dieguito I material does not appear to form a coherent assemblage either technologically or functionally. One possible explanation for this material is that it is quarry debris from the expedient exploitation of surface lithic material. Nodules encountered in the course of hunting or moving across the landscape would have been tested, leaving a number of pieces with one or two flakes removed, as well as some flakes scattered about the terrain. Pieces of good lithic material could have been further reduced to obtain useable flakes. The lighter flakes might have been removed leaving the heavier unifacial and bifacial cores behind. This scenario would result in just the kind of pattern of distribution noted by Rogers for the San Dieguito I–a scatter of rather crude tools on the ground surface over relatively large areas. Under this scenario the association of these items with cleared circles and other features would most often be fortuitous.

Figure 4. Bifaces (top row) and unifaces from Anza Borrego Desert State Park.

Second, we could see no clear distinction between San Dieguito II and San Dieguito III in the type collection. Warren (1967) has noted this on the coast and has suggested abandoning the phase designations and simply calling it all San Dieguito. In working on a site that Rogers characterized as San Dieguito II in the Carlsbad area of San Diego County, Gross also could not distinguish the material from San Dieguito III (Gross 2000:65-66). It seems appropriate to lump San Dieguito II and III in the desert areas, as well. As discussed above, San Dieguito I may actually represent the quarry and expedient tool use of more than one temporal period.

Third, the quarry debris scenario described above seems to fit almost all of the material in the Park collections. Also included in the Park collection are a number of items that at best are manuports, but do not bear definite evidence of modification or use by humans, although, as Begole points out, the items could have been used unmodified for some purposes. There is no way to classify these as artifacts at this point, however.

ACKNOWLEDGMENTS

Former District Archaeologist Rae Schwaderer not only asked the questions that led to the project, but she is also responsible for obtaining the funding. State Park Archaeologists Sue Wade has been patient with us in administering these contracts. Bob Begole provided important insights into his work at a visit to his house early in the project and has been an interesting and informative field companion on some of the site visits. We look forward to working with him in the future. Ken Hedges and Grace Johnson of the San Diego Museum of Man were valuable sources of information and were wonderful about allowing access to the collections and providing workspace in the rather crowded archaeology lab.

John Hildebrand handled the contracting details at Scripps Institution of Oceanography and has provided considerable encouragement. Jill Gardner administered the contract for Center for Archaeological Research, California State University, Bakersfield. Bonnie Bruce accompanied us in the field and performed the analysis of material from the Park collection.

Although we are grateful to all those named above, we, alone, are responsible for any errors or omissions in this paper.

REFERENCES

Begole, Robert S.
1973 An Archaeological Survey in the Anza-Borrego Desert State Park, 1972 Preliminary Report. *Pacific Coast Archaeological Society Quarterly* 9(2):27-55.

Brott, Clark W.
1966 How Stones became Tools and Weapons. In *Ancient Hunters of the Far West*, by Malcolm J. Rogers, pp. 139-193. Union-Tribune Publishing, San Diego.

Gross, G. Timothy
2000 *Archaeological Investigations on the Air Field Site, CA-SDI-13,327 (SDM-W-180), Carlsbad, California*. Affinis, El Cajon.

Haury, Emil W.
1950 *The Stratigraphy and Archaeology of Ventana Cave*. University of Arizona Press, Tucson.

Hayden, Julian D.
1967 Restoration of the San Dieguito Type Site to its Proper Place in the San Dieguito Sequence. *American Antiquity* 31:43-44.

Kaldenberg, Russell L.
1981 Pork Jerky Site Crescentic, Kane Springs Area, Imperial County, California. In *A Summarization of an Archaeological Sample of the Glamis/Dunes Area, Imperial Valley, California*, by Charles S. Bull. Recon, San Diego. Prepared for the Bureau of Land Management, Riverside, California.

Rogers, M.J.
N.D. Unpublished field notes. On file, San Diego Museum of Man.
1929 The Stone Art of the San Dieguito Plateau. *American Anthropologist* 31:454-467.
1939 Early Lithic Industries of the Lower Basin of the Colorado River and Adjacent Desert Areas. *San Diego Museum of Man Papers* No. 3.
1966 The Ancient Hunters–Who Were They? In *Ancient Hunters of the Far West*, by Malcolm J. Rogers, pp. 21-108. Union-Tribune Publishing, San Diego.

Warren, Claude N.
1967 The San Dieguito Complex: A Review and Hypothesis. *American Antiquity* 32:168-185.

Warren, Claude N. (editor)
1966 *The San Dieguito Type Site: M. J. Rogers' 1938 Excavation on the San Dieguito River*. San Diego Museum Papers No. 5. San Diego Museum of Man.

Warren, Claude N., and D.L. True
1961 The San Dieguito Complex and Its Place in San Diego County Prehistory. *Archaeological Survey Annual Report*, 1960-1961:246-337. University of California, Los Angeles.

JAY VON WERLHOF AND THE ARCHEOLOGY OF OCOTILLO WELLS STATE VEHICLE RECREATION AREA

Philip Hines, M.A.
Associate State Archeologist
Department of Parks and Recreation, State of California
Sacramento, California 93582

When Jay first moved to the Imperial County in the early 1970s to teach at Imperial Valley Junior College, his colleagues warned him there was very little archeology in the Colorado Desert. Less than several hundred sites had been recorded in Imperial County at the time. Today there are well over 6,000 sites. For the past several years, Jay has worked with state parks archeologists monitoring and adding to the list of recorded archeological sites at Ocotillo Wells State Vehicle Recreation Area (OWSVRA). OWSVRA is located in the Colorado Desert 100 mi northeast of San Diego and near the western shores of the Salton Sea. The unit covers 78,000 acres and includes state and federal lands (Bureau of Land Management).

Figure 1. Jay examining artifact at site in eastern acquisition unit.

Above, Jay examines a flake at a sparse flake scatter along the southern edge of an east/west trending tabular sandstone outcrop (Figure 1).

Jay brought with him the desire to work with local Native Americans just as he had done in many other areas of California. Jay's awareness of local customs has undoubtedly helped him identify features in the field such as geoglyphs that are frequently missed during cultural resource inventories. This paper will focus on these types of features.

During the third week of October 2003, with temperatures reaching 108 degrees, we began updating records for the 20 sites in the old eastern acquisition unit (20,000 acres) west of Pole Line Road. These sites were first recorded in the late 1980s during the original cultural resource inventory. The sites range from isolated hearths to large seasonal camps associated with the various shorelines of Lake Cauhilla. This area of OWSVRA is classified as an open riding area meaning vehicles can travel anywhere. A rigorous methodology was employed during the current project to re-record all of the previously recorded sites. Two-meter-wide transect intervals were used. All features, diagnostic artifacts, and concentrations of artifacts were mapped with pin flags and mapped to within 2 m accuracy with a Trimble Geo-Explorer III.

During the recordation of Imp-6313, an extensive lithic and pottery scatter, Jay walked a couple of hundred meters north of the site to a low terrace he saw in the distance. The terrace turned out to be an old mesquite dune anchored by the remains of a trunk from a long-dead mesquite bush. Hundreds of artifacts including lithic tools and debitage as well as a variety of pottery sherds surrounded the mesquite trunk. The site had not been recorded during previous surveys. The site was given the temporary designation of Jay's Site in his honor. During the final stages of recordation, Jay made an additional discovery: three rather large granitic boulders, 40 to 50 cm across were placed in a linear fashion 20 m apart across the site at a bearing of 89/179 degrees. The boulders would have been carried in from a wash 20 m south of the site.

The survey strategy was modified as a result of Jay's find to include re-surveying low terraces in areas that had been previously surveyed. The earlier surveys had concentrated on existing mesquite dunes and neglected older dunes that no longer had growth on them. We searched all of the low terraces between Tarantula Wash and IMP-6313, approximately 200 acres. Four previously unidentified sites were recorded, two having cremated human remains, a small scatter of artifacts including an almost complete pigment palate, a bifacial mano made from white quartzite, and a sparse flake scatter. Two extensive flake scatters were identified on low terraces along the northeast side of Tarantula Wash 700 m northeast of Imp-6313. Several wonderstone cores and flakes were found at these sites.

While walking out to re-record Imp-5184 along Tule Wash, a site recorded in 1989 as having three tabular sandstone enclosures, several flakes, and pot sherds, Jay found a rock-lined oval measuring 130 cm x 100 cm made with granitic rocks measuring 3 to 7 cm across. A single granitic rock in the center of the oval measures 13 x 26 cm. Jay described the features as a "prayer circle." Jay began scanning the area and picked out three granitic boulders running out from the oval at a bearing of 129 degrees (true). The boulders measured 30 to 40 cm across. The boulders were out of place among the tabular sandstone formations along the northern terrace of Tule Wash. They were likely to have been carried up onto the terrace from further up Tule Wash. The two easternmost boulders are partially buried in old lake sediments suggesting that they were originally placed near an active shoreline of Lake Cahuilla. The three tabular sandstone enclosure described for Imp-5184 lie 54 m northeast of the "prayer circle" and rock alignment (Figure 2). The features were added to the site record for –5184.

Figure 2. "Prayer circle" at Imp-5184.

While surveying and recording four previously unrecorded sites several miles northeast of the previously described features, the project geologist found a single granitic boulder measuring 60 centimeters across in old lake sediments that lay between two prehistoric lakeshore campsites situated on tabular sandstone formations. The boulder was out of place. By this time we had learned from Jay to scan areas for materials that might be out of place. We found three additional boulders of the same approximate size. The four granitic boulders are laid out in a "diamondlike formation" similar to those painted on the rocky knoll near the Imperial and Riverside county line at Travertine Point. The likely source for the boulders is a cobblefield 0.75 km southwest of their present location. The cobblefield is downslope of the diamond feature meaning the boulders had to be carried in and placed in the lake sediments.

During the final rotation of the 2004 spring season, Jay found another rock alignment (Figure 3). This time he was very animated. I was several hundred meters to the south recording a site when Jay began jumping up and down like a kid. My first thought was that he had walked across a sidewinder. That was not the case. He had discovered a rock alignment in an area that had been previously surveyed by two different crews in 1989 and by our crew during the current project. I probably walked across this alignment several times and did not notice while recording two seasonal shoreline campsites on nearby tabular sandstone formations. The reason for Jay's excitement was that this was the first example to his knowledge of this particular type of alignment to be identified on the west side of the Colorado Desert. He had recorded similar alignments along the Colorado River some 70 miles to the east of the project area. The alignment was made by placing 70 softball size or smaller globular and flat sandstone rocks in a linear fashion that measures 320 m in length. The feature is oriented true east/west. A second alignment extends to the southeast (150 degrees) from near the western end of the feature.

Working with individuals like Jay von Werlhof often makes us move away from the cultural resource management approach that dominates archeology today. We learn to take a little more time recording sites and increase our knowledge about the past cultures we are studying. Undoubtedly Jay's presence on the OWSVRA survey team has resulted in the recordation of sites and feature that would have been missed.

Thank you Jay.

Figure 3. Jay recording 320-m-long rock alignment

A TRIBUTE TO MY GOOD FRIEND, JAY VON WERLHOF

E. Henry James
Master Trail Finder
Mojave River Valley Museum
Barstow, California 92311

I met Jay and Sherilee von Werlhof through Russell Kaldenberg and Judyth Reed. In 1977 my wife Edith and I went to Ocotillo to help Jay and Emma Lou Davis excavate a huge rock cairn. The relationship has lasted all this time. We sort of took to each other because he was interested in trails and intaglios and geoglyphs as was I.

In 1986 a man came into the Mojave River Valley Museum and told us about five intaglios that he had found near Dunn, south of Interstate 15 near the Mojave River. He led us out to the site and showed them to us. We were amazed to find figures like that so close to us that we knew nothing about. We discussed it among ourselves as to why we had not found them during the past years. At first we accepted them as being genuine ancient geoglyphs.

We took Pat Barker, then BLM's Resource Area archaeologist, to the site. He immediately labeled them as recent. We then took Carol Rector, Curator of Anthropology at the San Bernardino County Museum to look at them. Carl labeled them as recent. Barker pulled out his aerial photos and found a good photo that showed that the terrace of the river and the geoglyphs were not in the photos which were taken in the late 1960s.

In order to verify that the intaglios were recent, I went trail hunting. I found other things that I could identify, but the geoglyphs did look new to me. I then met with Pat and Carol separately and asked both Pat and Carol why they had labeled the figures as recent. They both had the same answer. One of the figures is a mountain sheep that has feet on it. The old people did not put feet on their figures.

I called Jay von Werlhof about the recent discoveries, being excited about them. Jay was going to come up to Barstow and document them. He had heard from someone that they were authentic and was quite taken aback by my discussion with Pat and Carol. I asked him to hold off for a while.

Knowing that the figures were false, I remembered things that I should have thought about. First, Beth Pinnell and her family have an airplane that has flown all over the desert looking for ground figures and other historic sites. She has flown at a very low elevation looking for archaeology. When I first began mapping trails, she found them easily from the air. She remembers flying over the Dunn area at less than 300 feet elevation, and she does not remember seeing anything like that out there. I contacted Jay again, and he began sounding more convinced.

Finally, I received an anonymous letter stating that someone had faked the geoglyphs to play a trick on archaeologists. The unsigned letter says they were faked a few years back and went on to say that if I documented the geoglyphs I would eventually be embarrassed as having written up something modern and portrayed them as ancient.

I got back with Jay, a wonderful advisor. He said, "indeed, you have all of the evidence you need. Document your thought processes and that is all you need to do."

That is the type of relationship I had with Jay. He is always willing to help amateurs like me without making me feel dumb. It has been a real pleasure working with him.

E. Henry James

Postscript: Henry passed away recently. He was a master trail finder until his last days. The geoglyphs he was concerned about turned out to be a different set than Jay described along the north flank of the Mojave River in his famed presentation at the 1999 Kelso Conference.

JAY VON WERLHOF: HE WALKS, HE TALKS, HE READS AND WRITES ARCHAEOLOGY

Russell L. Kaldenberg, M.A. RPA
Command Archaeologist
China Lake Naval Air Weapons Station
China Lake, California 93555

Abstract

At the age of 83, Jay is still full of boundless energy. He has been doing California archaeology since the end of World War II. His pursuit of knowledge has taken him to every county in California. He has inspired two generations of archaeologists sharing his knowledge with them and fighting to help save California's precious past.

Jay's Bronze Rat

Jay walked into a curio shop in San Francisco as he often does when he visits the place he called home from 1954-1959. While looking around at the exotic merchandise, he noticed a very lifelike, life-sized, bronze statue of a rat. It had no price tag, but was so striking that he decided he must have it for his collection.

He took it to the old shop owner and asked, "How much for the bronze rat?"

"Aaaahhhh, you have chosen wisely, it is $12 for the rat and $100 for the story," said the old shopkeeper.

Jay quickly pulled out $12 for the rat. "I'll just take the rat, you can keep the story." Jay told him.

As he walked down the street carrying his bronze rat, Jay noticed that a few real rats had crawled out of the alleyways and sewers and had begun following him down the street. This was a bit puzzling so he began to walk faster. A couple of blocks later, he looked behind him and saw that to his puzzlement the herd of rats behind him had grown to hundreds, and they had begun squealing. Sweating now, Jay began to trot towards the bay.

Again, after a couple more blocks he looked around only to find that the rats now numbered in the MILLIONS and were squealing and coming towards him faster and faster.

Panicking, Jay ran to the edge of the bay and threw the bronze rat as far as he could into the bay. Amazingly, the millions of rats all jumped into the bay after the bronze rat, and were all drowned. Tired as he was, Jay walked back to the curio shop.

"Ahhhh," said the owner, "You have come back for the story."

"No sir," said Jay, "I came back to see if you have a bronze Republican."

In The Beginning there was No Beginning

This philosophy permeates Jay's life. Everything for him has always been an adventure that started sometime in the past. Just when is hard to tell because the quest for knowledge was always there. Until last year he was even uncertain of his birth year. Jay was born in Red Bluff, California, in 1923, the son of Victor Harry von Werlhof, a WWI test pilot who was also a lay preacher as well as newspaper editor for both the *Portland Telegram* and the *Sacramento Union*. Shortly after he took the job as editor of the *Sacramento Union* he was taken by spinal meningitis combined with double pneumonia and passed away when Jay was only 3. Jay has no memory of him except through stories. His mother, Bess Emily von Werlhof, took care of Jay and his three older brothers. She made a living by dressmaking and taking in sewing. Upon the death of Jay's father, his mother received a $1,000 life insurance policy, which she used to buy the old railroad stationmaster's house where Jay grew up.

Jay remembers his grandmother well; she had been a fortuneteller and his grandfather a bear hunter. One of the earliest memories he has of his family is of his grandmother divorcing his grandfather and marrying Tom Clegorn, an Indian who greatly influenced Jay. Jay feels strongly that this man may be the Yahi, Snowflake of the Ishi story. Jay cannot remember a time when he was not interested in anthropology, history, Indian cultural history, or other intellectually stimulating subjects.

He spent much of his early life in Red Bluff battling various illnesses, including rheumatic fever. He was pretty much of a loner even though he was elected class president. He was in high school when he was drafted into the Army. He was 20 years of age. As he was leaving for basic training in Texas, the local manager of the Red Bluff J.C. Penney store told him that when he came home from the war he would have his own J.C. Penney store in Willows, California. Penney's loss has been anthropology's gain (Figure 1).

Paper presented at the Society for California Archaeology Annual Meeting, March 20, 2004, Riverside, California. Session honoring the contributions of Jay von Werlhof. (Figure 1 courtesy of Sherilee von Werlhof, Figures 2, 3, 5, 6, 7, & 9 were taken by the author, Figure 4 was taken by Gary Breschini and Figure 8 is courtesy of the Archaeological Conservancy.

Figure 1. Jay upon being discharged from U.S. Army hospital in Denver, Colorado, waiting for General Eisenhower.

The Army

It was in the trenches in Europe during World War II that Jay made the decision that would lead him to history and finally anthropology. He was a sergeant in the 114th Signal Radio Intelligence Company (see Figure 1), G2, U.S. Army, much of the time behind enemy lines intercepting intelligence for detecting enemy troop movements. As Jay himself tells it, "Linguistics in the TA unit could handle any written European language or dialect. Sgt. Jeeter McPheeter, for example had a Ph.D. in European languages at University of North Carolina. Joseph B. Rounds, who occupied the 2nd half of my pup tent, had set up a library for the League of Nations, was on leave as Head Librarian of Grovsner Library in Buffalo, and spoke fluent French and German. Others in the unit covered the Slavonic languages. One of the members of our TA Platoon was a personal friend of Walt Disney who agreed to design the logo for our reorganized company. The original is in the National Archives" (von Werlhof 2002: 2).

With bombs literally bursting in air all around him, Jay intercepted German intelligence traffic and decoded messages while building order-of-battle maps. His wartime companions and their knowledge of history intellectually stimulated Jay. He was at the Battle of the Bulge where the intelligence gathering of his unit was paramount. If some of the advice Jay provided his superior officers been used, he feels that some of the lives of the 160,000 American and German casualties might have been spared. Jay's story of the Battle of the Bulge is riveting.

Those of you who know him know that he is a kind and gentle man. He tells of a snowstorm in Belgium during the war where a native woman was panicked that her child was lost. Jay organized a search party who found the youngster alive. As a token of appreciation, the woman and her friends stayed up all night making biscuits for Jay's company.

College, Jay, not J.C. Penney's

It was Joe Rounds who urged Jay to go on to college on the GI Bill. So, when Jay returned home to Red Bluff he worked at Penney's half time. He applied to college at Berkeley and was accepted. But he was on probation. He had been accepted provisionally based upon passing three courses. He passed the provisional courses via correspondence school and in June 1946 became a student in good standing ready to enroll in fall 1946.

Jay, without a high school diploma, enrolled in the exciting History Department. But he was interested in anthropology so his classes were eclectic. They included American Indian classes, religious philosophy, and California history and archaeology. Jay met A.L. Kroeber and was greatly influenced by his teachers who included, as he characterizes them, a cigar-chomping Herbert Bolton, the brilliant civil war specialist Kenneth Stamp, the physical anthropologist Ron Olsen, and, of course, his friend Robert Heizer. Jay developed the greatest esteem for Heizer, whom he worked for and with for the first decade after graduating from Berkeley. He assisted Nancy Heizer in keeping Michael Heizer interested in high school.

Jay began his archeological work in 1950 with surveys on Wintu sites, including archaeological sites that he located when he was in high school. He mapped Wintu sites on the west side of Blue Tent Creek, which had huge residential house pits. Heizer had quickly recognized Jay's artistic ability and had Jay drawing site maps for the University of California Archaeological Survey. His first archaeological report was written on the Wintu sites on Tent Creek and were written in 1950 with a supplemental site record written in 1951.

Jay spent until 1954 in graduate school, leaving UC Berkeley with an MA in history. He began his ethnographic work in 1954 with the Tachi on the Santa Rosa Rancheria north of Coalinga. He interviewed the last of the traditional Tachi chiefs and published significant ethnographic information regarding the Yokuts.

His Early Career

While Jay was at graduate school at UC Berkeley, Heizer asked him to prepare a class syllabus for the San Francisco School of Fine Arts integrating history/anthropology and art. Heizer said that they were looking for someone to implement the curriculum—why not the author? Jay taught from 1954-1959 at the School of Fine Arts. It was during this time that he became increasingly interested in Native American art, particularly in rock art. He began his Sierran Rock Art surveys during this time period (Figure 2). His notes are meticulous.

Figure 2. Jay meticulously drawing his sites.

In 1959, without his PhD, Jay was unable to get tenure at the School of Fine Arts, so he began the journey that ultimately led to him do archaeology in each of California's 57 counties. He and Robert Heizer went out for a hot dog, and Heizer asked him if he was interested in a new type of appointment. Jay thought the job would be interesting because it did not limit him to a single location. So, Heizer appointed him as an Archaeologist-in-non-Residence. Jay and James Davis were the only two ever to occupy such a position.

During this time, Jay received an American Philosophical Society grant to record rock art in the Owens Valley, California, and at the Valley of Fire, Nevada. This resulted in some of his best early professional work. It resulted in the publication of *The Rock Art of Owens Valley* (University of California Archaeological Survey Reports 65:1965).

He taught at the College of the Sequoias during this period also, but was fired from that institution for insubordination. He refused to wear a belt on his slacks so the college dean labeled him "a trouble causer." This trouble causer, though, conducted excavations at Hospital Rock for the National Park Service and on the Kaweah River for the Army Corps of Engineers.

In 1960-1961, he was able to obtain a grant from the National Science Foundation to document obsidian quarries in northern California. He also expanded his survey knowledge and conducted surveys in Tulare and Kern Counties, a Supplementary Report on the Tachi Indians with J. Vierhus, his work on Rocky Hill, his works on the petroglyphs of Fresno and Tulare counties, annotating notes on various Tulare and Kern County sites, and documentation of aboriginal trails of the Kaweah Basin, Tulare County. His excellent work on the Mouse Tank, Picnic, and Atlatl Rock sites was also done during this period of time.

From 1961-1963 Jay was an Archaeological Research Assistant at the Robert Lowie Museum. It was here that he got his first good experience with working with the various publics that he would spend the rest of his life with. He conducted experimental summer field school classes for Lassen and Siskiyou counties and taught archaeology at high school in Plumas and San Luis Obispo counties. Part of his job when he was teaching in Orleans was to be the assistant school librarian. The librarian asked him if he knew how to shelve books and he said, "Yes, I do, what system do you use, the Dewey Decimal or some other?" The librarian replied that they used another system; books were filed as "Easy To Read, Hard To Read, and Harder To Read."

He continued recording and excavating sites with his high school students, taught eighth grade English and published reports and monographs in the University of California Archaeological Survey series.

Jay Goes Back to Graduate School

Believing that he needed a doctorate, he went back to school, this time to the University of California at Los Angeles in the Department of History. He spent from 1964 to 1967 completing his requirements and his dissertation on intellectual history. Due to a change in the college administration and to the philosophy of one member of his committee, Jay left UCLA with his dissertation 2/3 signed, never looking back, he would say. He had made good friends at this time with Clem Meighan who told him about a temporary job open at Chico State University.

Mid-Career Challenges

Jay taught at Chico State in 1967 and 1968 and then at California State Polytechnic University, San Luis Obispo, from 1968-1971 where he became the director of the San Luis Obispo County Archeological Society. It was at this time that Paul Shumaker invited him to be one of the founding members of the Society for California Archaeology.

Again, without the doctorate, though, he was unable to get tenure. This time he moved to Cuesta Community College, where he taught from 1971-1973. It was at this time that he established the museum at the Hollister Adobe on campus, and created the Chumash Indian Research Center. He enjoyed tremendous support of the community. One of his more interesting projects was going to the Planning Department and getting a map of all vacant lots in San Luis Obispo then having his students survey every vacant lot. Every vacant lot had evidence of coastal shellfish. He then had his students study shellfish remains as an important cultural ecofact. He also worked with Roberta Greenwood and did his first work on the Carrizo Plain, looking at the subsistence patterns of the Chumash, Yokuts, and Salinan.

109 Recorded Sites

He met Sherilee (Figure 3), his partner for the last 30 plus years, at that time. Sherilee's family was from El Centro. Her mother heard that Imperial Valley College was in the market for someone to teach anthropology. Jay interviewed and excited the resident of the college about teaching at San Diego State University (Imperial Campus) also as well as developing a local program involving the community. At the time of his arrival in 1973, there were 109 sites recorded in Imperial County. He talked with Heizer about this change, and Heizer asked him if he thought that he would really be happy with a place that had so few archaeological sites. Jay thought that he might organize a survey class and find a few more.

The rest, as they say, is history. From 1973 until his retirement in 1992, Jay taught anthropology, Indian studies, and archaeological survey and supervised the museum. He also taught classes at Arizona Western Community College, University of Redlands, and Northern Arizona University (Yuma campus). He served as the director of the Imperial Valley College Desert Museum and was instrumental in constructing the museum as it is today in Ocotillo.

He has surveyed most of the county. There are a few holes out there but he feels that he has found all of the geoglyphs and rock art in the county. More than 10,000 sites have now been recorded in the county. He has conducted about 4,000 separate field surveys and has completed 185 field projects as a result of CEQA, NEPA, and research projects. He also developed and continues to maintain his interest in the archaeology of Baja California.

Figure 3. Jay and Sherilee von Werlhof in 2003.

Achievements

Among Jay's prestigious achievements are receiving the Society for California Archaeology's covetous Mark R. Harrington Award for Conservation Archaeology in 1982 and the society's Lifetime Achievement Award in 1996. He also received the Imperial Valley College Student Body Teacher of the Year Award in 1991 and The California Bureau of Land Management's State Director's Volunteer of the Year Award in 1993 (Figure 4).

His enthusiasm is contagious. At 81 years of age he is just now resigning as Director or the Museum so he can concentrate upon his writing and his summary of his accomplishments. His volunteers will carry on to completion what he has started.

Figure 4. Jay receiving the prestigious Mark R. Harrington Award for Conservation from the Society for California Archaeology. Presented by Dr. Paul Ezell and Russell Kaldenberg 1982.

From childhood Jay has always been interested in archaeology and religion. His work on rock art has contributed to a better understanding of the belief systems of local native peoples, wherever he has chosen to work. He has taken a great deal of pain and effort to document, make public, and understand the meaning and sensitivity of the ground figures of the California Desert, particularly where he uses ethnography to attempt to understand those along the Colorado River. Jay's *Spirit of the Earth, Volume 1 and 2* will stand as monuments to his ability to bring to life the prehistoric past of our desert area. Those of you who have surveyed with Jay, particularly on a geoglyph survey, know that when he is out among the ancient evidence of humans, he takes on another presence. It is almost as if he is walking an inch above the ground, floating above the sites so they will not be disturbed (Figures 5 and 6).

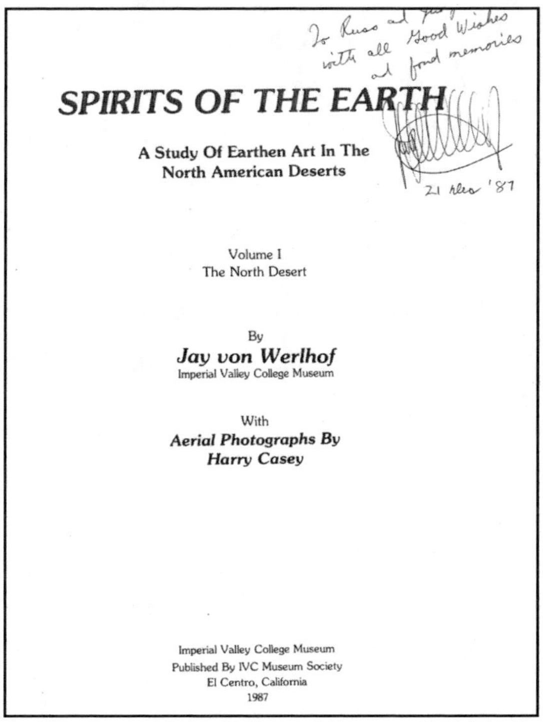
SPIRITS OF THE EARTH

A Study Of Earthen Art In The North American Deserts

Volume I
The North Desert

By
Jay von Werlhof
Imperial Valley College Museum

With
Aerial Photographs By Harry Casey

Imperial Valley College Museum
Published By IVC Museum Society
El Centro, California
1987

Figure 5. Cover page from *Volume I Spirits of the Earth.*

He was greatly influenced by Emma Lou Davis in getting to know the makers of the sites they left. Interpretation and the use of ethnography have been paramount in his ability to understand and assign meaningful contexts to lines on the ground. Jay was one of the first landscape archaeologists (Figure 7).

Of course Jay, is interested in the peopling of North America, as are most archaeologists. Early on in his desert career, he worked to understand what he thought to be early manifestations of archaeology in Imperial Valley in the form of the Malpais tradition. Jay has come to understand the Malpais not as a culture but as work production centers of perhaps the San Dieguito people. His feeling is that due to the constant inundation of the landscape that any early sites would have been removed or mixed by the successions of the Gulf of California intruding into the Imperial Valley. He feels that the first occupation of the area was probably about 12,000 BP but still longer for older sites; he accepts dates of about 6,000 years for the oldest recognizable sites in the Imperial Valley.

THAT THEY MAY KNOW AND REMEMBER

Vol. 2: Spirits of the Earth

By Jay von Werlhof

Aerial Photographs by Harry Casey

Figure 6. Cover page from *Volume 2 Spirits of the Earth.*

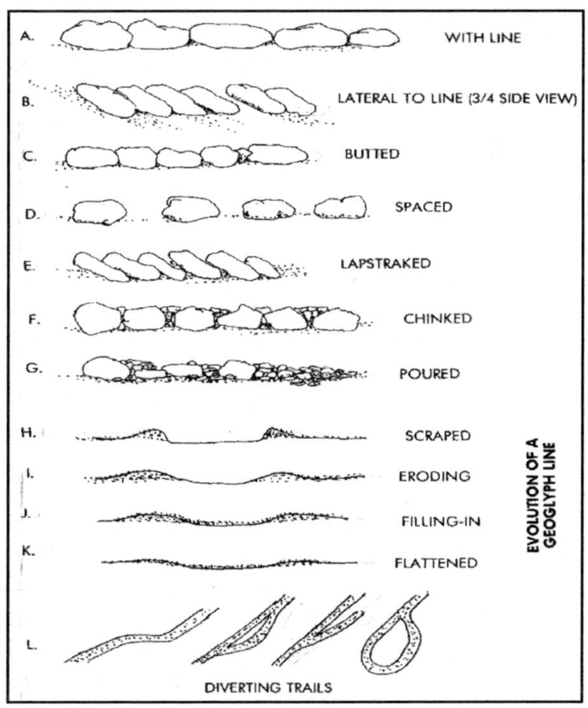

Figure 7. Jay's concept of alignment, geoglyph and trail construction.

Influences

While Jay was a student of Heizer and Bolton, he was a contemporary of Paul Ezell. Like Paul, he also had great admiration for Malcolm Rogers, an early archaeologist in San Diego County who died before Jay could meet him. After moving to El Centro, he and Paul worked closely together, sharing students and devoting time to developing a feel for public archeology.

Along with Bob Heizer, Jay greatly admired Paul Ezell and worked with Paul as much as possible. Others who have been particularly influential are James Moriarity and Emma Lou Davis, Aden Treganza, James Bennyhoff, Frank Fenenga, Fritz Riddell, and Martin Baumhauf. Native Americans Lorey Cachora, Preston Arroweed and Joe Homer, Carmen Lucas and Laura Cline have been special to Jay in his attempt to document the past.

Harry Casey, an avocational archaeologist, helped Jay for his first ten years or so in the Imperial Valley with photographing geoglyphs from a slow-flying airplane.

Jay has few regrets in his life. He would do most things just as he has. Not having the PhD, he feels, has made him more flexible and less prone to stay in one place. He has never looked back and said that he should have dotted the i's for the one committee member who did not like his topic.

Jay has always been interested in the fish traps associated with Ancient Lake Cahuilla. A few years ago the Archaeological Conservancy featured him on its cover examining fish traps that eventually were purchased for an addition to Anza Borrego State Park (Figures 8 and 9).

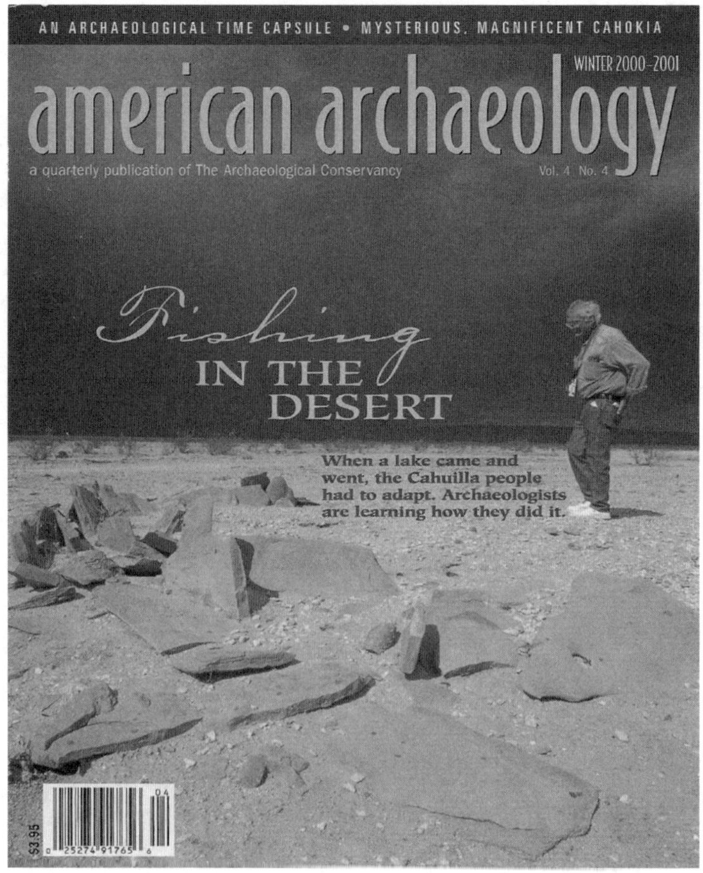

Figure 8. Jay on the cover of *American Archaeology*, Winter 2001-2001.

He feels that his most important contributions to the discipline are his study of geoglyphs and his ability to add people to the equation. His greatest fear is that these fragile resources will be destroyed out of complacency; there are many more ground-disturbing activities than people to prevent them. He feels that the Anza Borrego Desert State Park is the best protector of the resources and that BLM and the Forest Service are the worst because they lack the funding and sometimes the management wherewithal. Jay sadly recounts that he has seen too many archaeological sites disappear due to recreation needs and agency indifference. He wants more protection for the resource.

I remember he proposed that the Yuha Desert Area be set aside as a cultural park. Jay developed an energetic plan to have this cultural park created out of a need to save some of the sites for perpetuity. This plan is still exciting. Unfortunately, while the concept of cultural parks is boldly identified in the 1992 revisions to the National Historic Preservation Act, no implementing regulations have ever been written.

Jay is always thinking out of the box. I am holding here a bunch of dowels. Jay mailed them to me and said to try them to measure bedrock mortars. I used that method at a site in the Temblor Mountains; it really worked well, and then I set it aside forgetting that I had them for the last 20 years. I found them as I was looking for something in the garage recently. They had been stuck in a map rack. This is another example of Jay, the artist archaeologist thinking outside of the box. Most of us ask *why* when confronted with a puzzle, Jay thinks *why not* and goes on to attempt to solve the mystery.

The archaeologist he never met but would have enjoyed meeting the most is Francois Bordes. Jay would ask him about cultural evolution related to development of cost-saving lithic tool productions such as the Levallois technique of stone tool reduction.

He wants to be remembered for his work with geoglyphs. I think he should also be remembered for his community museum that sits at the Ocotillo off ramp between San Diego and Yuma. His last major goal now that he has published his Volume 2 of *Spirits of the Earth* is to open his museum. He does not need much money, just a little. Lets all help him to meet that goal by working to help him open a room at a time.

Figure 9. Jay with volunteers examining a fish trap in Imperial County.

I asked Jay if he had any regrets at all. "Well," he smiled; "I wish that I had minded my money a little better. I have always been broke." I think to all of us in this room and to everyone Jay has touched with his enthusiasm, he is indeed a very wealthy man. He has more friends than he knows what to do with. To quote this remarkable man, "In the Beginning there was no Beginning. " His enthusiasm and scholarship had no beginning. They were always there as has been his willingness to share information, thoughts, ideas, and always with his pleasant smile…none of these will have an end. Jay, through his notes, his writing, and his enthusiasm, will always be there waiting for us to ask just the right question and borrowing from his various approaches to find a solution.

Some Key Publications

1950 Report on Wintun Indian Sites around Blue Tent Creek near Red Bluff, Tehama County.

1951 Supplemental Report on Wintun Indian Site around Blue Tent Creek near Red Bluff, Tehama County.

1958a Rocky Hill, Tulare County, California. Petroglyphs in Fresno and Tulare Counties. On File at the Phoebe Hearst Museum, University of California, Berkeley.

1958b "Granite Galleries" in Pacific Discovery Volume XI - Number 4, July-August pgs. 149-156.

1959 Survey Report and Supplementary Report on Tachi Indians (with J. Vierhus). On File at the Pheobe Hearst Museum, University of California, Berkeley.

1960 Archaeological Investigations at Hospital Rock, Tulare County, Californian. San Francisco: Report to the National Park Service.

1961a Aboriginal Trails of the Kaweah Basin. San Francisco: Report to the National Park Service.

1961b What We Found at Hospital Rock. San Francisco: Report to the National Park Service.

1965 Rock Art of the Owens Valley, California. Berkeley: University of California Archaeological Survey Reports 65.

1979 With Sherilee. Archaeological Examinations of the Magma Site, East Mesa, El Centro: Imperial Valley College Museum.

1987 Spirits of the Earth, Volume 1. El Centro: Imperial Valley College Museum.

1988 An Archaeological Park: A proposal to Bureau of Land Management of the Preservation and Public Education of Geoglyphs on the Yuha Mesa, Imperial County, California.

2002 One Man's Decision: Why SHAEF Failed to Halt the Battle of the Bulge.

2004 That They May Know and Remember. Vol. 2: Spirits of the Earth. Ocotillo: Imperial Valley College Desert Museum Society.

And, hundreds of gray literature archaeological reports on file at colleges and universities throughout California.

"A KIND AND GENTLE PERSON": SOME PERSONAL REFLECTIONS ON JAY VON WERLHOF

Jeff LaFave, Esquire
Site Steward
233 First Avenue, #201
San Diego, California 92101

I first met Jay von Werlhof on October 30, 1999, while attending a training class for the California Archaeological Site Steward Program, held in El Centro. Jay was directing the training program. That was a very significant day for two reasons: 1) I became a site steward; and more importantly, 2) I started a friendship with one of the most interesting, kind, and intelligent people I know.

Prior to the meeting, I was excited about meeting Jay. Since I was someone who had done a lot of reading about California archaeology, it was impossible to be unaware of many of his contributions. The funny thing was that I wondered if I was going to meet Jay von Werlof, "Junior." I wondered this because I had seen dates on his publications going back to at least the early 1950's and I "did the math" and doubted that author would be the person running my site steward program.

Within a few minutes of meeting Jay, I knew I had met one of those rare people in life who really makes an impact on you. I was impressed by his knowledge of archaeology and history revealed by his speaking, but I was more impressed by the sense I got from him as a person. A few days later, I met Jay again in Ocotillo so that I could receive my field training regarding the site I was going to monitor. Having a couple of hours of Jay's time all to myself was great, especially since we visited one of his favorite rock alignment sites. It didn't take long before I started thinking, "Jay has forgotten more about earthen art and rock alignments than I will ever know." The time I spent that day and many other days with Jay, especially in the field, are memories that I will always cherish. Jay has also really impacted my thinking about archaeology because he taught me to try and be aware of things such as earthen art, rock alignments, trails, cleared circles, cairns, and other often subtle features that I previously might have missed.

One of the things that has most impressed me about Jay has been his kindness towards others and his respect in treating people without an attitude of superiority. Many archaeologists I have encountered, and professionals of any type, have a tendency to look down on nonprofessionals in their areas of expertise. I am probably guilty of the same thing regarding my own profession of law. Not Jay. He immediately treated me with respect and acted as though I were his equal regarding archaeology, though nothing could be further from the truth. Jay has always been very encouraging about my site steward efforts, my later efforts at recording sites, and eventually my presentations and publications. Jay has asked me to accompany and help him with field work and even called me to ask my input on ideas he had. I remember nights when he would call me to ask questions or commend me after reading my site reports or memos I would send him about discoveries I had made. His encouragement and respectful treatment have been something that motivated me to be more dedicated as a site steward and to get much more involved in archaeology and especially rock art (even though Jay has never fully understood my obsession with rock art over other aspects of archaeology).

Jay was also responsible for introducing me to two of my best friends. Jay knew them because of their discovery of an impressive rock alignment near his home in Ocotillo. He suggested that I meet them because he was sure we would become good friends. He gave me a name and a phone number and told me to call. I was a little bit hesitant at the prospect of this "blind date," but Jay was right and I thank him for his suggestion.

One of the other things that comes to mind about Jay is his overwhelming dedication to the protection of archaeological sites and the education of the public about this important goal. He has been actively involved in conservation and education through teaching, training, writing, site recording, activism, and public outreach. Jay was the leading force in the preservation efforts for the geoglyph near Ocotillo and the ongoing efforts to create a Museum of the Desert on the same property. He has also been a powerful force in trying to preserve numerous archaeological sites including the Yuha Mesa earthen art sites and the precious resources in the Indian Pass area which have been threatened by mining. Whether it be teaching classes, lobbying politicians, working with Native Americans, directing programs, or leading weekend field trips, Jay has spent a considerable amount of time fostering public understanding of archaeology and promoting conservation and documentation of sites.

Whenever I speak to Jay I learn something new. To me, Jay is something of a "time capsule," a holder of knowledge that very few people on this earth have. Whether it is stories of Robert Heizer or Theodora Kroeber, his involvement in World War II, or his fondness for earthen art, Jay always amazes me with facts, experiences, and perspectives that I would not otherwise be exposed to. Jay himself is a real treasure, and I feel very lucky to be his friend.

ROCK FEATURES, QUARRYING, ROCK ART, AND OTHER PREHISTORIC ACTIVITIES AT THE PISGAH CRATER LAVA FLOW IN THE MOJAVE DESERT

Meg McDonald, Ph.D.
Natural Resources and Environmental Division, Marine Air Ground Task Force Training Command, Marine Corps Air Ground Combat Center
Twentynine Palms, California 92278
and
Daniel F. McCarthy, M.A.
San Jacinto Ranger District, San Bernardino National Forest,
54270 Pinecrest
Idyllwild, California 92549

ABSTRACT

A complex of 124 sites at Pisgah Crater includes pavement quarries and lakeside camps with a connecting trail. Among these sites are a petroglyph site and large, complex rock alignments, including a 50-m-diameter circle, an alignment of four parallel branched lines, a closed-cell feature, and a curvilinear alignment. Abundant artifact assemblages are indicative of daily subsistence activities. Other sites in the complex also have structural features, including small rock rings, rock cairns, rock clusters, and cleared circles with berms and borders. General assemblage characteristics indicate Gypsum period (3,500-1,500 B.P.) occupation, but some evidence suggests general use may span 8,000-1,500 B.P.

INTRODUCTION

Pisgah Crater and Lavic Lake are located in the northwest section of the Marine Corps Air Ground Combat Center (MCAGCC), Twentynine Palms, California (Figure 1), in the central Mojave Desert. Topographically, the Mojave Desert represents the southwestern portion of the Great Basin and in turn belongs to the Basin and Range physiographic region (Fenneman 1931; Warren and Crabtree 1986). Mojave Desert basins average 2,000 ft above mean sea level (AMSL) and the mountains reach heights of over 10,000 ft. At Lavic Lake, the shoreline is 1,900 ft AMSL and the Bullion Mountains to the south rise to nearly 4,700 ft. Interspersed among the Mojave Desert ranges are a series of lava flows, the two largest of which, Amboy and Pisgah, are located in the northwest and northeast corners of the MCAGCC. The Mojave River, an important water source and transportation corridor across the Mojave Desert, is located about 30 km northwest of Lavic Lake. The area exhibits characteristic plant groups of the creosote scrub plant community, and small mammals and reptiles, including the desert tortoise, are abundant. Mule deer and desert bighorn sheep are more common in the mountainous areas, as were pronghorn prior to the turn of the century.

Lavic Lake is one of five remnant playas of major Pleistocene and early Holocene lakes at MCAGCC. Its sediments consist of well-sorted clay, silt, and fine sand to a depth of 160 ft (Gawarecki 1968:10-4). There are no distinct beach ridges or sand dunes at Lavic Lake, and drainage into the playa is internal. It is believed that Lavic Lake was connected to Troy Lake by northward drainage during the Pleistocene and this drainage connection was dammed by the flow of basalt from Pisgah Crater (Gawarecki 1968:10-4).

On the average, the area receives about 9 cm of yearly precipitation, and annual temperatures range from lows around 25° F to as high as 120°F. Daily temperature fluctuation also can be drastic. In the winter, it is not uncommon to have daily high temperatures in the 80s and lows in the 20s and 30s. Short-term climatic patterns are also quite variable and have undergone several major shifts during the Holocene period. These patterns, plus the general paucity of permanent water sources and seasonal availability of water at the playas, were key factors influencing prehistoric settlement patterns in the Mojave Desert.

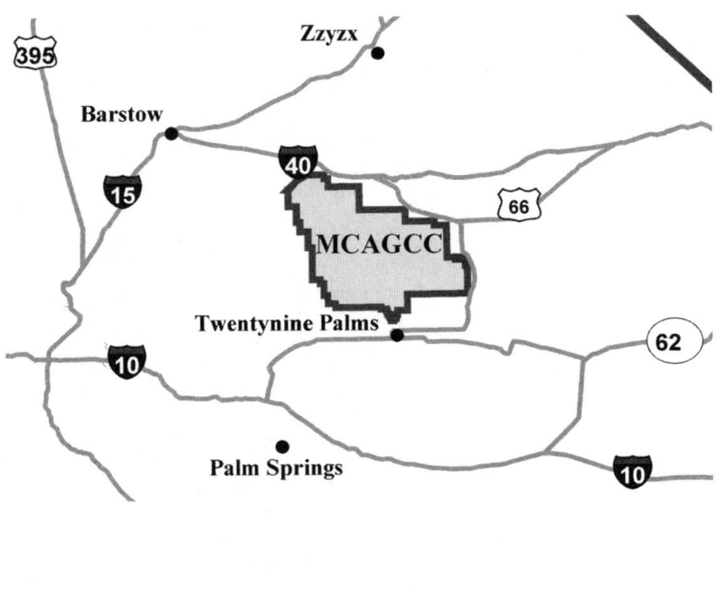

Figure 1. General location of the Marine Corps Air Ground Combat Center.

A dramatic moon-like landscape was created when lava flows from the three Pisgah Crater eruptions flowed south to Lavic Lake sometime in the Quaternary period. Erosion has not altered the thick lava from the second and third flows, and ripple marks and pillow formations still remain clearly defined from the viscous lava rivers. The flow surface is 6 m higher than the alluvial surface in areas, and coves were formed along the margins of the flows. As the lava solidified, great blocks broke off at the margins and interior tubes collapsed, leaving many depressions of wildly varying shapes and sizes. The larger depressions within the first lava flow now support a mixed alkali scrub vegetative community of saltbush (*Atriplex* spp.), grasses, and large acacia trees. The playa embayments also support atriplex and grasses. Almost all prehistoric sites in the survey area were discovered in and near the playa coves, and sites on the lava terrace above the playa were found close to the larger tubes.

The cultural history of the Mojave Desert region has been the subject of numerous chronologies and models. A general framework from Warren (1984; Warren and Crabtree 1986) and others (Donnan 1964; Drover 1979; Rector et al. 1983; Schneider 1987, 1988; Sutton 1996; Wallace 1958; Yohe 1992) is presented here (Table 1), with supplemental data from various Fort Irwin studies (Basgall 1991, 1993; Hall 1992; Hall and Basgall 1988; Jenkins and Warren 1986; McGuire and Hall 1988).

Rock Features at Pisgah Crater and Lavic Lake

Nine sites in the Pisgah Crater and Lavic Lake survey areas have large and small rock rings, cleared circles, and a large circular rock alignment (Table 2). Several of these are found as isolated features, but most were found as part of larger, more complex sites or in association with other site types. Other rock features (not included herein) include large linear rock alignments and rock cairns.

Table 1. Mojave Desert Chronological Sequence

Period	Site and Settlement Attributes
Lake Mojave 12,500-7,500 B.P.	• Sites on remnant Pleistocene surfaces • Small/medium animals dominate faunal remains • Silver Lake and Lake Mojave points, scrapers, choppers, hammerstones, crescentics, and casual milling tools
Pinto Period 7,500-3,500 B.P.	• Relatively few sites, typical of temporary camps • Role of environment de-emphasized given the persistence of essentially modern populations of flora and fauna since the end of the Pleistocene • Diagnostic artifacts include Pinto points, a flake industry, and rarely, milling tools • Highly mobile, used a diverse range of areas and resources
Gypsum Period 3,500-1,500 B.P.	• Socioeconomic and technological diversification • Episodic playa lakes may have formed more often and endured longer • Sites more variable in size and location, including intensively occupied valley sites • Gypsum, Humboldt, and Elko series point types • Rose Spring points indicating introduction of the bow and arrow at the end of the period • Artifacts and features indicating northern Arizona influences and Pacific Coast trade • Exploited a wide range of resources within a generalized, mobile subsistence strategy
Saratoga Springs 1,500-800 B.P.	• Gypsum Period technological and economic developments persist and expand, including Anasazi and Yuman influences • Arrow points, milling equipment (including mortars and pestles), and ornamental and ritual objects common • Major habitation areas, temporary camps, and processing stations found in all areas of the desert; site occupation both more lengthy and regular • Animal and plant subsistence resources diverse, local resources exploited more intensively
Protohistoric Period 800 B.P.-contact	• Anasazi influences disappear and are replaced by Shoshonean connections • Three distinct linguistic groups emerge; archaeological correlations less definitive • Desert Side-notched and Cottonwood Triangular points persist from previous period • Ceramics introduced; brown wares north of the Mojave River related to historic Paiute ceramics, buff ceramics south of the river related to Yuman influences
Historic Period	• Father Francisco Garcés follows the Mojave River route across the desert in 1776, meeting Chemehuevi, Vanyume, and Mohave Indians along the route (Coues 1900) • Chemehuevi, possibly Serrano, use area until forced to missions (Bean and Smith 1978) • Chemehuevi mention "spirits" at nearby Amboy Crater, and "when the world was young they said that the mountain used to smoke and throw out lava" (Kelly MS)

Table 2. Rock Feature and Site Descriptions

Site (CA-SBR-)	Feature Size and Construction	Setting and Site Constituents
Small Rock Rings		
7890	1.2 m x 1.6 m, >30 rocks (-10-20 cm) "butted and loosely arranged, deeply embedded in lava Apavement"	Large site with four large rock alignments (circular, closed cell, and linear) located in the center of the terrace overlooking Lavic Lake. Small rock ring and rock cairns are located inside the large circular rock alignment. Abundant flaked stone debris and discarded bifaces include a notched Coso obsidian dart point.
7910	-1.0 m x -1.0 m, 14 rocks (-10-60 cm) 95 cm (outside diam.) / 66 cm (inside diam.) Placed end-to-end, open to the west	Site is a desert pavement prospect or quarry located on the hills east of the lava flow and northeast of the lake.
7941	0.8 m x 1.5 m (inside diameter), -20 rocks (-10-30 cm) Loose semi-circular arrangement with natural ledge forming one side	Site consists of a rock ring positioned over a short lava tube (approximately 10-12 m long) located in the interior of the lava flow in the second flow area. No artifacts found at this site.
7958	1.4 m x 1.5 m, 33 rocks (-10-30 cm) Rocks are placed end-to-end, are deeply embedded	Rock ring is located in front of a small shelter near a "cove" along the northwestern margin of the lava flow. Pottery and bone are scattered 2-3 m southeast. Feature may be a hearth but has no visible charcoal. Only rock ring site in survey area with late prehistoric artifacts in association.
7965	1.5 m x 1.3 m, 23 rocks (8-58 cm) Rocks in Feature 1 are loosely arranged 1.5 m outside diameter, 38 rocks (8-48 cm) Rocks in Feature 2 are loosely arranged and scattered	Located on small lava finger in a large cove at the northern lakeshore and western margin of lava flow. The two features are 75 cm apart. A rhyolite core fragment was found near these features. A Gypsum Contracting Stem dart point and a Bristol Mountains obsidian flake (-5.0μ; no chronological estimate) were recovered.
Cleared Circles		
7898	1.5 m x 2.6 m Two conjoined cleared circles with gravel berms (Feature 1) 1.0 m x 1.0 m Single cleared circle with gravel berm (Feature 2)	Boulder-strewn hillside overlooking a large wash on the eastern margin of the lava flow. The two features are 2 m apart and each has associated flaked stone debris. Abstract, well-patinated petroglyph images on small boulders are the main site constituents.
Large Rock Rings (Simple and Stacked)		
7918	If one feature, 3.5 m x 5.2 m, -84 rocks (5-50 cm) If two features, 2.2 m x 2.1 m, -38 rocks (5-30 cm) and 2.2 m x 5.2 m, -46 rocks (5-50 cm) Scattered rock cluster (one or two rock circles; Feature 1), presumably a simple rock ring 2.6 m x 3.8 m, -40 rocks (-10-25 cm) Scattered rock cluster, appears to have been a simple rock circle of about 3-m diameter (Feature 2).	Features appear to be two or three rock circles at a trailside site in an enclosed "plaza" surrounded by higher lava from second flow. Trail leads from pavement quarries to lakeside sites. Few flakes noted within area of rock features and the only other site constituent is a lithic reduction area located at the other end of the "plaza."
7930	2.4 m outside diameter / 1.65-2.0 m inside diameter >277 rocks (-5-25 cm) Stacked rock circle open to the east (Feature 1) 2.7 m x 3.4 m (outside) / 1.4 m x 2.8 m (inside) -100 rocks (-5-30 cm) Stacked rocks on the south, scattered rocks in the center and on the north side (Feature 2)	Large, complex site with dense lithic scatters located in the center of the terrace overlooking Lavic Lake. Dense lithic scatters are found inside each of these rock rings and a metate is associated with Feature 2.
Tamped Circles		
7940	-1-m diameter circles Three connecting circles tamped in desert pavement	Open hilltop site overlooking pavement quarries and lake has a small amount of flaked stone debris.
Large Circular Rock Alignment		
7890	-50-m diameter circle End-to-end, stacked, or well-separated basalt rocks and natural lava flow features incorporated in feature construction	Large site in center of terrace overlooking lake. Small rock ring and rock cairns located inside large circular alignment. Other large rock alignments located outside. Flaked stone debris and discarded bifaces abundant. Notched dart point midsection probably dates to Early or Middle Holocene

Small Rock Rings

Five sites in the Lavic Lake survey areas have small rock rings, usually between 1.0 and 1.5 m in diameter and constructed of a small number of rocks either placed end to end or loosely arranged. The number of rocks in these features ranges from 14 to 38, and most are relatively small in size, generally 10 cm to 60 cm across their largest dimension.

Three sites with small rock rings (CA-SBR-7941, CA-SBR-7958, and CA-SBR-7965) are located along the western margin of the lava flow in an area relatively isolated from other concentrations of sites. In particular, CA-SDI-7941 is isolated, and the rock ring is the only cultural attribute. There are no artifacts or other features near the rock ring or inside the small lava tube that the rock ring is positioned over (Figure 2). The rock ring itself is too small to have functioned as a sleeping area, but the lava tube is large enough to have been used as a shelter if necessary or may have been used as a cache with the rock ring left as a marker.

Figure 2. Small rock ring positioned over lava tube at CA-SBR-7941.

At CA-SDI-7965, two small rock rings (Figure 3) are perched on a small lava finger extending into a large "cove" along the margin of the lake. The two features are 75 cm apart, and a large rhyolite core was found between the two. The site has a sparse deposit of lithic debris on and around the lava finger, and test excavations at this site recovered an exhausted white chalcedony Gypsum Contracting Stem dart point and an obsidian flake with a relatively large hydration rind of 5.0 F from the nearby Bristol Mountains source. Although few hydration rinds have been reported and no hydration rates have been proposed for Bristol Mountains obsidian, it may have been used over a long span of prehistory. It has been assumed to be most useful for manufacturing arrow points because it occurs in nodules averaging 4-5 cm in size, but it has been found at sites with dart points and arrow points. CA-SBR-7965 appears to have been used for camping and possibly "retooling," given the discarded dart point, but may not have been used on more than one occasion, probably prior to the introduction of arrow points to the region.

Figure 3. Rock ring features at CA-SBR-7965.

CA-SBR-7958 is located in a sandy area adjacent to the lava flow. This site is the only one of the five with any late prehistoric artifacts, and pottery, bone, and a hearth are located near the rock ring, which is positioned in front of a small rockshelter. The rock circle is deeply embedded in the sand (Figure 4) and may be a deeply buried stacked rock feature rather than a simple rock ring (see Hatley [1982] for excavation photos of a similar feature in Stoddard Valley). Although this site is located in the same area as the previous two sites, it is clearly a small campsite and is not as isolated as the others.

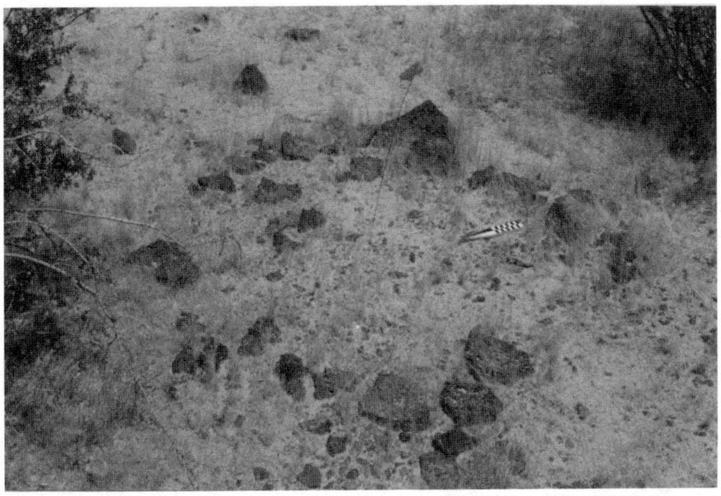

Figure 4. Embedded rock ring at campsite (CA-SBR-7958).

CA-SBR-7910 is a pavement quarry site on the eastern margin of the lava flow. The rock ring at this site is small, constructed with only 14 rocks placed end to end to create a 1-m-diameter circle with an opening to the west (Figure 5). Most of the rocks are deeply embedded in the pavement, but a few have been recently removed and turned over. No charcoal or any particular artifacts are associated with this feature, but an anvil rock with a dense flake scatter surrounding it are located nearby.

Figure 5. Open small rock ring at pavement quarry site (CA-SBR-7910).

The final small rock ring is found in a more unusual situation, as it is found within a much larger circular rock alignment that is part of a complex of linear and closed cell rock alignments. This small rock ring is constructed of at least 30 chunks of lava that are either abutted or loosely arranged, forming a scattered circle of about 1.2 m by 1.6 m in size. The site is located on the older lava "terrace" overlooking Lavic Lake, amongst a group of large, complex sites with abundant habitation debris, linear rock alignments, and other rock features.

Small rock rings (generally less than 1 m in diameter) have been interpreted variously as basket rests (Goodyear 1977), hearths, shamanic circles (Davis 1980), and vision quest sites (Begole 1974). An early description (Smith 1964) of Lavic Lake archaeological sites interpreted the stacked rock rings on or near the lakebed as "tortoise roasters." These particular stacked rock features are no longer apparent at Lavic Lake. Similar but larger features in extreme southern California were used as supports for large acorn granaries.

The small rock rings at Lavic Lake are fairly consistent in their construction methods, using small rocks with loose or abutted placement and no stacking, but vary widely in site context. They range from the isolated location of the single feature at CA-SBR-7941, without artifacts or other features, to a large archaic campsite with complex rock alignments overlooking the lake.

We can presume from their construction, lack of contents, and settings that most of these features probably were not used as basket rests, tortoise roasters, or hearths. These same attributes make it easier to assume that these small features served a religious or similar function, especially rock rings such as the isolated one at CA-SBR-7941. In truth, the function of these features is as yet unknown, and the few that have been excavated (e.g., Hatley 1982; Hevly 1982) are as enigmatic as the ones that have not been examined.

Cleared Circles

Two cleared circle features with gravel berms (Figure 6) are located at the only petroglyph site in the survey area. CA-SBR-7898 is located just above a wash at the southeast base of the large hill where the pavement quarry sites are found. Features include 55 petroglyph boulders with 63 images, a cleared oval area that appears to be two intersecting cleared circles (1.5 m by 2.6 m), and another cleared circle (1 m diameter) located within 2 m of the first. Each of these features has a gravel berm and is bordered by larger stones ranging from 5 cm to about 60 cm in size. Pavement is reforming in the interiors of these two features. These appear to be relatively old, as compared to a cleared circle with a dusty interior at the end of a set of tire tracks at a nearby site. Debitage and other flaked stone artifacts are not abundant, presumably because of the steep slope, and consist primarily of red jasper. A large block metate with a very rough surface is the only other artifact type found at this site.

The petroglyphs at CA-SDI-7898 are all heavily patinated abstract images and appear to best fit into the Western Archaic style tradition (Schaafsma 1980:18). Bisected circles (atlatl images?), meandering lines, rakes, rectangular grids, interlocking triangles, and other images (Figure 7) are found on the panels at this site. Cupules and incised images are also included, but there are no images of the bow and arrow. Darker image patination, the absence of images commonly associated with late prehistoric sites, and the general Gypsum period context of sites at Lavic Lake point toward the early origin of the petroglyphs at this site, possibly restricted to the Gypsum period.

The function of these cleared circles, like that of the small rock rings, is enigmatic, but their association with abstract, darkly patinated petroglyphs and their pavemented interiors testify to their relative antiquity and possible religious or ritual use. Similar cleared circles are found in other contexts in the southern California deserts, such as along trails in the McCoy Mountains of eastern Riverside County (McCarthy 1993:105).

Figure 6. Cleared circles at petroglyph site (CA-SBR-7898).

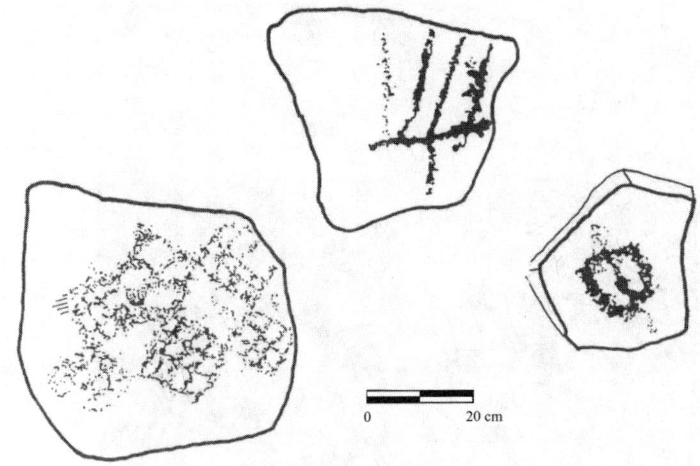

Figure 7. Selected petroglyph elements from CA-SBR-7898.

Large Rock Rings

Larger rock rings and cleared circles in the California deserts have often been considered to be "sleeping circles," used as temporary brush shelters. Others have suggested possible uses that include storage areas or winnowing platforms (Begole 1974; Whalen 1976). Large concentrations of rock rings have been reported in the Colorado Desert and western Arizona, but it appears that fewer have been described in the literature of the Mojave Desert. Rogers (1966) thought that early rock rings were constructed with several courses of rocks and are found along extinct watercourses, and that "newer" features are simple rock-outlined rings with openings to the south or east.

The features at CA-SBR-7918 appear to be scattered remnants of simple rock-outlined rings. The site consists of features and artifacts associated with a prehistoric trail (CA-SBR-7917; Figure 8) that links the pavement quarry area to the lowest "terrace" of the lava flow overlooking Lavic Lake where the most complex occupation sites are located. The trail traverses very rough lava terrain and crosses two open "plaza" areas before reaching the terrace area. The features and artifacts of CA-SBR-7918 are situated in the two plazas. A few jasper flakes are scattered along the trail, but the only large concentration of artifacts is a dense lithic scatter of core reduction debitage found at the southern end of the trail where it emerges onto the lower terrace created by the oldest flow surface.

Two loosely scattered arrangements of rocks (Figure 9) are situated where the northern portion of the trail crosses the northeastern "plaza," but very few artifacts (only a few jasper flakes) are found in this area. At the time of the survey, these features were recorded as disturbed rock rings, each 2-3 m in diameter, that may have served as the foundations for temporary brush shelters (McDonald and McCarthy 1994:83), but scaled drawings show little discernable arrangement to the rocks.

Two other large, stacked rock rings (Figure 10) are constructed at a large campsite in the central area of the terrace, 250 m north of the lake. CA-SBR-7930 is located just south of three large, deep, collapsed lava tubes that expose the old lakebed surface. Six of nine lithic reduction stations at the site are clustered at the southern margin of one lava tube, one is in the center of the same tube, and the other two are located about 100 m south in two small, sediment-filled depressions in the lava. Two dense lithic scatters were more or less contained within the two stacked rock rings. CA-SBR-7930 and other sites in the same area of the lava flow have fewer features, less variety in artifact types, and a more discrete spatial distribution of artifacts when compared to CA-SBR-7890, the site with several large rock alignments. No temporally significant artifacts were noted at CA-SBR-7930, but the proportion of biface fragments to flakes in each lithic scatter is relatively high compared to other nearby sites. It appears that sites on this terrace of the lava flow may represent different activity areas or different levels of activity and use.

Tamped Circles

Three interconnecting circles found at CA-SBR-7940 are interesting and unique. These features are not cleared circles, but merely circular outlines lightly "tamped" into the surface of the pavement (Figure 11), somewhat like a trail surface in desert pavement, but smaller in width and less convex in cross section. They are very regular in size and appearance and actually appear as if heavy circular objects such as large drums or tires were set on the pavement; however, during the time of the survey there was no disturbance on the surface of the site that would indicate that Marine Corps, mining or other modern activities at the site created these features. Similar features have been recorded at CA-SBR-4547. Each features measures about 1 m in diameter, and they are adjacent to one another. They are located in smooth areas of pavement devoid of larger rocks. Except for their size, they are similar to features described as "dance circles" that measure several hundred feet in diameter (Bourke 1889; Johnson 1985, 1994). Several large dance circles occur in eastern Riverside County as sites associated with major trail systems on desert pavements along the California side of the upper Colorado River terraces. In size, these rings are similar to small cleared circles at other Lavic Lake and Mojave Desert sites that have been interpreted as shamanic circles (Davis 1980) and vision quest sites (Begole 1974).

CA-SBR-7940 is located on a saddle of the large hill where the petroglyphs are located and overlooks Lavic Lake to the southwest and the pavement quarry sites on the northeast. The site is small and isolated, with a few artifacts that could have been deposited in one or a few episodes of lithic reduction. CA-SBR-7940 appears to be an almost ideal location for a "vision quest" or similar solitary pursuit. Basalt flakes are more common than jasper flakes and more deeply embedded in the pavement; this may indicate that use of the site dates to the Lake Mojave Period, but this relationship is scant chronological evidence.

Figure 8. Trail through second lava flow connecting pavement quarries to lakeside sites.

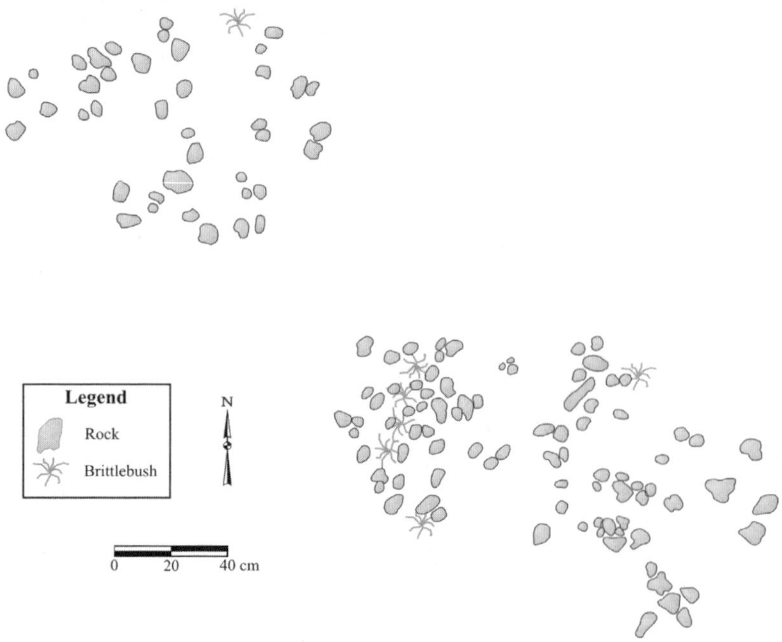

Figure 9. Scattered rock rings at CA-SBR-7918.

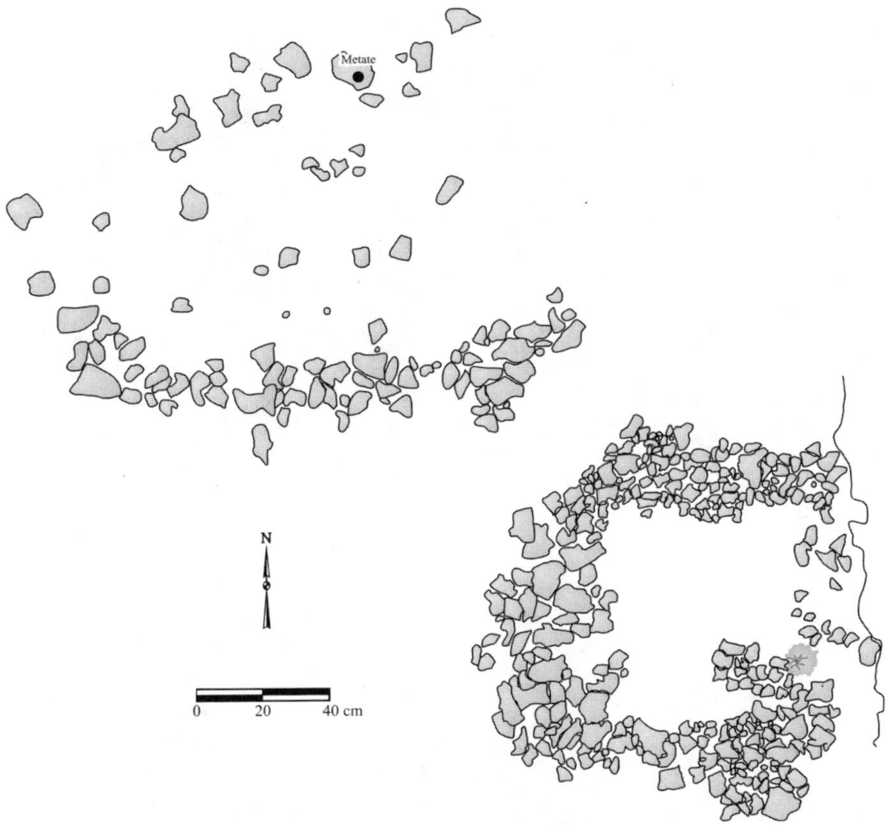

Figure 10. Rock ring (Features 1 and 2) at CA-SBR-7930 (plan view).

Figure 11. "Tamped" circles at CA-SBR-7940.

Large Circular Rock Alignment

CA-SBR-7890 is located on the oldest flow of the lava field, on the terrace overlooking Lavic Lake. In addition to several rock alignments, the site has numerous concentrations of debitage, bifaces and cores, and several milling stone fragments. The largest and most impressive alignment is a large circle approximately 55 m in diameter (Figures 12 and 13). In some areas, it is constructed of several courses of rock and is up to three rocks wide (Figure 14), other areas are simply rocks spaced several centimeters apart, and a part of the circumference is formed by the fracturing and natural "extrusion" of volcanic rocks between two lava "plates" (Figure 15). This circular alignment is unusual because of its size and preservation and is reminiscent of "dance circles" found near the Colorado River (cf. Johnson 1985:Figs. 43, 82, 83), but the function and possible symbolism of such a rock alignment in this region and on a lava flow are without precedent. Since this site was recorded, a similar rock ring has been recorded at a site about 50 mi from Lavic Lake (R. Laska, personal communication 1997).

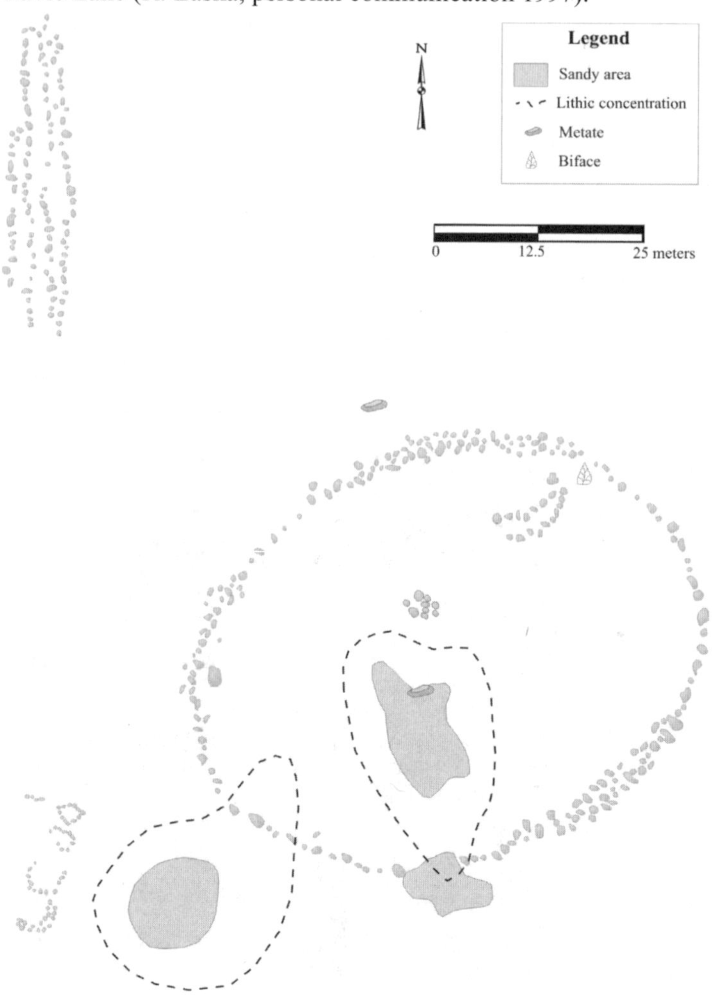

Figure 12. CA-SBR-7890 site map.

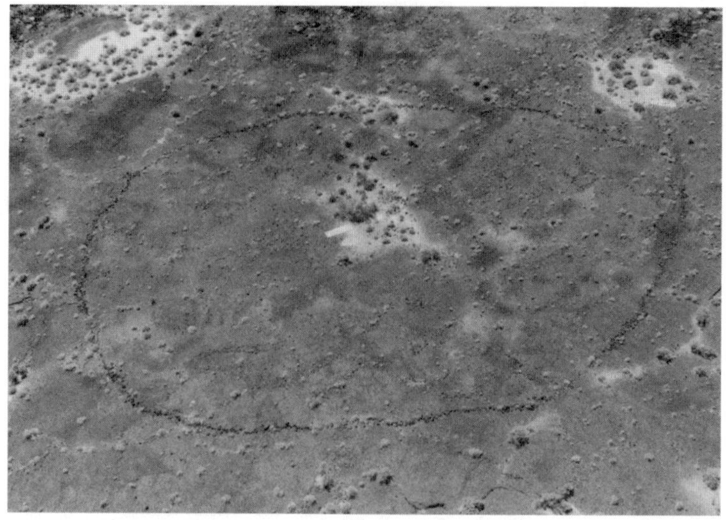

Figure 13. Large circular rock alignment at CA-SBR-7890.

Figure 14. Stacked rock construction of large circular rock alignment at CA-SBR-7890.

Johnson (1985:12) noted that circular rock alignments (excluding habitation sites) are "usually found in association with trails, dance pattern features, or linear rock patterns . . ." The circular rock alignment on the Pisgah flow is associated with both a linear rock feature and a compartmentalized rock feature, an association that seems to be rather unusual. These linear and compartmentalized features are quite similar to many found in Panamint Valley (cf. McCarthy 1986; von Werlhof 1987) and other Mojave Desert areas, but those illustrated by von Werlhof are not associated with large circles.

Three other rock alignment figures, all closed figures (cf. von Werlhof 1987), also occur at CA-SBR-7890 (Figure 14). One that resembles a meteor (Figure 16) is found within the large circular alignment, as is the small rock ring previously described, another is a closed cell arrangement (Figure 17); and the third has four parallel lines (Figure 18). Large, dense lithic scatters, numerous utilized flakes, flaked stone tools, and mano and metate fragments at this site are indicative of daily subsistence activities in addition to whatever social or ceremonial uses the rock alignments may have had.

Figure 15. Natural "extruded" areas used in construction of large circular rock alignment at CA-SBR-7890.

Solari and Johnson (1982:425-426) discussed various functional interpretations of intaglios (and related large rock alignments); these include trail markers, power sources to travelers, boundary markers, ceremonial constructions (as enlarged sand paintings) possibly for use at annual rituals, fertility rituals, or pure art. Intaglio sites along the Colorado River generally have little or no habitation debris, but habitation sites are usually found nearby (Solari and Johnson 1982:426). Johnson (1994:4) described a complex of images inside the Black Point intaglio (near the Colorado River) as "teaching stations for a number of different lessons about times and places along the path of the great journey [of life]."

The large circular rock alignment at Lavic Lake and the related alignments inside and nearby may have served a similar ceremonial function, may have been astronomical functions, or may have been a dance circle. Waterman (1910:281) described Diegueño (Kumeyaay) dance circles as being surrounded by a "wall of brush . . . held in place by large rocks." Johnson (1985:17-18) summarized historic references to Chemehuevi and Kumeyaay dances as follows: "The most common dance style was for people to line up shoulder to shoulder (sometimes holding hands) in a circle and revolve counterclockwise by moving one foot after another in a sideways shuffle. The dance was led by a drummer and song leader, who usually sat somewhere in the center area of the dance circle." The close association of this rock alignment with habitation debris appears to be unusual; systematic analysis and further comparison to similar sites in the Mojave and Colorado deserts may provide more information and aid better interpretation of this site.

Figure 16. Linear "meteor" rock alignment at CA-SBR-7890.

Figure 17. Closed cell rock alignment at CA-SBR-7890.

Figure 18. Parallel linear rock alignment at CA-SBR-7890.

Pavement Quarries and the Lithic Reduction Sequence

Two main resource use areas were delineated by the Lavic Lake surveys (McDonald and McCarthy 1994; Schaefer and Pallette 1991), one complex of large, temporary camps along the shoreline of Lavic Lake and on the lava terrace overlooking the lake, and another complex of lithic prospecting sites on older pavements containing tabular jasper nodules located northeast of the playa. A trail cleared through the rough terrain of the second Pisgah Crater lava flow connects the two areas.

Rejected and broken bifaces and unmodified debitage from nodules obtained from the lithic prospecting sites are abundant at the Pisgah Crater/Lavic Lake sites. Replicative systems analysis of the flaked stone artifacts at 12 of these sites (McDonald et al. 1996) determined that the reduction strategy was primarily directed towards initial core reduction and percussion bifacial thinning of prepared core nuclei, although core reduction (Stage 1) is representative of the quarry sites, and early percussion biface thinning of those cores took place at occupational sites overlooking the lake.

Nodule size and shape have repeatedly been shown to be a determining factor in the choice of reduction methods, but it is equally obvious that the desired end product is also a determining factor. Arrow points can be made from relatively small flakes, but as summarized by Basgall et al. (1988:308-309), "technological requirements demand that dartpoints pass through a 'biface' stage more frequently than arrowpoints." Dart points can be easily manufactured from bifaces made on flake blanks (e.g., Elston and Zeier 1984) or directly from bifacial reduction of cobbles, but are unlikely to be made from raw source materials that are very small in size. Extensive excavations of Gypsum period sites at Fort Irwin have illustrated that sites from this period usually contain abundant debitage and discarded bifaces. Use of bifacial cores for the production of spear and dart points appears to have been a common toolmaking strategy used during the Gypsum period, resulting in site assemblages typical of those at Lavic Lake, dominated by flaked stone tools, bifaces, cores, and debitage (cf. Basgall and Hall 1994; Hall and Basgall 1994). Although it is conceivable that the same reduction strategy could have been used during late prehistoric times when arrow points were the desired end-product, the evidence suggests that arrow points were not an intended end-product; however, it is not unlikely that late prehistoric people using the area may have recycled flakes or used the same reduction strategy for arrow point manufacture, at least to some point along the continuum.

Discussion

Given the relatively small number of temporally sensitive artifacts and sites with these artifacts, it is difficult to draw inferences regarding the chronology of use of the Lavic Lake area and how resource use in the area may have changed over time. However, some points can be made regarding the distribution of certain artifact types, features, and site assemblages. Only two of ten ceramic-bearing sites were found away from the margins of the playa, and only one of these sites (CA-SBR-7958) also has a rock feature, a rock ring located in sandy soils. Other artifacts (*Olivella biplicata* wall disk beads, a Desert Side-notched arrow point, and possibly Bristol Mountains obsidian) typical of the ceramic phase of the Saratoga Springs period were found only at sites with ceramics. Despite a small ceramic sample, the predominance of buff wares of sedimentary clay origin suggests late preshistoric populations whose territorial range probably remained in the desert area and did not extend to the Peninsular or Transverse Ranges where residual clay brown wares would be produced.

It appears that the nonceramic site assemblages at Lavic Lake most closely resemble Gypsum period assemblages excavated from Fort Irwin sites (Basgall and Hall 1994) in that projectile points are few, bifaces and waste flakes of locally available microcrystalline quartzes are abundant, and few milling tools are found on these sites (Table 3). These comparisons assume that settlement and subsistence practices during earlier time periods (i.e., Lake Mojave

Table 3. Comparison of Lavic Lake Assemblage Characteristics with Middle and Late Holocene Sites at Fort Irwin

Assemblage Components	Lavic Lake Assemblages (McDonald and McCarthy 1994; McDonald et al. 1996)	Lake Mojave Assemblages (Basgall and Hall 1994)	Newberry/Gypsum Assemblages (Hall and Basgall 1994)
Settlement/ Subsistence Patterns	Recurrent occupation of area probably focused around infillings of Lavic Lake; frequency & duration of lakestands unknown	Presumed Pleistocene lacustrine focus probably an artifact of differential preservation of surfaces & sites	Areas (not sites) recurrently occupied, possibly connoting more regularized subsistence-settlement routines
General Assemblage Characteristics	Assemblages dominated by tremendous amount of bifacial reduction debitage, rejected & broken bifaces & bifacial cores, few milling tools, no flake tools	Assemblages dominated by tremendous numbers of bifaces, flake tools, cores/core tools, & few milling tools	Assemblages dominated by flaked stone tools, bifaces, cores, & flakes
Projectile Points	Very few dart & spear points recovered, appear to have long retention	Moderate numbers; usually basalt or related materials	Very few recovered, mostly "spent" tools; prolonged retention
Bifaces	Abundant bifaces made from locally available jasper & chalcedony, some bifaces made from locally available igneous materials	Abundant quantities; usually basalt or related materials	Early-stage bifaces local materials; late-stage bifaces extra-local materials
Formed Flake Tools	None recovered; presumably very low proportion or none in site assemblages	Abundant; usually high-quality, heat-treated microcrystallines	
Casual Flake Tools	None recovered; ready supply of suitable tool blanks from copious debitage, presumably very low proportion or none in site assemblages	Less common than formed flake tools	Minimal wear noted on flake tools; ready supply of suitable tool blanks from copious debitage [implying greater use of casual flake tools?]
Flake Cores/ Core Tools	Few flake cores recovered; bifaces used as general, "carry-around" core forms	Some; bifaces used as general, "carry-around" core forms	Served as principal, carry-around artifact form for tool/flake manufacture
Other Tool Types		Crescentics & drills/borers few in number	
Milling Tools/ Battered Stone	Small but regular component; typically casual, unshaped handstones & unifacial block metates	Small but regular component; typically casual, unshaped handstones & unifacial block metates	Frequent but not abundant component; shaped handstones & thin, portable, well-used slabs
Predominant Toolstone	Microcrystallines abundant due to local source; fine-grained igneous also available & used, but not as abundant. Igneous debitage appears to be older	Fine-grained igneous common; microcrystallines rare except as finished points	Virtually exclusive use of microcrystalline silicates for all flaked tool classes; exhausted, discarded tools mostly extra-local materials, new production on-site of local/near-local materials
Debitage	Copious unmodified debitage	Copious unmodified debitage	
Fauna	Sporadic occurrence, but dominated by tortoise & medium/small fauna (rabbits/hares, rodents) with some artiodactyl bone	Sporadic occurrence, but dominated by medium/small fauna (rabbits/hares, rodents, birds, lizards) with some artiodactyl bone	Situational exploitation of a wide range of animal resources dependent upon locality (artiodactyl/large mammal at upland sites, smaller fauna & tortoise at lowland sites)
Plant Resources	Milling tools at sites near lakebed suggests plant foods, especially those associated with a marshy environment, were a regular diet component		Composite flaked stone to ground stone tool ratios suggest much greater emphasis on vegetal processing at Drinkwater Basin rather than Tiefort Basin

and Gypsum periods) were widespread over the Mojave Desert. A cursory overview of the literature suggests that this may be the case, as archaeological remains around Pleistocene lakes in the Mojave Sink and smaller, ephemeral lake playas are plentiful and diverse. Geologic evidence from other Mojave Desert lakes indicates that shallow lakestands of short duration (several years) or longer duration (one to several hundred years?) occurred during the late Holocene (Enzel et al. 1992). Major lakestands and even minor seasonal infillings producing a marshy environment would have provided an attractive seasonal base for the exploitation of local plant and animal resources.

Rock rings and cleared circles in the Mojave Desert often occur on geologically older surfaces, are often embedded in the surface and heavily varnished, or have some new pavement formation in the cleared area and have few or no diagnostic artifacts, possible indications of their origin prior to the Saratoga Springs or Protohistoric periods (>1,500 B.P.). For the most part, the same is true of the features at Lavic Lake.

The major rock alignments at Lavic Lake are similar to rock alignments constructed throughout the Mojave Desert, particularly in areas further north such as Avawatz Valley (Kaldenberg 1981), Panamint Valley, and Death Valley. The functions of these types of alignments are unknown, but commonly have been assumed to be astronomical or ceremonial in nature, and a general lack of associated temporally diagnostic artifacts has led previous researchers to conclude that these constructs are relatively ancient in origin (Davis and Winslow 1965; von Werlhof 1996). The Lavic Lake rock alignments are associated with occupational debris that probably dates from the Gypsum period or earlier, judging by the correspondence of assemblage characteristics to similar Gypsum period sites at Fort Irwin (see Table 3), and it appears that these rock alignments were constructed during the same period of time that the sites were occupied. The most striking of the primary elements in the rock alignments is a large circular alignment with associated smaller alignments and features, reminiscent of dance circles or "ceremonial pathways" constructed by the Mohave people near the Colorado River. Construction methods are understandably different, given the different environmental settings, but the overall milieu is similar—on a terrace overlooking the lake or the river. The general age of the Lavic Lake sites suggests an association with ancestral Yuman peoples prior to the Numic expansion.

Data from excavated sites at Fort Irwin suggest that there were "highly specific tendencies in the use of raw material types to produce certain tool classes" and that most early and middle Holocene bifaces were made from fine-grained igneous stone such as basalt (Basgall and Hall 1994:71). If this lithic material acquisition and use pattern at Fort Irwin is typical of early and middle Holocene Mojave Desert assemblages, limited use of igneous materials at the Lavic Lake sites supports other lines of inference that suggest the Lavic Lake sites were used predominantly during the Gypsum period (ca. 3,500-1,500 years B.P.), and to a lesser degree during earlier periods. However, this line of inquiry does not take into account the effects of raw material morphology and availability, their relationship to the desired end product, and how these affect choice. It may be unreasonable to assume that the high quality, readily available, easily worked, and relatively large nodules at the Lavic Lake quarry sites were much less desirable than the less abundant igneous materials found in the same and adjacent pavements.

Overall, combined but circumscribed evidence from survey and excavation indicates intensive use of the area during the Gypsum period, and limited use of the area during the middle Holocene and late prehistoric periods also. Minor evidence of middle Holocene and late prehistoric occupation in the area may be indicative of a "tight" occupation period, ranging only from the transition periods on either end of the Gypsum period (3,500-1,500 years B.P.), but this cannot be definitely asserted from the data at hand, and general use of the area may range from ca. 8,000 years B.P. to less than 1,500 years B.P. The findings at these sites have wide-ranging implications for dating similar rock alignments and rock art sites in the Mojave Desert where artifacts are generally not found in association. Nonetheless, the limitations of these data should be considered carefully in view of the fact that the assemblages are derived primarily from surface contexts and are lacking more tangible chronological indicators such as radiocarbon dates, temporally diagnostic artifact types, or a large sample of obsidian hydration data. In spite of these limitations, many of these sites are significant to our understanding of the regional prehistory, especially viewed in the context of the abundant lithic resources and likely bountiful lacustrine resources available during seasonal infillings of Lavic Lake. Although probably of inadequate duration and reliability for year-round occupation, Lavic Lake would have fit well into the opportunistic and flexible settlement strategy characteristic of southern California hunters and gatherers and other hunter-gatherer societies worldwide.

ACKNOWLEDGMENTS

This paper is adapted from a paper delivered by the senior author at the Stone Circle Symposium, 56th Annual Plains Anthropological Conference, Bismarck, North Dakota, October 1998. The information is derived from two surveys (McDonald and McCarthy 1994; Schaefer and Pallette 1991) and a test excavation project (McDonald et al. 1996) conducted in the Lavic Lake area for the U.S. Army Corps of Engineers, Los Angeles District and the Natural Resources and Environmental Affairs Division, Marine Corps Air Ground Combat Center, Twentynine Palms. The aerial photographs in Figures 13 and 19 are courtesy of Mike Gorman, Landmark Aerial Photography, Riverside, California.

REFERENCES

Bard, Robert Charles
1972 Settlement Pattern of the Eastern Mojave Desert. Unpublished Ph.D. dissertation, University of California, Los Angeles.

Basgall, Mark E.
1991 The Archaeology of Nelson Basin and Adjacent Areas, Fort Irwin, San Bernardino County, California. Report prepared by Far Western Anthropological Research Group, Inc., Davis, California, for to the U.S. Army Corps of Engineers, Los Angeles District.
1993 Early Holocene Prehistory of the North-central Mojave Desert. Unpublished Ph.D. dissertation, University of California, Davis.

Basgall, Mark E., and M. C. Hall
1994 Perspectives on the Early Holocene Archaeological Record of the Mojave Desert. In Kelso Conference Papers 1987-1992, G. Dicken Everson and Joan S. Schneider, editors, pp. 63-81. California State University, Bakersfield, Museum of Anthropology, Occasional Papers in Anthropology No. 4.

Basgall, Mark E., M. C. Hall, and William R. Hildebrandt
1988 The Late Holocene Archaeology of Drinkwater Basin, Fort Irwin, San Bernardino County, California. Report prepared by Far Western Anthropological Research Group, Davis, California, for the U.S. Army Corps of Engineers, Los Angeles District.

Bean, Lowell J., and Charles R. Smith
1978 Serrano. In California, edited by R. F. Heizer, pp. 570-574. Handbook of North American Indians, vol. 8. Smithsonian Institution, Washington.

Begole, Robert S.
1973 An Archaeological Survey in the Anza-Borrego Desert State Park: 1972 Preliminary Report. Pacific Coast Archaeological Society Quarterly 9(2):30-55.
1974 Archaeological Phenomena in the California Deserts: Desert Cairns and Rock Alignments. Pacific Coast Archaeological Society Quarterly 10(2):51-69.
1981 1978-1980 Investigations in the Anza-Borrego Desert State Park. Pacific Coast Archaeological Society Quarterly 17(4):1-38.

Bettinger, R. L., and R. E. Taylor
1974 Suggested Revisions in Archaeological Sequences of the Great Basin and Interior Southern California. Nevada Archaeological Survey Research Papers 5:1-26.

Coues, Elliot
1900 On the Trail of a Spanish Pioneer, the Dairy and Itinerary of Francisco Garcés. 2 vols. Francis P. Harper, New York.

Davis, E. L.
1970 Archaeology of the North Basin of Panamint Valley, Inyo County, California. Nevada State Museum Anthropological Papers 15:83-141.
1980 Shamanism, Little Stone Circles, and the Calico "Hearth." In Evaluation of Early Human Activities and Remains in the California Desert, edited by Emma Lou Davis, Kathryn H. Brown, and Jacqueline Nichols, pp. 284-296. Bureau of Land Management, California Desert District, Riverside.

Davis, E. L., and Carol Panlaqui (editors)
1978 The Ancient Californians: Rancholabrean Hunters of the Mojave Lakes Country. Los Angeles County Museum of Natural History, Science Series 29:4-152.

Donnan, C. B.
1964 A Suggested Culture Sequence for Providence Mountains (Eastern Mojave Desert). University of California Archaeological Survey Annual Report, 1963-1964:1-26. Los Angeles.

Drover, Christopher E.
1979 The Late Prehistoric Human Ecology of the Northern Mojave Sink, San Bernardino County, California. Unpublished Ph.D. dissertation, Department of Anthropology, University of California, Riverside.

Ebling, Walter
1986 Handbook of Indian Foods and Fibers of Arid America. University of California Press, Berkeley.

Elston, G. Robert, and Charles D. Zeier
1984 The Sugarloaf Obsidian Quarry. Department of the Navy, Naval Weapons Center, China Lake, California.

Enzel, Yehouda, William J. Brown, Roger Y. Anderson, Leslie D. McFadden, and Stephen G. Wells
1992 Short-Duration Holocene Lakes in the Mojave River Drainage Basin, Southern California. Quaternary Research 38:60-73.

Fenneman, Nevin M.
1931 Physiography of Western United States. McGraw-Hill, New York.

Gawarecki, Stephen J.
1968 Infrared Survey of the Pisgah Crater Area, San Bernardino County, California: A Geological Interpretation. In Earth Resources Aircraft Program, Status Review; Vol. 1, Geology, Geography, and Sensor Studies, Section 10, pp.

1-26. U. S. National Aeronautics and Space Administration, Manned Spacecraft Center, Houston.

Goodyear, Albert C., III
1977 HECLA II and III: An Interpretive Study of Archaeological Remains from the Lakeshore Project, Papago Reservation, Southcentral Arizona. Arizona State University Anthropological Research Paper 9.

Hall, M. C.
1992 Final Report on the Archaeology of Tiefort Basin, Fort Irwin, San Bernardino County, California. Draft report prepared by Far Western Anthropological Research Group, Inc., Davis, California, for the U.S. Army Corps of Engineers, Los Angeles District.

Hall, M. C., and Mark E. Basgall
1988 Late Quaternary Environments. In The Late Holocene Archaeology of Drinkwater Basin, Fort Irwin, San Bernardino County, California, by Mark E. Basgall, M. C. Hall, and William R. Hildebrandt, pp. 33-53. Report prepared by Far Western Anthropological Research Group, Davis, California, for the U.S. Army Corps of Engineers, Los Angeles District.
1994 Lessons to Learn from Newberry/Gypsum Period Archaeology of the Mojave Desert. In Kelso Conference Papers 1987-1992, edited by G. Dicken Everson and Joan S. Schneider, pp. 82-94. California State University, Bakersfield, Museum of Anthropology, Occasional Papers in Anthropology No. 4.

Hatley, M. Jay
1982 Archaeological Investigations of the Sidewinder Well Quarry Workshop and of the Stoddard Valley Stone Ring Formations, San Bernardino County, California. In The Archaeology of Owl Canyon and Stoddard Valley, Mojave Desert, California, by William T. Eckhardt and M. Jay Hatley, pp. 28-99. Cultural Resource Publications, Anthropology - History, Bureau of Land Management, California Desert District, Riverside.

Hevly, Richard H.
1982 Palynological Studies of an Archaeological Stone Ring: Stoddard Valley, California. In The Archaeology of Owl Canyon and Stoddard Valley, Mojave Desert, California, by William T. Eckhardt and M. Jay Hatley, App. B. Cultural Resource Publications, Anthropology - History, Bureau of Land Management, California Desert District, Riverside.

Hunt, Alice P.
1960 The Archaeology of the Death Valley Salt Pan, California. University of Utah Anthropological Papers No. 47.

Jenkins, Dennis L., and Claude N. Warren
1986 Test Excavation and Data Recovery at the Awl Site, 4-SBr-4562, A Pinto Site at Fort Irwin, San Bernardino County, California. National Park Service, Interagency Archeological Services, San Francisco, Fort Irwin Archaeological Project, Research Report No. 22.

Johnson, Boma
1985 Earth Figures of the Lower Colorado and Gila River Deserts: A Functional Analysis. The Arizona Archaeologist No. 20, Phoenix.
1994 Earth Figures of the Lower Colorado River Valley. Paper presented at the annual meeting of the Society of American Archaeology, Anaheim.

Kelly, Isabel T.
MS Las Vegas Southern Paiute Field Notes. (Notebook of the Las Vegas Band, Southern Paiute Field Notes.) University of California ARC microfilm CU-23.1. University of California Museum of Anthropology Ethnological Documents #138,2.

Kubly, Dennis M., and Gerald A. Cole
1979 Limnologic Studies on the Desert Playas of Southern California. Report on file at the Bureau of Land Management, Riverside District Office.

Lanning, E. P.
1963 The Archaeology of the Rose Spring Site (Iny-372). University of California Publications in American Archaeology and Ethnology 49(3):237-336.

Leakey, L. S. B., R. D. Simpson, T. Clements, R. Berger, J. Witthoft, and participants of the Calico Conference
1972 Pleistocene Man at Calico: A Report on the International Conference on the Calico Mountains Excavations, San Bernardino County, California. San Bernardino County Museum, Redlands.

McCarthy, Daniel F.
1986 Preliminary Report on the Rock Alignments of Panamint Valley, California. American Indian Rock Art 11:33-41. American Rock Art Research Association, El Toro, California.
1993 Prehistoric Land-Use at McCoy Spring: An Arid-Land Oasis in Eastern Riverside County, California. Unpublished Master's thesis, Department of Anthropology, University of California, Riverside.

McDonald, Meg, and Daniel F. McCarthy
1994 Prehistoric and Historic Land Use of the Pisgah Crater Lava Flows and Lavic Lake Area, Marine Corps Air Ground Combat Center, San Bernardino County, California. Report prepared by Brian F. Mooney

Associates, San Diego, for the U.S. Army Corps of Engineers, Los Angeles District.

McDonald, Meg, J. Jeffrey Flenniken, and Daniel F. McCarthy
1996 Evaluation of Prehistoric Resources at Pisgah Crater Lava Flows and Lavic Lake, Marine Corps Air Ground Combat Center, San Bernardino County, California. Report prepared by ASM Affiliates, Inc., Encinitas, California, for the NREA Directorate, Marine Corps Air Ground Combat Center, Twentynine Palms, California.

McGuire, Kelly R., and M. C. Hall
1988 The Archaeology of Tiefort Basin, Fort Irwin, San Bernardino County, California. Report prepared by Far Western Anthropological Research Group, Inc., Davis, California, for the U.S. Army Corps of Engineers, Los Angeles District.

Payen, Louis Arthur
1982 The Pre-Clovis of North America: Temporal and Artifactual Evidence. Unpublished Ph.D. dissertation, Department of Anthropology, University of California, Riverside.

Rector, C. H., J. D. Swenson, and P. J. Wilke (editors)
1983 Archaeological Studies at Oro Grande, Mojave Desert, California. San Bernardino County Museum Association, Redlands.

Rogers, Malcolm J.
1966 Ancient Hunters of the Far West. Edited by Richard F. Pourade. The Union-Tribune Publishing Company, San Diego.

Schaafsma, Polly
1980 Indian Rock Art of the Southwest. School of American Research, Santa Fe, and University of New Mexico Press, Albuquerque.

Schaefer, Jerry, and Drew Pallette
1991 Prehistoric and Historic Land Use and Settlement Patterns at Lavic Lake, Twentynine Palms Marine Corps Air Ground Combat Center, San Bernardino County, California. Report prepared by Brian F. Mooney Associates, San Diego, for the U.S. Army Corps of Engineers, Los Angeles District.

Schneider, Joan S.
1987 Archaeological Investigations at Afton Canyon (CA-SBr-85), Mojave Desert, San Bernardino County, California. Unpublished Master's thesis, Department of Anthropology, University of California, Riverside.

1988 Late Prehistoric Times in the Central Mojave Desert: Some Problems. Pacific Coast Archaeological Society Quarterly 24(1):30-44.

Smith, G.
1964 Archaeological Site Survey Record, SBCM-315 [SBR-420]. Site record on file at the San Bernardino County Archaeological Inventory, San Bernardino County Museum, Redlands. [Narrative attached to the site record contains information about the Lavic Lake area given to G. Smith by Sylvia and Slim Winslow.]

Solari, Elaine Maryse, and Boma Johnson
1982 Intaglios: Synthesis and Recommendations for Management. In Hohokam and Patayan: Prehistory of Southwestern Arizona, edited by Randall H. McGuire and Michael B. Schiffer, Appendix A. Academic Press, New York.

Sutton, Mark Q.
1996 The Current Status of Archaeological Research in the Mojave Desert. Journal of California and Great Basin Anthropology 18(2):221-257.

von Werlhof, Jay
1987 The North Desert. In Spirits of the Earth, A Study of Earthen Art in the North American Deserts, vol. 1. Imperial Valley College Museum Society, El Centro, California.

Wallace, William J.
1958 Archaeological Investigation in Death Valley National Monument. University of California Archaeological Survey Reports 42:7-22.
1962 Prehistoric Cultural Development in the Southern California Deserts. American Antiquity 28(2):172-180.
1977 A Half Century of Death Valley Archaeology. The Journal of California Anthropology 4(2):249-258.

Wallace, William J., and Edith S. Taylor
1978 Ancient Peoples and Cultures of Death Valley National Monument. Acoma Books, Ramona, California.

Warren, Claude N.
1984 The Desert Region. In California Archaeology, by M. J. Moratto, pp. 339-420. Academic Press, Orlando.

Warren C. N., and R. H. Crabtree
1986 Prehistory of the Southwestern Area. In Great Basin, edited by Warren L. d'Azevedo, pp. 183-193. Handbook of North American Indians, vol. 11. Smithsonian Institution, Washington.

Waterman, Thomas T.
1910 The Religious Practices of the Diegueño Indians. University of California Publications in American Archaeology and Ethnology 8:271-358.

Whalen, Norman M.
1976 An Archaeological Survey in Southeastern Imperial County, California. Pacific Coast Archaeological Society Quarterly 12(2):25-50.

Yohe, Robert M., II
1992 A Reevaluation of Western Great Basin Cultural Chronology and Evidence for the Timing of the Introduction of the Bow and Arrow to Eastern California Based on New Excavations at the Rose Spring Site (CA-INY-372). Unpublished Ph.D. dissertation, Department of Anthropology, University of California, Riverside.

GRANITE GALLERIES

COURTESY OF CALIFORNIA ACADEMY OF SCIENCES

Reprint from Volume XI - Number 4

Pacific Discovery

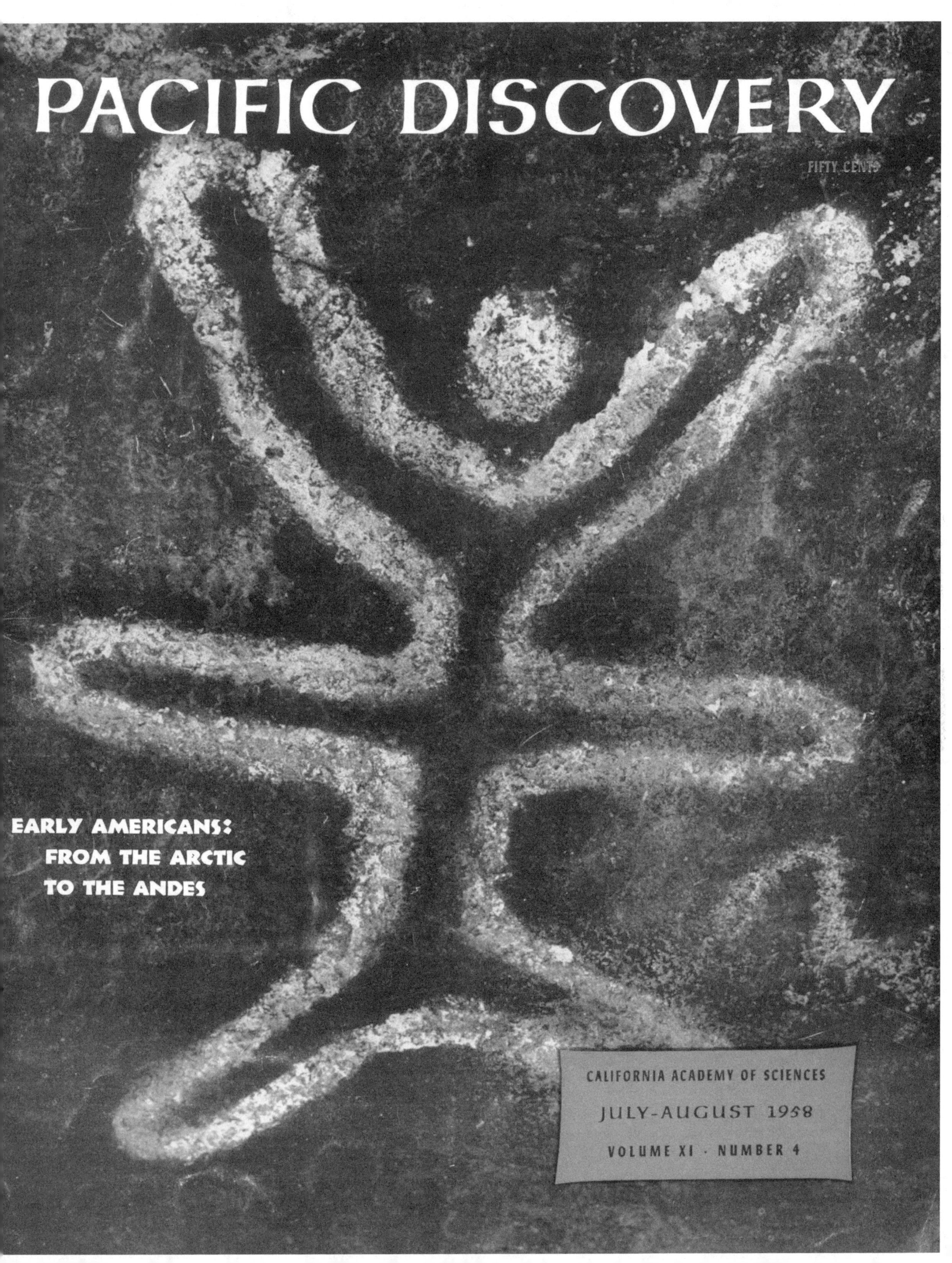

PACIFIC DISCOVERY

FIFTY CENTS

EARLY AMERICANS: FROM THE ARCTIC TO THE ANDES

CALIFORNIA ACADEMY OF SCIENCES

JULY-AUGUST 1958

VOLUME XI · NUMBER 4

VOLUME XI · NUMBER 4
JULY-AUGUST 1958

PACIFIC DISCOVERY

PRE-DISCOVERY

In the history of an institution such as the California Academy of Sciences, or of a part of one, such as this magazine, times come when those directing things must think of growth and improvement. Despite gratifying comments from many readers who seemed happy with *PD* just as it was, the truth has always been on the editorial mind that nothing is ever so good it cannot be made better. It does not take long, however, to find out that it is humanly impossible to make each one just a little better than the last – indefinitely! Eyes must therefore be cast over the long run to some attainable objective. As for getting there, it is like a multi-stage rocket. In order to keep it moving another charge must be fired off at the right moment. *PD* has never stopped accelerating, but we think the time has come to shoot for higher goals.

Our editorial following–"A New Pacific Discovery"–we hasten to assure our constant readers, does not foreshadow a completely different magazine. Just as the institution grows by means of periodic physical additions and internal reorganization to fulfill its destiny, so must its official voice, *Pacific Discovery*, reach out for wider audiences through timely changes of emphasis and increases of volume. Our recent look around the Pacific has confirmed the already expressed belief that wider audiences are waiting to be reached. Dedicated as it is to public service and enlightenment through science, the Academy in all its parts has an obligation to grow with the expanding times. People are moving Pacificward from both sides of the earth; the center of gravity is shifting. The Pacific world is discovering itself. For the present, then, we promise more "Pacific" in *Pacific Discovery*. We shall accelerate towards our long-established goal of increasing knowledge and understanding of nature and man in the Pacific World. Our plans will not materialize overnight. Evolution, not revolution, is to be the process of growth. More than ever, we shall feel the need of communication *from* as well as with our readers. Finally, our great need, as always, is for *more* readers; or, to put the shoe on the right foot, we must find more readers who need *Pacific Discovery!*

PD'S AUTHORS

Did the University of California Press have a tip-off on the Alaska statehood vote? We're just asking. . . . Their *Landscapes of Alaska* happened to come out only a week or two before the news broke. Under the editorship of **Dr. Howel Williams,** University of California professor of geology, the book greatly enlarges our view of the vast chunk of land soon to be Number 49. The Introduction he wrote with **John C. Reed** is presented here almost entire, with the publisher's kind permission. . . . ¶The second of the "Eskimo Hunters" series, by **Richard D. Taber** of the Montana State University School of Forestry faculty, makes a pointed observation on why men shoot birds–Eskimo men, that is (any suggested resemblance to more southerly American males is purely a suggestion). . . . ¶**Annette Richards Parent** of Tucson, Arizona, again pleads for support of the program of the National Park Service, this time in behalf of the many and sadly run-down archeological national monuments in the Southwest. . . . ¶A combined faculty-student project of the California School of Fine Arts of San Francisco, under the auspices of the California Archeological Survey, has resulted in a most unusual pictorial presentation. On the faculty side, **Jay C. von Werlhof** (who studied anthropology at the University of California) wrote the story; the photographer, **Pirkle Jones,** also teaches his art. Three of the photos are by student participants. . . . ¶**Philip Ferry** recently paid an extended visit to South America–merely a chaser to the trip around the world he'd just tossed off. . . . ¶Morrison Planetarium lecturer **Charles F. Hagar** believes that if enough people ask a certain question, you get the answer into print–in *PD* of course. This saves a lot of correspondence; and besides, more people will have to buy *PD*–see? D.G.K.

IN THIS ISSUE

THE COVER

Untitled, by Unknown California Artist. Indian painters were busy with brush and palette long before the "artists' colony" idea came into California with the promoter breed, as this Tulare County composition shows. Photo by Pirkle Jones *(see pages 16–22).*

Pacific Discovery is published bimonthly at Gillick Press by the California Academy of Sciences. Publication office: 2057 Center Street, Berkeley 4. Editorial, Advertising and Circulation offices: Golden Gate Park, San Francisco 18. Annual subscriptions: U.S., $3; Foreign, $3.50. Single copies, 50c. Members of the Academy subscribe through their dues. Entered as second-class matter, February 17, 1948, at the Post Office, Berkeley 4, California, under the act of August 24, 1912.

A JOURNAL OF NATURE AND MAN IN THE PACIFIC WORLD

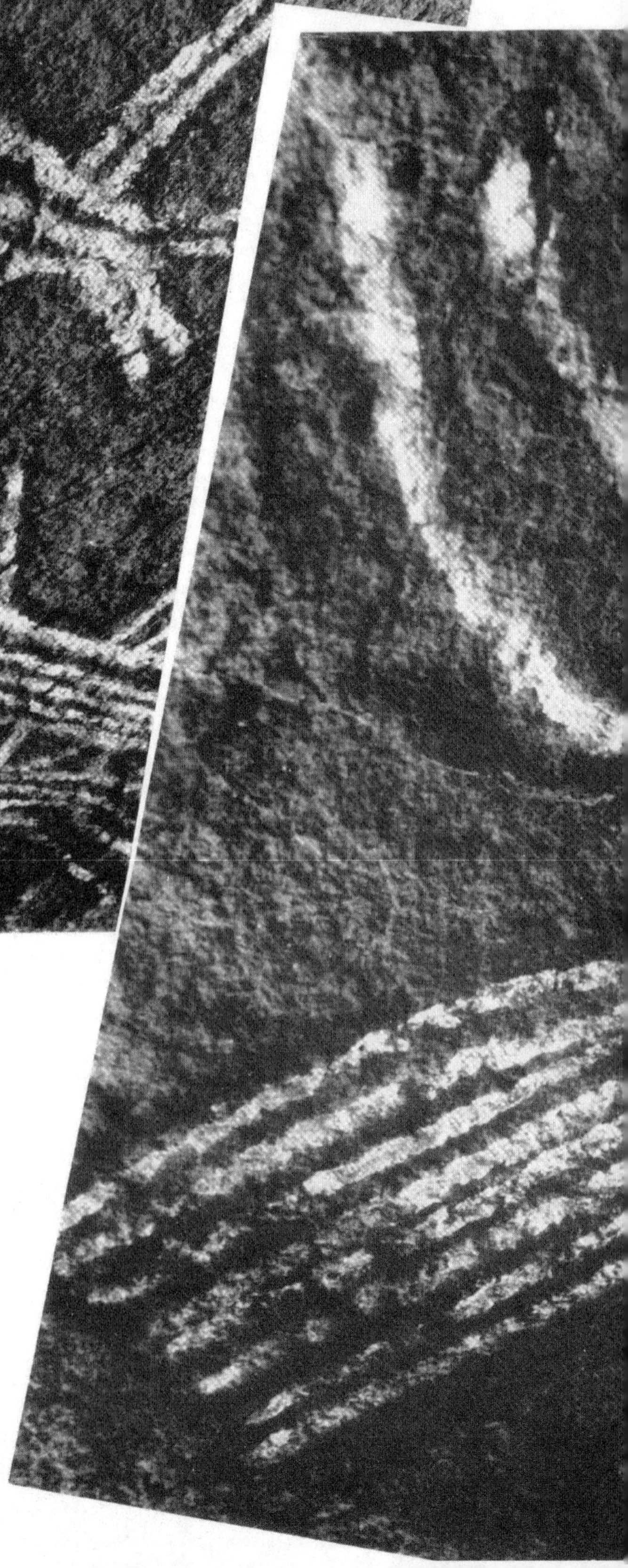

JAY C. VON WERLHOF

THE California Archeological Survey, of the University of California, has been extending its efforts the past few years to record petroglyphs and pictographs throughout the state. Some of the drawings, paintings, and carvings by California Indians had been known locally for some time. Little had been done, however, to record and study them. Few persons until recently took the works of art seriously. While the spectacular cave paintings near Santa Barbara

16

With stiff brushes of animal hair bound around a stick, Tule-Kaweah Indians made strong lines on the rough surfaces of granite and sandstone boulders. (Photographs by Pirkle Jones)

In caves and beneath jumbles of boulders California's earliest artists left their paintings which are now being recorded, before their inevitable destruction, by a survey team from the California School of Fine Arts. (Photograph by Fran Conor)

ANITE GALLERIES

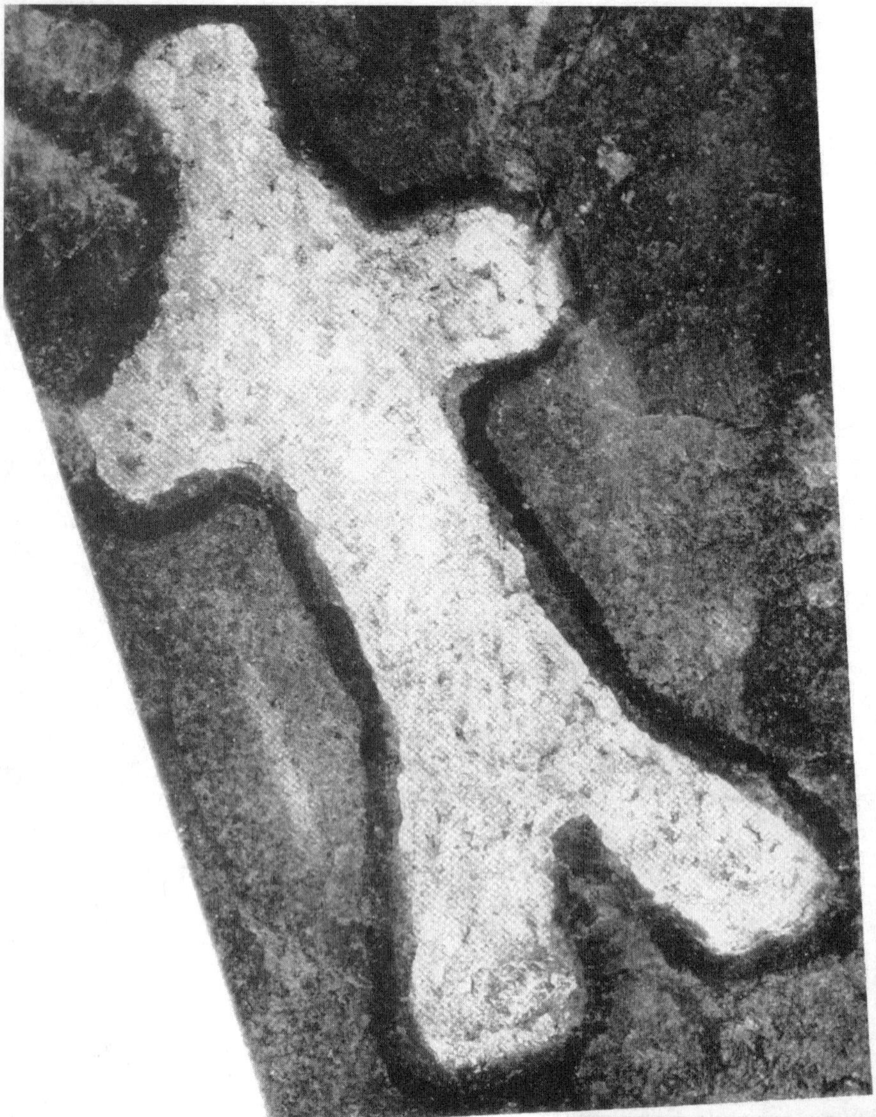

have been publicized and visited by thousands, the art of painting had not been recognized as a meaningful expression of California natives.

In contrast to a written language the interpretation of symbols is not the most fruitful result of art study. Of more significance is the degree of technical mastery in painting or carving. Also to be given consideration is the aesthetic sensitivity felt in the work. And, perhaps the most important of all, the study of art affords another means for tracing intercultural influences. By charting recurring designs or techniques used in art, migrational movements or intertribal contacts can be followed more definitely.

The photographs shown here were taken at Rocky Hill, near Exeter in Tulare County. The California School of Fine Arts in San Francisco, to enhance its own interest in Primitive Art, organized the expedition with the coöperation of the California Archeological Survey. The site is actually made up of six areas along the lower slopes of Rocky Hill. The paintings were found in caves and the undersides of large rock formations. Only a few of the works were fully

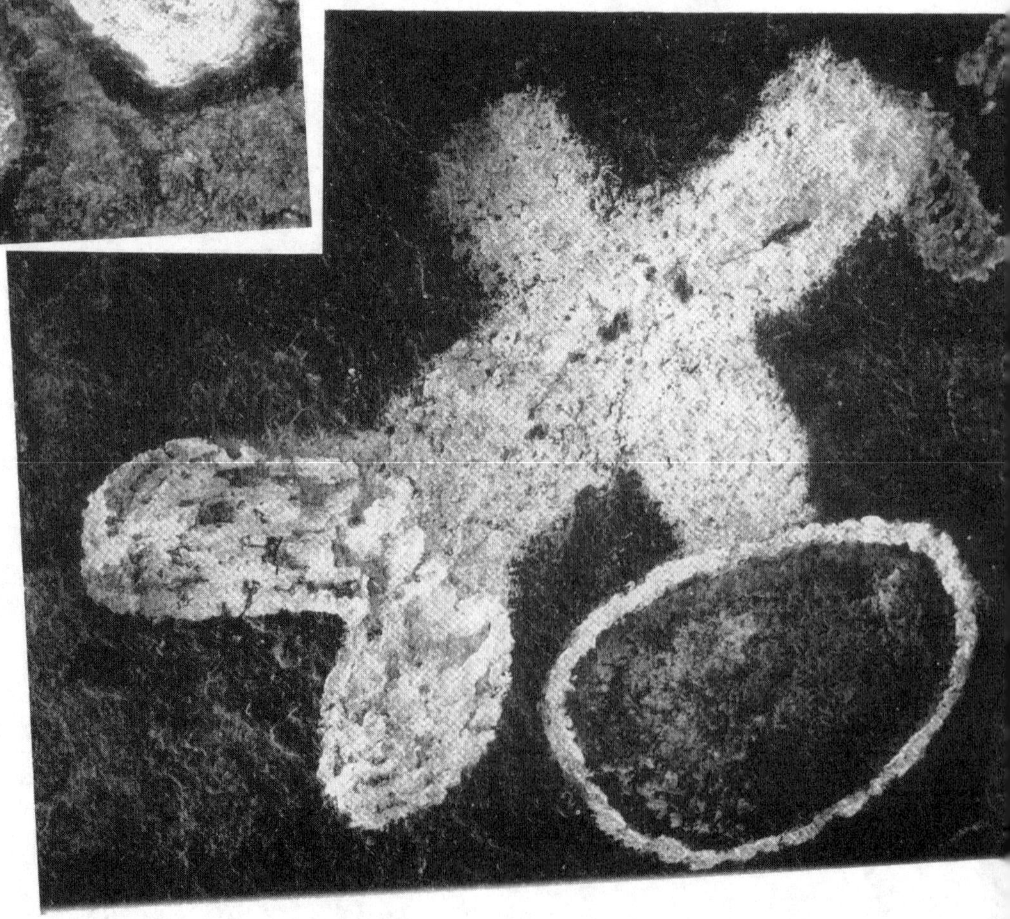

↑ *This figure was painted with a white pigment, outlined in black. The Indian painter daubed two coats of thick pigment on the white area with his crude brush.*

→ *Left unfinished, this painting's general form was quickly defined with undercoating. The final coat, intended to smooth out the lines and give solid body, was only begun at the lower part of the figure and around the circle. The inside of the circle was blocked in with red which also was used for the line protruding behind the head.*

PHOTOGRAPHS BY PIRKLE JONES
(unless otherwise noted)

exposed on outside walls, lessening greatly the danger of destruction by erosion or vandals that already has ruined so many sites. Paintings, as at Sunflower Valley near Avenal, Orosi, and Primavera, were quite distinct twenty years ago but by now are too weathered to accurately record. Other sites, as the flamboyant series at Hospital Rock, have been damaged beyond recognition by vandals within the past five years. This summer the California School of Fine Arts will conduct field experiments with ultra-violet light in an attempt to reveal work that has been eroded or mutilated.

The symbols painted at Rocky Hill were drawn from natural forms, and are representative enough to be recognized as man, plant, or animal. A few designs are too abstract to identify counterparts in nature. However, these are done well enough technically to permit the assumption that the designs were intentionally abstracted instead of being poor attempts at creating likenesses to nature.

The paintings are products of the Tule-Kaweah tribe. These Indians formed a part of the Yokut nation which occupied the bulk of the San Joaquin Valley from Fresno to Bakersfield, and from the lower slopes

↑ A wide span of time appears in the painting of this surface. An element marred by age barely survives at the left; a younger one at the top is quite distinct. A third element is almost bright in contrast but cruder (compare with the cover photograph as to quality of line, proportion, and connecting of lines).

→ Pigments came from the earth: red was hematite or iron oxide, plentiful in lower Sierra slopes; white was prepared from Cretaceous foothill strata, the chalk broken into small bits and mixed with animal fat and human saliva as binding agents; black was soot or powdered charcoal. Pigments were mixed in a crucible or on a stone slab with the liquid ingredients. Some plants were chewed for their saliva-producing effect. Colors turned out with an even and opaque consistency.

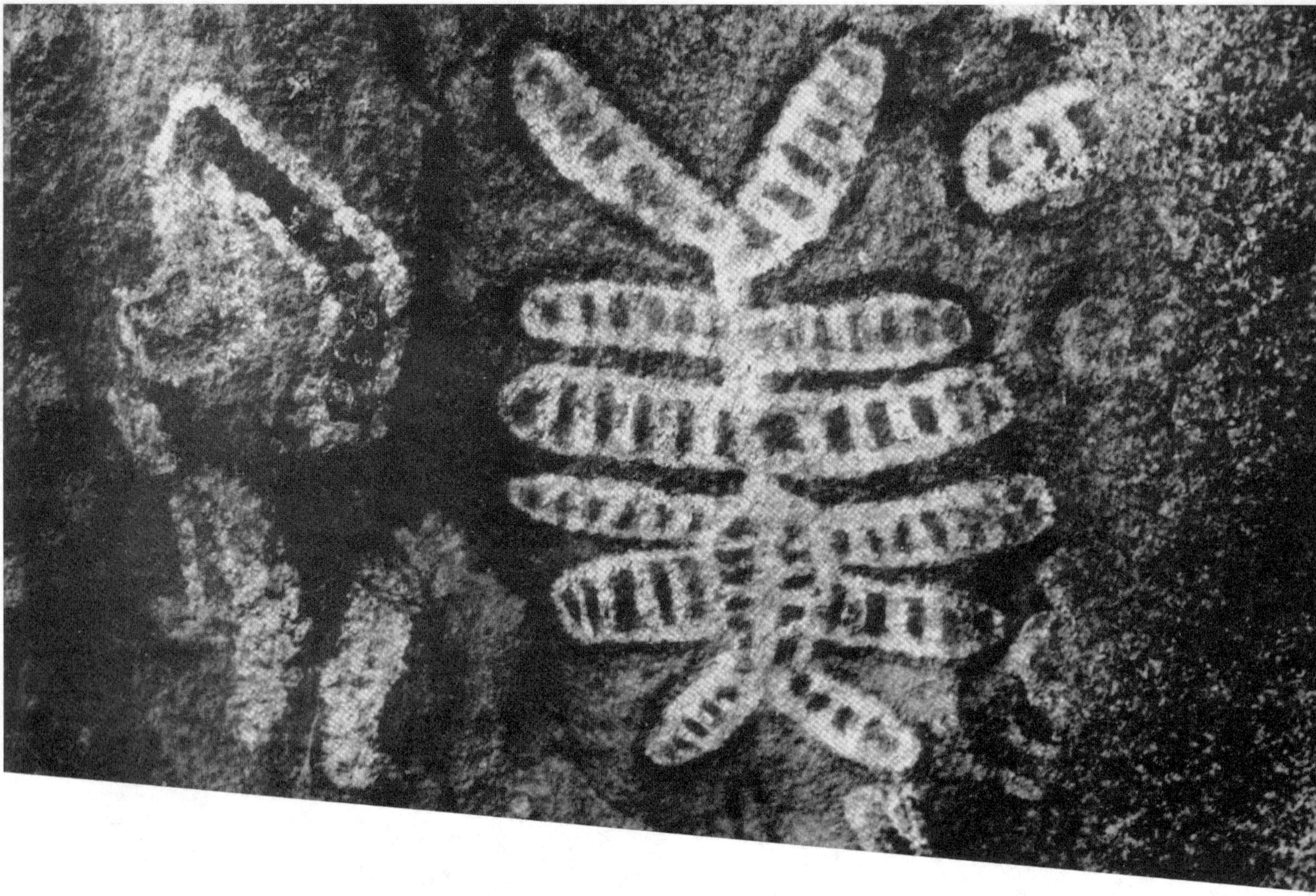

of the Sierra to those of the Coast Range. The Yokuts were known for their peaceful ways, and the various tribes that made up the nation kept close to their own territorial limits. Within their own borders each tribe found a self-sufficient life that seldom necessitated jealous glances to the lands of neighbors. They actively traded across the valley, and with a few of the foreign tribes that surrounded the Yokut peoples.

The Tule-Kaweah tribe occupied the central sector of the eastern Yokut territory. Geographically their land was quite varied. From the valley floor at Exeter into the 1,500-2,000-foot levels of the Sierra the In-

Looking westward over San Joaquin Valley from the southern slope of Rocky Hill. At the large boulder's base is a series of bed-rock mortar holes for grinding seeds, grasses, and other edible plants into meal. Nearby were found large patches of Jimson (loco) weed which the Yokuts used to make medicinal and ceremonial brews. A natural spring near the fenced trough (center right) was the only permanent water source in the area. Paintings were found on all sides of the boulder. Several test holes at this site yielded pot sherds, pit ash, bird bones, and flint.
(Art Adams)

20

← *Plant shapes — white, dividing lines red, the whole outlined in black. (Fran Conor)*

dians found a diversified supply of plant and animal food. Natural springs in the hill regions provided them with water during the months when the creeks were dry.

The Tule-Kaweah were hunters and not farmers. Their animistic religion, tribal organization, codes of ethics and morals reflected their dependence upon accepting the world as it seemed to exist. This was so in their total way of life.

Near the center of their area stood Rocky Hill. Appropriately named, the hill is a chaos of boulders, and rises above dozens of nearby hills that are barren as dunes. The hill was a religious site for ceremonial

→ *A complicated series on the inside walls of a cave opening shows some elements badly preserved, others obviously overpainted on earlier ones — example, the lower procession of horned, tailless animals painted over parts of the turtle. In some cases the poor condition may be owing to the painter's using only a single coat. The colors are white, red, and black.*

↓ *This group is at the opposite end of the same cave, away from wind and sunlight — it is very well preserved. The white and red areas are very distinct. One of the elements is a series of near-circles, an element appearing in only one other example at Rocky Hill.*

gatherings held several times each year. At these ceremonies, which ushered in changes of seasons or brought forth initiates into the tribe, dances and feasts were held. On the southwestern slope is a huge boulder (see photo) around which a camp was kept during festivities. The boulder is broken in two, and at some time the halves were painted with murals. Earth movements later brought the pieces close together again, and today only a few inches separate them. The opening is too narrow to observe the full scene, and only the elements nearest the outside can be accurately viewed.

Shamans, or "medicine men," presided over the festivities because of the religious significance inherent in each celebration. The shamans were the interpreters of nature which was so full of mystery and power. They claimed to communicate with the spirits, demons and ghosts which they felt inhabited every object and act within the natural world. As interpreters and communication experts the shamans were also the tribal artists.

The specific meaning of many paintings cannot at the moment be determined. And there are no Indians in the Yokut area today who can say anything positive about the work.

Ho-En-Nick, a former chief of the Tachi (another of the Yokut tribes), told this writer that paintings were made "long time back, way back," but not during his 85 years or even in the lifetime of his father. It is assumed that paintings still visible today are between 150-300 years old. Those that have been well protected are perhaps as old as 500 years.

Of certainty, the shamans did not paint for the aesthetic experience. Also, there are at least two distinct kinds of work. One is enshrouded in religious meaning, the other has a documentary or communicative purpose. Little is actually known about either.

The religious paintings show evidence of subjects that were meaningful to the tribe, and involve descriptive and explanatory matters pertaining to people, animals and plant life, and spirits. The documentary and communicative work has only recently begun to be studied. Maps of cemeteries and camp sites, and perhaps other physical features of the tribal area, seem to have been drawn for purposes of record or communication. The shaman was an important figure as interpreter and recorder to the natives. Women did most of the handicrafts in tribal life, and developed an acute sense of spatial relationships. The men, however, who painted on rock walls were not so interested in composition. With rather large surfaces on which to work, the paintings were apparently unplanned. They seem to have been spontaneously created, element by element, and made over long periods of time. Many elements had been placed over older ones, and in several instances the quality of work and painting technique shows considerable variation.

The Survey keeps a chart of the individual elements recorded. By this manner intercultural influences become obvious, as do later introductions of elements into a painting by malicious or ignorant passers-by.

As more and more people turn to the lures of the outdoors and chance upon Indian rock art the defacement of sites becomes more frequent. This defacement, coupled to deterioration by natural elements, seriously limits the life of the art work. For this reason too, the California Archeological Survey and California School of Fine Arts are extending efforts to record petroglyph and pictograph sites in California.

Jay von Werlhof Photos

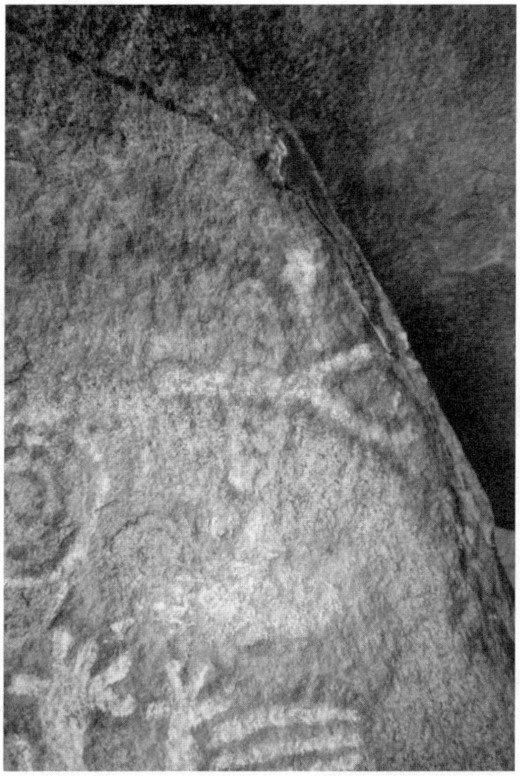

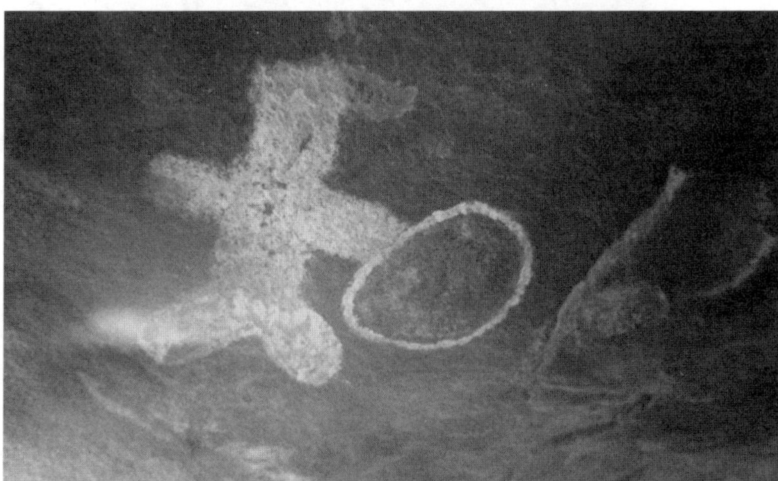

Photographs of Rocky Hill pictographs
by Russell Kaldenberg, 2001

Jay von Werlhof Photos

Photographs of Rocky Hill pictographs
by Russell Kaldenberg, 2001

LATE PLEISTOCENE/EARLY HOLOCENE ARCHAEOLOGY OF THE SOUTHWEST SEARLES LAKE BASIN, KERN COUNTY, CALIFORNIA

Judyth Reed, M.A.
Bureau of Land Management
Wyoming State Office
Cheyenne, Wyoming 82009

ABSTRACT

In 2001 BLM archaeologists learned that an organized competitive off-highway vehicle event had traversed an existing route inside the Christmas Canyon Area of Critical Environmental Concern (ACEC) in the southwest portion of the Searles (dry) Lake basin. Inventory of the route resulted in location of 16 prehistoric sites that had been impacted by the event and ultimately led to inventory of more than 2000 acres in and adjacent to the ACEC. More than 40 prehistoric archaeological sites were recorded, including lithic quarries/workshops, rock shelters, rock alignments, stacked stone features, and milling features. Dating of some features and limited testing have been carried out. This paper reports on the results of these recent studies and implications for study of Late Pleistocene/Early Holocene archaeology in the Mojave Lakes country.

INTRODUCTION

In the spring of 2001, inquiries into the status of archaeological resources in the Christmas Canyon Area of Critical Environmental Concern (ACEC) in the Searles Lake basin, and an off-highway vehicle event that went the wrong way, led Bureau of Land Management (BLM) archaeologists to take a new look at a portion of the southwest Searles Lake basin. In addition to worked lithic materials deeply embedded in variously formed desert pavements and suspected to be quite old, which had been known about for decades, recent investigations have identified a large and complex archaeological resource spread over a very old landscape. Thus far, rock alignments; rock cairns and other similar features; rock shelters with midden deposits, at least one nearly a meter deep; trails; open-air habitation sites with lithic reduction and milling; quarries; and other types of sites have been documented. Inventory of portions of the Searles Lake basin adjacent to this area but within the China Lake Naval Air Weapons Station has resulted in identification of the same kinds of resources in abundance. With over 3,000 acres inventoried and much still to be done, it is clear that a very widespread archaeological complex exists along the southern Searles lakeshore. Evidence also indicates that some of this complex is very old.

SEARLES LAKE BASIN

The southwest Searles Lake basin lies in the northwestern corner of San Bernardino County (Figure 1). The Christmas Canyon ACEC, which falls along a portion of the ancient lake shoreline, contains "low hills, pediments, and dry channels, and in late Pleistocene times, a succession of large lakes filled the basin in Searles Valley. Shorelines, beaches, and bars are exposed on the upper levels of the valley. At the last high lake stand, which occurred 10,000 years ago and reached the 2,250-foot contour level, the northern part of the ACEC was under water" (USDI, BLM 1988:1). Searles Lake was one of a system of lakes that existed throughout the Great Basin during the Pleistocene (Figure 2).

> Reconstruction of the history of Searles Lake, which is based primarily on the work of Smith (1979), suggests that the lake reached its highest elevation by 16,000 to 15,000 years ago, and remained high until after 14,000 years ago, when it began to decline sharply. This part of the chronology correlates well with the high stands and declines of Lake Lahontan and Lake Bonneville to the east. The subsequent history of Searles Lake is unique in that it rose again about 13,000 years ago and remained high until about 11,300 years ago when it began to decline again. A final rise in the level of Searles Lake occurred between 11,000 and 10,500 years ago at the same time as a rise in Lake Bonneville (Grayson 1993:102). In the Late Pleistocene and continuing into the Middle Holocene, Teagle Wash, which crosses Searles Valley within the

Figure 1. Searles Lake Basin

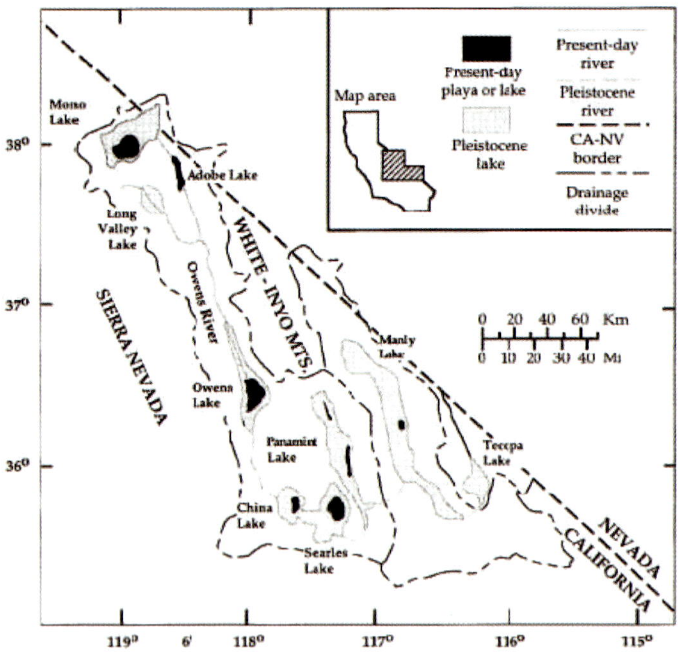

Figure 2. Map of Great Basin Pleistocene lakes system

Study area, was part of the lake and river system, connecting Searles Lake with China Lake (Wells 2003:4).

The Christmas Canyon ACEC is located in a sub-basin of Searles Valley … Between roughly 19,000 and 12,000 YBP (years before present), when these lakes were at their apogee, unusual hydrological and geomorphological conditions resulted in the creation of a tombolo or barrier spit (see Figures 3, 4, 5) across the northeastern end of the Christmas Canyon Basin, thereby separating it from Searles Valley as a whole. This unusual circumstance had two consequences. First, it contributed to the preservation of an intact Late Pleistocene and Early Holocene landscape; one that has not suffered from the subsequent erosion and degradation experienced by the vast majority of the California Desert. Second, and due to this first fact, this intact ancient landscape contains a dense, very well-preserved and extremely important concentration of Late Pleistocene and Early Holocene archaeological sites; that is, sites dating between 7,000 and 11,000 YBP, if not potentially earlier (Whitley n.d.(a):1-2).

Background to Current Research

The Federal Land Policy and Management Act (FLPMA) of 1976 established the California Desert Conservation Area (CDCA) on 12 million acres of public lands in the Colorado and Mojave Deserts of southern California and directed the Bureau of Land Management to create a land management plan for the new conservation area. The California Desert Conservation Area Plan (USDI, Bureau of

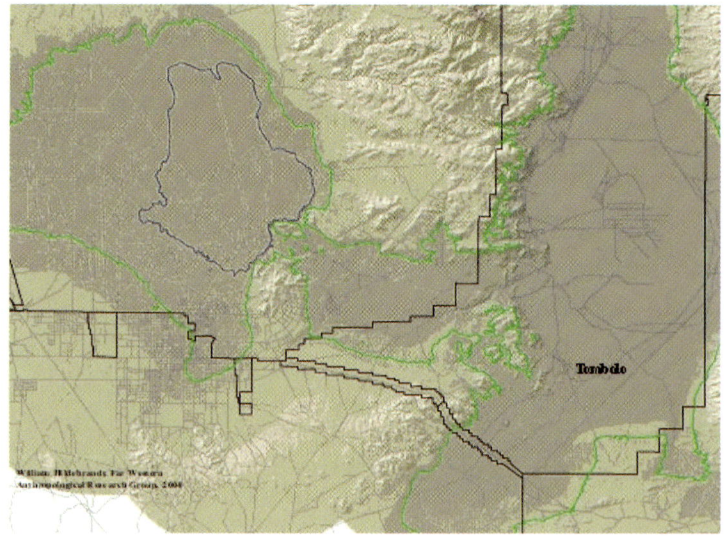

Figure 3. Map showing location of tombolo or barrier spit. (William Hildebrandt, for Western Anthropological Resources Group (2004)

Figure 4. View of back side of tombolo showing beach lines

Land Management 1980) was signed in 1980. The plan broke the desert into units and established various levels of land use on each unit, with differing kinds of uses and activities allowed within each level. For lands not inside wilderness, those areas that contain the most significant or most endangered resources were identified as Areas of Critical Environmental Concern (ACEC) and were singled out for special management that was outlined in individual management plans for each ACEC. The goal of most of the ACEC management plans was to identify activities that were threats to the resources, to implement actions to reduce or eliminate those threats, and to take measures to ensure the permanent protection of the resources. Preparation of the plans often required extensive inventory within the ACECs. At the opposite end of the spectrum, a number of off-highway vehicle (OHV) open areas, in which four-wheel-drive vehicles, motorcycles, dune buggies, quad runners, and other such vehicles may go wherever their drivers can make them go, were created. Competitive OHV events are also permitted within the open areas. In a feat of planning difficult to understand now, the desert planners managed to meld these two extremes by designating the Christmas Canyon Area of Critical Environmental Concern for cultural resources and designating the Spangler Hills OHV Open Area in such a way that the ACEC is contained almost entirely within the boundary of the open area.

Figure 5. Top of tombolo

At the time the Christmas Canyon ACEC was proposed, information on the resources there was limited and problematic. Sylvia Winslow, a local artist who took great interest in archaeological materials that she and her husband found around Searles and Panamint dry lakes in the 1960s, provided verbal information to BLM archaeologists about sites she had seen in the Searles lakeshore area. She had found what she called hand axes within archaeological sites at a number of locations, and she thought they were very old. By the time the ACEC management plan was being written, Winslow had become reticent about her activities at Searles Lake and did not offer much information to the archaeologist who was writing the management plan (USDI, BLM n.d.) In an undated paper (Winslow and Winslow n.d.), the Winslows discussed their observations and some conclusions about artifacts they had seen in the Searles basin. In referring to the locations of "hand axe sites," the Winslows said:

> In the Searles Lake basin, sites are located well above the old shorelines. They are also to be found on high terraces, above ancient river channels, or on the banks of streams which once carried water into Pleistocene lakes. Out of thirty sites, only one was located near a spring; others have been found more than twenty miles from the nearest source of water at the present time. . . Rather than being concentrated . . . the hand axes and associated implements are scattered over a considerable part of the terrain. Sometimes they are found embedded in the desert pavement, sometimes half-buried in erosion gullies, or merely lying on a hillside. . .the larger implements are made of basalt or chert. . . Belonging in the complex are the massive choppers, scrapers and other tools which are usually associated with hand axe sites in other countries (Winslow and Winslow n.d.:2-4).

A note at the end of the paper states that "only stone implements from 30 sites investigated by the authors have been referred to in the proceeding (sic)" (Ibid.:7). It appears certain that the Winslows did collect artifacts from the Christmas Canyon ACEC area but the whereabouts of those artifacts now is unknown.

G. I. Smith, a geologist with the U.S. Geological Survey, who studied the Searles basin extensively, had seen the Winslows' artifacts and knew where they had been collected. In a 1976 letter to the California Desert District Manager, he wrote:

> Your staff would, I am sure, like more information on the artifacts found by Mr. and Mrs. Winslow in Searles Valley. At least some of them were found lying on the surface in an area of dark colored, boulder-rich gravels of pre-Wisconsin age that crop out west of a small playa (elev. 1,939 ft) about 4 km SSW of The Pinnacles. Presumably, the tools were originally incorporated in younger, finer grained deposits that have been eroded away during subsequent lacustrine periods, leaving them as part of the lag-gravel. As I remember them, the tools were mostly large hand-axe types, made of felsitic volcanic rock from the northern Lava Mountains that are included in the gravels of that area, crudely worked (both unifacial and bifacial, I think), and visibly patinated. On the basis of the patination that characterizes dated deposits of alluvium and lake gravels, I would guess that the tools have been lying on that surface for the last 10,000 years. No points or imported rock (e.g. obsidian) were found with them. The lack of similarly old-looking material in the collections by your staff is not surprising because most of the "intensively evaluated" areas . . . are either on much younger deposits which would have concealed any older layers that contained similar tools, or are not near any of the possible "paleoshorelines" of the lake which seem to be among the more promising areas of artifact concentrations (Smith 1976:1-2).

The information provided by Smith would place these locations in the area of what became the Christmas Canyon ACEC.

During preparation of the ACEC management plan, some inventory was carried out within the ACEC and a number of sites were recorded to add to a few that had been recorded during inventory for the California Desert Conservation Area Plan. A total of 18 prehistoric sites had been recorded. "Most of the sites are associated with the procurement and reduction of chert, including an extensive quarry. The artifacts exhibit varying degrees of patination, and many are associated with desert pavement. Milling stones have been identified with some lithic scatters. About 7% of the ACEC has been inventoried for archaeological resources, and additional inventory is expected to yield many more sites" (USDI, BLM 1988:4). At the time that the management plan was being written, China Lake Naval Air Weapons Station had applied for a withdrawal of 4,200 acres of the ACEC to provide a security zone for the Navy's Mojave "B"/Randsburg Wash Test Range. This withdrawal was ultimately approved and reduced the original 7,540-acre ACEC to 3,340 acres. It also removed Christmas Canyon, the feature for which the ACEC was named, from the ACEC.

The greatest challenge in managing the ACEC lay in the fact that 90% of the ACEC fell within an area identified for intensive use, the OHV open area. Given that the ACEC was at the far eastern boundary of the open area, it begs the question, why didn't they just adjust the boundary of the open area to leave the ACEC outside the open area? In the long run, that would have been far easier at the time than what has since transpired. The management plan notes that "recreational use within the ACEC has increased in the last 2 years and is posing a conflict with the preservation and protection of archaeological resources" (USDI, BLM 1988:7). Management actions identified in the plan included requiring that competitive events be confined to existing routes, which might have been acceptable in other areas, but new "existing" routes crop up all the time in an area in which vehicle free play is allowed, and existing routes on the desert frequently run through archaeological sites. Other requirements of the plan, such as inventory and evaluation of sites, regular monitoring, fencing where needed, and increased patrol, don't appear to have been implemented at all.

In 1992 a management plan for the OHV open area was written (USDI, BLM 1992). It allowed for competitive events on existing routes inside the ACEC only after appropriate inventory and mitigation had been carried out. No record of any effort to inventory routes within the ACEC or conduct mitigation of impacts to sites on or adjacent to routes prior to recent efforts has been found.

In March of 2001, in response to an inquiry by a member of the public about resources in the Christmas Canyon ACEC (now minus Christmas Canyon) and how they had fared inside an open area, the author and her supervisor made a tour of inspection through the ACEC to examine the general state of health of the resources and landscape. Archaeology was found all over the place. We also observed what we considered to be evidence that OHV events had been run through the ACEC for some time and noted lithic reduction materials along and in one route that had clearly been damaged by vehicle use. Our findings were reported back. The following month, a "truck and buggy" event was permitted through the OHV open area. (This is the most destructive type of event permitted in the open area; vehicles go as fast as they can and cut large dips in the route, making it uncomfortable for other vehicles to use. Consequently, the route widens as other vehicles drive around the rough spots.) Although the officially approved route was not the one used, it is apparent from appearances on the ground that the participants used a route they had been using for some time.

When the deviation from the approved route was discovered, a BLM archaeologist inventoried the route through the ACEC and a motorcycle route he found in the same area. He recorded 16 prehistoric sites, mostly lithic reduction sites but with some milling, that had been damaged by use of the route for the competitive event (Scott 2001). The need to mitigate damage to these sites prompted the initial work reported here. As understanding of the landscape and resources within the ACEC grew, efforts to document the resources and improve management of the ACEC expanded.

Archaeology of the Christmas Canyon Basin

Mitigation of Effects of OHV Event

The first recent inventory in the ACEC was carried out after the spring 2001 event and covered only the portion of the event route that crosses the ACEC and a portion of a motorcycle route that lies near the event route and that had also affected prehistoric sites. This inventory was carried out by a BLM staff archaeologist out of the Ridgecrest office (Scott 2001) and covered 96.9 acres on an area of desert pavement near an observable shoreline. Sixteen sites, primarily lithic reduction areas, were recorded. Because this area had been noted and photographed during the March inspection tour, it was possible to compare the condition before and after the event. Although the route was well-used before this particular event, its condition was substantially altered by the event. Deep ruts were cut in the soil, particularly at turns. The route had been widened by race vehicles going wide around ruts created by competitors ahead of them on the course. The location of artifacts in the route was documented.

To begin mitigation of effects caused by the race event, W&S Consultants was hired to carry out more detailed recording of these sites, to micro-map each site and artifacts within the site, and to document the nature and amount of disturbance at each site. The setting was summarized by Whitley thus:

> The 16 sites examined during this study are located on a single landform consisting of a large and very ancient alluvial fan surface. This pre-dates the late Pleistocene and, because of this fact, it was repeatedly submerged and exposed by changing lake levels during this period and, thus, has been sculpted by terrestrial and marine processes.
>
> Four circumstances contribute towards making the geomorphological setting of the 16 study sites unusual and thereby potentially enhancing their significance:
>
> First, although the Pleistocene lake sequence for Searles has not yet been precisely and reliably determined, it is nonetheless clear that, as part of the larger eastern California lake system, its history is similar to that defined for other

lakes in this system. This being so, the landform containing the study sites was most likely submerged between roughly 16,000 and 12,500 RYBP (radiocarbon years before present). It therefore seems unlikely sites on the landform could predate about 12,500.

Second, and despite this fact, the landform is very well preserved, representing in effect a Terminal Pleistocene/Early Holocene setting; that is, one dating from about 12,000 to 8,000 RYBP.

Third, the landform abuts the mud playa at the bottom of the sub-basin. During the Terminal Pleistocene, this likely would have been a marshy environment containing substantial kinds of resources amenable to human exploitation.

Finally, the landform contains significant quantities of metavolcanic (especially rhyolite and basalt) and cryptocrystallate (especially jasper) cobbles and, in some cases, boulders. These are very usable for stone tool manufacture.

The combination of these circumstances and factors suggests that this landform has a high likelihood of containing well-preserved sites dating from the Terminal Pleistocene and Early Holocene Periods (Whitley n.d(b):1-2).

Whitley identified 67 features within the 16 sites, generally cairns and other types of rock structures or concentrated scatters of debitage, including intact chipping stations. "Typically (but not invariably) the debitage and artifacts making up the lithic scatters are embedded in the desert pavement" (Whitley n.d(b).:4). He also noted that all of the sites but one lie above the low stand of the lake and significantly below the high stand (Ibid.:4).

Site 16 contained 9 features, 8 of which are lithic scatters. The 9th feature is a rock structure (Figure 6) of undetermined function at this time, possibly a hearth or rock ring, or possibly something else. It is composed of "16 cobbles ranging from fist-sized to about 0.75 m in diameter. With one exception, the cobbles are all basalt. The cultural origin of this concentration of rocks is established by the fact that three of the cobbles are standing essentially upright; that is, with their long-axes upwards" (Ibid.:14). One cobble from this feature was removed and the carbonate rind on it was radiocarbon dated by Ronald I. Dorn. It yielded a date of 8440 ± 60 BP; in the Early Holocene (Ibid.:14). Among other management recommendations, Whitley recommended that some of the sites be subjected to sub-surface testing since they may not all be merely surface phenomena, and that they be nominated to the National Register of Historic Places because they have the potential to increase our knowledge of the earliest prehistory of the California Desert. These two things remain to be done, although the State Office of Historic Preservation has concurred that the archaeological complex in Christmas Canyon ACEC appears to be of National Register quality.

Figure 6. Rock feature in Site # 16

Inventory

The situation in the ACEC made it evident that additional inventory within the ACEC and in adjacent portions of the open area would be prudent. In June 2002 BLM archaeologists carried out inventory of several areas covering approximately 1,130 acres. This inventory covered a block adjacent to the routes in which the original 16 sites had been damaged, a block at the spillway where in-filling of the lake would have occurred, the tombolo or spit that created the Christmas Canyon sub-basin, a terrace next to a small remnant lagoon (dry) that was dubbed Trail Mesa, the entire periphery of the same lagoon, and 3 other areas scattered inside the ACEC. Twenty sites were documented in this survey. Fourteen other sites were flagged and recorded later by Ancient Enterprises (Scott 2002).

In 2002 and spring 2003, Ancient Enterprises, Inc. was contracted to inventory 2,000 acres inside the ACEC (Wells 2003; Walsh and Clewlow 2003). They recorded a total of 30 sites. The total number of recorded sites has now risen to 75 with 3,227 acres inventoried. Finally, with grant funding from the State of California Off-Highway Motorized Vehicle Division of California State Parks, ASM Planning and Research Collaborative (ASM-PARC) has been contracted to inventory 20,000 acres within the Spangler Hills OHV Open Area. Specific inventory blocks have been identified for this effort, including a transect along the 16.5-kya shoreline. This work is in progress and results are not yet available.

Results of Inventory

Table 1 presents the results of inventory by project. Table 2 lists the total number of site/feature types recorded. Lithic workshops/assay areas/quarry areas are ubiquitous, with 119 such areas recorded. One survey block approximately a quarter-mile square was found to contain lithic prospecting areas and workshops throughout its extent. The area contains abundant good-quality chert, rhyolite, and chalcedony, so these results were not unexpected.

Table 1. Recent Inventory within Christmas Canyon Basin

Inventory	Done by	Acres	No. Sites	Site Types
Inventory of Unauthorized OHV Event Route, Christmas Canyon ACEC	BLM	96.9	16	Lithic scatters consisting of assay areas, assayed cobbles, and flake scatters of chert, rhyolite, quartz, and chalcedony. One possible hearth.
Section 110 Inventory Within and Adjacent to Christmas Canyon ACEC	BLM	1130	20 (+ 14 recorded later)	A quarry/workshop area approximately ¼ square mile in extent; four prehistoric trail segments; temporary habitation sites containing flaked stone artifacts and debitage, milling, and rock rings; large rock rings (house rings).
Intensive Archaeological Inventory of 1000 Acres Adjacent to Ancient Lake Searles in the Christmas Canyon ACEC, San Bernardino Co., CA	Ancient Enterprises	1000	33	Components: 27 lithic scatters, 43 rock cairns, 6 rock shelters, 2 milling components, 3 fire-affected rock components, 2 hearths/roasting pits; 2 stacked stone features other than cairns; 2 geoglyphs; 2 quarries; 2 rock rings; 1 historic.
Intensive Archaeological Inventory of an Additional 1000 Acres Adjacent to Ancient Lake Searles in the Christmas Canyon ACEC, San Bernardino Co., CA	Ancient Enterprises	1000	10	Components: 11 lithic scatters; 1 cobble circle (hearth?); 10 rock shelters; 2 milling components; 3 stone alignments; 8 rock cairns.
Class II Archaeological Inventory of the Spangler Hills OHV Open Area	ASM Affiliates	20,000	In progress	
TOTALS		**23,226.9**	**79**	

Table 2. Sites/Features Recorded within Christmas Canyon Basin During Recent Inventory

Site/Features Recorded	Number of Occurrences
Lithic scatter/assay area	119
Rock cairn	40
Rock shelter	16
Trail segment	10
Groundstone	6
Hearth/roasting pit/fire pit	3
Fire-affected rock	3
Large rock ring/house ring	3
Trail break	2
Quarry	2
Circular rock arrangement/rock ring	2
Rock alignment/geoglyph	2
Rock wall	1
TOTAL	**209**

The presence of 16 rock shelters, along with several open-air temporary camps, indicates that at least short-term habitation took place in the basin. One of the shelters was tested and found to be over 70 cm deep (Wells 2003:32), which may indicate long-term occupation or repeated short-term occupation over a long period of time. All levels within the test unit were found to contain lithic artifacts, primarily of chert, and charcoal. The chert appears to all have come from local sources. Three multi-directional chert cores were recovered along with 3 obsidian flakes. Two of the obsidian flakes have been sourced to the West Sugarloaf source in the Coso Volcanic Field (Hughes in Wells 2003: Appendix A:2), which is 30 to 40 miles northwest of the basin. Obsidian hydration yielded measurements of 8.9, 7.5 and 3.3 microns (Origer in Wells 2003: Appendix B:2). "A commonly used formula for converting obsidian hydration measurements to years before present (ybp) has been developed by Basgall (1990). Using this formula, the age for Specimen CA-SBR-10836-1 is 3672 ybp or 1670 BC, the age for Specimen CA-SBR-10836-3 is 368 ybp (AD 1635) and the age for Specimen CA-SBR-10836-2 is 2469 ybp (467 BC). The amount of rodent disturbance encountered in the test unit could account for the occurrence in the first level of the youngest and oldest of the three artifacts. It is also possible that the measurement on Specimen CA-SBR-10836-3 may be from a post-depositional break" (Wells 2003).

More unusual features include 2 geoglyphs, one informally dubbed the "rattlesnake" (CA-SBR-10838) and the other the "running man" (CA-SBR-10839) for purposes of reference (Figures 7,8). The "rattlesnake" (Wells 2003:16) measures 203 cm long, 63 cm wide, and 40 cm high, and is situated on a low ridge. It is composed of approximately 20 medium to large (15-to 50-cm-diameter) cobbles of basalt and granitic composition. Although no other cultural features were found in direct association, the geoglyph is within ⅛ mile of the largest quarry and the largest rock shelter. "It is not possible to determine the age of this feature using survey data, but its location, which is not visible from the jeep and motorcycle trail, suggests that this rock feature is probably prehistoric. Geoglyphs with similar shapes have been found elsewhere in the Mojave Desert. Davis (1980a:309-310) argues that these represent snakes or serpents and that they are powerful religious symbols that have their roots in Paleolithic shamanism" (Ibid).

The "running man" is approximately 5.5 m long by 4 m wide and is situated on a ridge. It is composed of 40 to 50 granitic and basalt cobbles. A large pack rat midden has been built into the central portion of this feature. A cobble from the feature was dated by cation ratio dating to 900 ± 350 years before present.

A single large flake, dubbed the "Davey flake" (Figure. 9) in homage to Emma Lou Davis and found built into a stacked stone cairn, was dated by cation ratio dating to 6600 ± 1300 years before present. Table 3 summarizes dates obtained by various methods and on various materials collected within the Christmas Canyon ACEC.

Figure 7. "Running Man" rock alignment.

Dating the Archaeology of the Christmas Canyon Basin

Standard methods of dating archaeological materials such as radiocarbon dating, accelerator mass spectrometer (AMS), and obsidian hydration have been combined with experimental techniques such as cation ratio dating, radiocarbon dating of pedogenic carbonate, and rock varnish microlamination. The experimental techniques allow dating of materials that could not be dated using current standard methods, while reliance on the standard techniques provides the opportunity to compare and confirm or refute dates acquired using more experimental techniques. If such experimental techniques prove reliable, they will provide means of dating rock materials and features in desert archaeological sites. Rock is by far the most frequently occurring material in prehistoric sites in the desert, and except for obsidian hydration, no reliable techniques to date their use as cultural objects has been perfected.

Continued experimentation with such techniques until such time as their viability for archaeological use can be determined is critical for the scientific method. If any of these techniques are found to work reliably, they will be a valuable tool for archaeological research and may help to answer some of the most important questions in desert archaeology, including when people first lived there, which could have implications for when people first settled in the western hemisphere. The experimental techniques used here all aim at dating anthropogenic modifications to rock coating catenas resulting from movement of the rocks for construction of rock features or use in other cultural activities (Cerveny *et al.* n.d.[draft]).

Table 3. Dates obtained from archaeological materials within the Christmas Canyon basin.

Site Designation	Sample Design./Material	Age Before Present	Method
10834	Davey Flake	6600±1300	Cation Ratio
10835	Quarry Face 1	4800±1500	Cation Ratio
10835	Quarry Face 2	8800±2000	Cation Ratio
10835	Quarry Face 3	800±200	Cation Ratio
10839 "Running Man"	Beta 168080	4110±40	AMS
10836	6-1	3672	Obsidian hydration
10836	6-2	2469	Obsidian hydration
10836	6-3	368	Obsidian hydration
01-BLM-XMAS-17	Rock ring sample	7260±40	Pedogenic carbonate inversion (Beta 164603)
02-BLM-XMAS-16	Rock ring sample	8440±60	Pedogenic carbonate inversion (Beta 168081)
02-BLM-XMAS-16	Rock ring sample	3860±50	Pedogenic carbonate inversion (Beta 168711)
02-BLM-XMAS-14	Flake embedded in tufa	20,820±130	Radiocarbon (Beta 163526)
02-BLM-XMAS-14	Flake embedded in tufa	5900±50	Radiocarbon (Beta 167811)
Lagoon	Artifact	10,200±50	Radiocarbon carbonates (Beta 175484)
Lagoon	Artifact	11,160±50	Radiocarbon carbonates (Beta 175485)
Site # 1			
01-BLM-XMAS	Chert corer	2100±500	Cation Ratio
01-BLM-XMAS	Rhyolitic flake	2300±600	Cation Ratio
01-BLM-XMAS	Chert corer	8600±800	Cation Ratio
01-BLM-XMAS	Quartz flake	8000±900	Cation Ratio
Site # 13			
01-BLM-XMAS-3a	Quartz flake scar from a scatter	7100±1000	Cation Ratio
01-BLM-XMAS-4a	Quartz flake scar from a scatter	6700±1200	Cation Ratio
01-BLM-XMAS-5a	Quartz flake scar from a scatter	6000±1000	Cation Ratio
Site # 11			
01-BLM-XMAS-27a	Refitted flake	800±250	Cation Ratio
01-BLM-XMAS-28A	Refitted flake	950±250	Cation Ratio
Site # 2			
01-BLM-XMAS-6a	Quartz flake scar from a scatter	2100±600	Cation Ratio
01-BLM-XMAS-7a	Quartz flake scar from a scatter	1600±300	Cation Ratio
01-BLM-XMAS-8a	Quartz flake scar from a scatter	1800±600	Cation Ratio
01-BLM-XMAS-9a	Quartz flake scar from a scatter	1500±200	Cation Ratio
01-BLM-XMAS-10a	Quartz flake scar from a scatter	2000±600	Cation Ratio

Our new strategy to understand rock features derives from prior research establishing changes to catenas in biogeochemical landscapes (Ferring, 1992; Perel'man, 1980). Prior landscape geochemistry (Perel'man, 1966, 1980) and geoarchaeology (Bettis III & Mandel, 2002; Ferring, 1992; Fredlund, Bousman, & Boyd, 1988; Goldberg, Lev-Tadum, & Bar-Yosef, 1994) research has long interpreted and mapped environmental changes from analyses of soils, regoliths, and Quaternary sediments. Even in the extraordinarily arid western deserts of South America, extensive environmental history can be extracted through analyses of changes to rock coatings (Jones, 1991) and weathered debris and precipitates, enabling an interpretation of changes from high water tables with lacustrine settings, submergence, re-emergence, and then drying (Cabrol & Bettis III:7810-7811) (Cerveny *et al.* n.d.[draft]:5).

Creation of rock features such as rock alignments, rock rings, and hearths alters the position of the rocks in question and affects the rock coatings on their surfaces, which are formed over time. "The new portion of a rock fully exposed to the atmosphere will form black varnish on top of the prior rock coating. Alternatively, if a former black surface varnish is thrust deep enough into the soil, into the Bk horizon, pedogenic carbonate forms over the black varnish" (Ibid:5). The varnish on the newly exposed surface can provide information on age through use of rock varnish microstratigraphy.

"The study of varnish microlaminations (VML) (Cremaschi, 1996; Dorn, 1990), refined by extensive independent calibrations of layering patterns (Tanzhuo Liu, 2003; T. Liu, Broecker, Bell, & Mandeville, 2000; T. Liu & Dorn, 1996), permits assignment of correlative surface ages into broad time classes that correspond with regional climatic events" (Ibid:10). At present, the calibration available for the study area only discriminates Holocene from terminal Pleistocene and older ages (Ibid:11). The carbonate formed over the original black varnish can also be dated. To use this method it is necessary to determine that old carbonate has not contaminated the age of newer pedogenic carbonate. Radiocarbon dating of the outermost (newest) carbonate deposits on materials in this study did not show evidence of contamination from ancient carbonate (Ibid:8). Because sample sizes were too small for conventional radiocarbon dating, accelerator mass spectrometry (AMS) was used. For various reasons (Ibid:13), the pedogenic carbonate dates should be interpreted as minimum dates for construction of the features dated, which included 2 rock rings, a chert flake, and a rock alignment. One rock ring dated to 7260± 40 BP using radiocarbon dating of pedogenic carbonate. Varnish microlamination chronology on this rock is consistent with this date. Two dates on pedogenic carbonates on the second rock ring came in at 8440±60 BP and 3860±50 BP. The chert flake dated to 5900±50 BP. The rock from the alignment dated to 4110±40 BP. Varnish microlamination study of a boulder from a rock cairn on a ridge overlooking the high stand of the sub-basin, using Liu's calibration, "suggests that the cairn boulder was rotated sometime in the latest Pleistocene, before the 11,000 BP microlaminae and after the 14,000 BP layer that is not present" (Ibid:15). Varnish microlamination study of a projectile point near the cairn indicates that "the point's varnish initiated between 11, 000 and 10, 500 BP" (Ibid:15). (For details on all dating procedures, see Cerveny *et al.* n.d. [draft]). The cairn and point were within a prehistoric site containing other lithic materials and at which volcanic (ancient) soil was observed at the surface. The complete suite of dates indicates use of the basin over a very long period of time, from terminal Pleistocene into recent times. The presence of terminal Pleistocene archaeology is strongly indicated.

CONCLUSIONS

Recent research in this sub-basin of the Searles Lake basin has shown the presence of a diverse prehistoric complex spread over a large and unusually intact landscape that still contains undisturbed terminal Pleistocene landform features. The presence of terminal Pleistocene occupation is strongly indicated, if not proven. Many areas have not yet been inventoried, and in-depth study has not been carried out at many of the recorded sites. The complex has the potential to provide information on the earliest occupation of the California desert and possibly the earliest human occupation of North America. It must be ranked as among the most important archaeological resources in the California desert, if not the arid west. Continued research here would seem to be singularly important to questions of the earliest prehistory of the region and perhaps a wider area.

Although the sub-basin offers a fragile and rare example of arid archaeology in the western United States, and has been designated as an Area of Critical Environmental Concern for these resources, it is under immediate threat and great jeopardy of severe damage or complete destruction by use of this same area as an off-highway vehicle open area. While "multiple use" is an acceptable management strategy in general, off-highway vehicle use is not compatible with preservation of these fragile, irreplaceable, unique, and scientifically important resources. It is imperative that management of the area be turned from unfettered consumptive use to preservation of resources that will have national and even worldwide importance if they address questions surrounding the peopling of the Americas.

ACKNOWLEDGMENTS

This paper is based upon work carried out by BLM archaeologists Russell Kaldenberg, Jack Scott, Joan Oxendine, Sally Cunkleman, John Murray, Wanda Rasckow, and Cheryl Foster-Curley; Ancient Enterprises, Inc.; ASM-PARC; W & S Consultants; and Ron Dorn and Niccole Cerveny. Their dedicated work is deeply appreciated.

REFERENCES

Cerveny, Niccole Villa, Russell Kaldenberg, Judyth Reed, David S. Whitley, Joseph Simon, and Ronald I. Dorn
n.d. A New Strategy for Analyzing the Chronometry of Constructed Rock Features in Deserts (draft). Manuscript in possession of author.

Davis, Emma Lou, Kathyrn H. Brown, and Jacqueline Nichols
1980 Evaluation of Early Human Activities and Remains n the California Desert. BLM, California Desert District Cultural Resources Publications, Eric Ritter, General Editor.

Hughes, Richard E.
2002 Geochemical Research Laboratory Letter Report 2002-77. In: Wells 2003, Appendix A.

Origer, Thomas
In Wells 2003.

Scott, David J.
2001 Christmas Canyon ACEC, Survey of Unauthorized OHV Race Courses and Partial Assessment of Damage to Sites. On file, BLM Ridgecrest Field Office, Ridgecrest, California.
2002 Section 110 Inventory for 2002. On file, BLM Ridgecrest Field Office, Ridgecrest, California.

Smith, George I.
1976 Letter dated November 11, 1976 to Gerald Hillier, BLM California Desert District Manager. On file, BLM Ridgecrest Field Office, Ridgecrest, California.

USDI, Bureau of Land Management
n.d. Christmas Canyon ACEC. Notes and information on file, BLM, Ridgecrest Field Office, Ridgecrest, California.
1980 California Desert Conservation Area Management Plan. On file, BLM, California Desert District Office, San Moreno, California. Revised and reprinted 2000.
1988 Management Plan for the Christmas Canyon Area of Critical Environmental Concern (CA-06-ACEC-23). On file, BLM Ridgecrest Field Office, Ridgecrest, California.
1992 Spangler Hills Off-Highway Vehicle Area Management Plan. On file, BLM Ridgecrest Field Office, Ridgecrest, California.

Walsh, Michael R., and C. William Clewlow, Jr.
2003 Intensive Archaeological Inventory of an Additional 1000 Acres Adjacent to Ancient Lake Searles in the Christmas Canyon ACEC, San Bernardino County, California. Illustrated by Clarus Backes. Prepared for Bureau of Land Management, California State Office, Sacramento, California. On file, BLM Ridgecrest Field Office, Ridgecrest, California.

Wells, Helen
2003 Intensive Archaeological Inventory of 1000 Acres Adjacent to Ancient Lake Searles in the Christmas Canyon ACEC, San Bernardino County, California. Illustrated by Clarus Backes. C. William Clewlow, Jr., Principle Investigator. Prepared for Bureau of Land Management, California State Office, Sacramento, California. On file, BLM Ridgecrest Field Office, Ridgecrest, California.

Whitley, David S.
nd(a) Assessment of Christmas Canyon ACEC Boundaries, Inyo County, California (Draft). On file, BLM, Ridgecrest Field Office, Ridgecrest, California.
nd(b) Report on Mapping 16 Archaeological Sites in the Christmas Canyon ACEC, Inyo County, California (Draft). On file, BLM, Ridgecrest Field Office, Ridgecrest, California.

Winslow, Sylvia and Alvan K.
nd Some Unusual Sites in the Mojave Desert. On file, BLM Ridgecrest Field Office, Ridgecrest, California.

PETROGLYPHS AND MILLING-STONE PRODUCTION AT A BASALT OUTCROP IN NORTHWESTERN ARIZONA

Joan S. Schneider, Ph.D.
California State Parks
Colorado Desert District
200 Palm Canyon Drive
Borrego Springs, California 92004
and
Department of Anthropology
University of California, Riverside
Riverside, California 92521

Loretta Jackson
Department of Cultural Resources
Hualapai Tribe
P.O. Box 310
Peach Springs, Arizona 86434

ABSTRACT

The Hualapai Valley, near Kingman, Arizona, has a low basalt hill that recently has been identified as an important traditional cultural-use site. The hill has numerous petroglyphs as well as abundant debris of stone quarrying and manufacture associated with the production of metates. Based on traditional knowledge, technological observations, and comparative research, the functional and ritual nature of this site has been confirmed and the understanding of it enhanced. The pattern of dual use at this site is further examined in light of other sites with similar patterns at locations in Arizona and California.

The Hualapai Valley, in northwestern Arizona near Kingman, is part of the ancestral lands of the Hualapai Nation (Figure 1). The Hualapai have geographical affiliations and territorial claims and use areas that include lands from the Colorado River on the north and west to the Bill Williams Ford of the Santa Maria River to the south (Euler 1958), and the San Francisco Peaks to the east. The affiliations and claims arise out of the oral traditions of the Hualapai creation account. The affiliations and claims are further supported by tribal and other historical sources, population movements, trade and exchange, external and internal group interactions, archaeological evidence, burial sites, and ceremonial and religious activities held to be significant by the Hualapai and others (Casebier 1980; Dobyns 1954, 1974; Dobyns and Euler 1976, 1998; Euler 1958; Ewing 1903; Kline and Coder 1993; Kroeber 1935; Malach 1975; Manner 1957, 1975; McGuire 1993; Watahomigie 1981; U. S. Congress, Senate 1936).

The word "Pai" means "the people." According to traditional oral transmission, all Pai bands consider themselves as "one ethnic group, the only true human beings on earth." Tribal beliefs indicate a creation within the riparian world of the Grand Canyon, with the gift of canyon and plateau lands from the Creator. In accepting the gift, the Pai regard the land with all its associated resources—water, animals, plants, landforms, sediments, and minerals—as important.

Two prehistoric Hualapai bands, the Cerbat Mountain and Red Rock bands, lived in the Cerbat Mountains; they were known to their relatives in other bands as *Witoov Mi'uka Ba'a,* the Middle Mountain People, because the primary range of these bands was the Cerbat Mountain-White Hills chain between Hualapai Valley on the east and Sacramento and Detrital valleys to the west. The main band

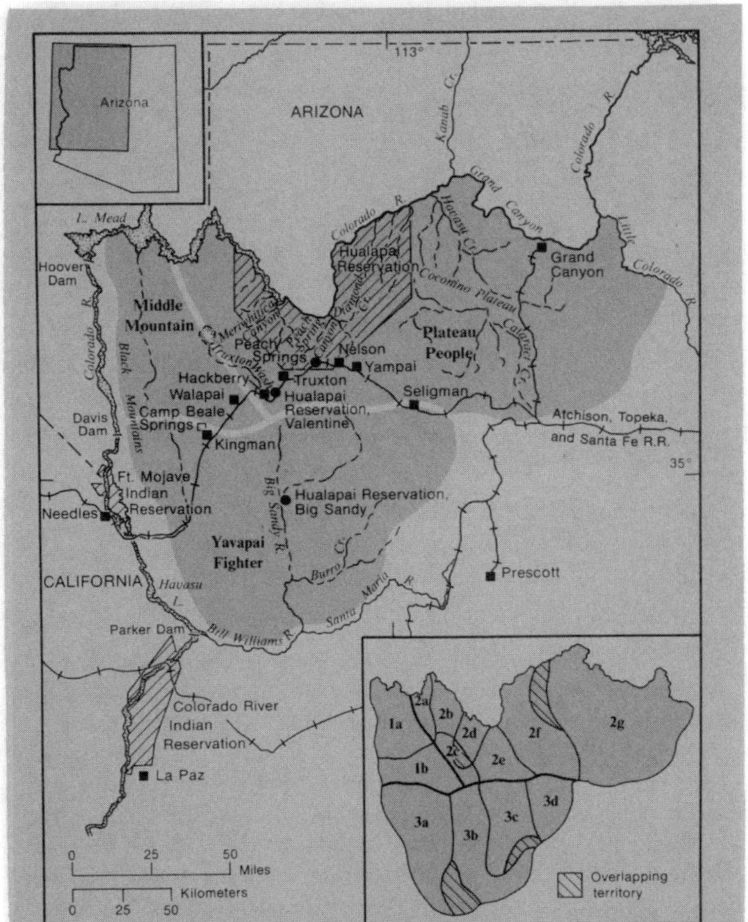

Figure 1. Location of the Hualapai Valley in northwestern Arizona and traditional locations of major Hualapai bands in the area of Metate Hill.

living in proximity to Metate Hill, the subject of this paper, was the *Ha' Mte' Ba'a,* the White Rock (Clay) Water People. Kniffen (in Kroeber 1935) described the band as follows: "This group held the Cerbat Mountains and the country west to the Colorado. *Hamte'* was the principal village, near a spring at the eastern base of the mountains. Most of the other larger villages lay to the north or south of *Hamte'* in similar situations. Several villages of lesser importance were established to the west across the Detrital-Sacramento Valley, along the crest and eastern margin of the Black Mountain Range".

Metates were important tools for processing subsistence items, as well as being a basic item in every Hualipai household. The typical Hualapai word for metate is *Pi'.* The handstone (mano) is *Va'cha.* Metates are revered and well-cared-for in Hualapai homes and are considered family heirlooms. Even today, there are important uses for what was a prehistoric implement (Figure 2). In prehistoric times, metates were left in harvesting camps, upside down, as markers, so the tools could be relocated during return visits to the camps during successive harvesting seasons. Not only was it easier to leave a large metate at the site where it was used, but the metate also served as a personal or group marker of the camp and nearby subsistence resources.

Figure 2. A Hualapai elder passes on to a Hualapai child, her traditional knowledge of the proper way to use a metate (photo courtesay of Lucille Watahomigie).

With the encroachment of Euroamerican settlers in the late 1800s and the early 1900s, modern metal tools replaced the traditional use of metates (as well as other stone tools) for processing and cooking. Traditional beliefs of the Hualapai people, however, encourage the continuing presence of the metate in the Hualapai home. If a whole metate is discovered in the countryside, it means that it is to be carried home for family use by the person who found it. Finding a complete metate is a sign of good fortune. Broken metates are avoided, since they are a sure indicator of trouble. A broken metate may be associated with the general practice of destroying a person's belongings at their death.

Metate Hill: The Site

In 1969, during his dissertation research in the Hualapai Valley, R. G. Matson (1971) noted a low basalt hill with many petroglyphs (Figures 3 - 5). Matson initially realized that some other activity had taken place at the hill, but was not sure what it was. Much later, after reading a published account of the Antelope Hill Milling-Implement Quarry on the Gila River (Schneider 1996), Matson contacted Schneider with the suggestion that the hill in the Hualapai Valley was another metate production site. Schneider, with Matson's directions, visited the hill and confirmed that hypothesis. She, in turn, contacted John Rose, Bureau of Land Management (BLM) archaeologist in Kingman, and Loretta Jackson of the Hualapai Cultural Resources Office in Peach Springs. Three visits to the site have resulted in the preliminary findings that are the subject of this paper.

Figure 3. View of a "grid" lineal abstract petroglyph on a boulder on the south-facing slope of Metate Hill.

The entire low hill is composed of vesicular basalt but only certain outcrops have been exploited for milling-implement stone. Although other basalt outcrops are available (Figure 6), the metate-production activity is limited to those portions of the hill where the stone has the following characteristics: (1) a jointing pattern consisting of widely spaced cracks resulting in relatively large sizes of separated blocks, and (2) textural qualities of basalt are such that it is homogeneous and has very small vesicles (Figure 7). A preliminary survey of other portions of the hill indicates that metate production is limited to the eastern portion of the hill, near and below the outcrops on the southern face (Figure 8).

Figure 4. View of a quadriped-motif petroglyph on a boulder on the south-facing slope of Metate Hill.

Figure 5. Portion of petroglyph display at Metate Hill. View to the north; figure for scale. Note petroglyphs on left side of photograph.

Figure 6. Outcrop on the southern aspect of Metate Hill showing the large vesicles in the rock and the heterogeneity of the texture.

Figure 7. An outcrop at the eastern end of Metate Hill showing an exploited outcrop and its jointing pattern. Note that the debris from working a block of this material (in the foreground of the photograph) has freshly exposed surfaces showing the small vesicles and the homogeneity of the texture.

Archaeological Deposits of Metate Production

Percussion flakes lie on the southern face of the hill, some accumulations forming the characteristic circular features of production workshops (Figure 9). In many cases, however, debitage resulting from the reduction of large blocks of stone and shaping of metates lies randomly scattered along the slope (Figure 10). Interspersed with tumbled blocks of stone and debitage are blocks that have been tested, i.e., had a few flakes removed, and others that were abandoned at a stage of production that was further advanced (Figure 11). Within the circular production workshops and in the general scatter of debris are hammerstones of materials different from the stone of the hill itself (Figure 12 a, b). Most show battering on their margins and at the apexes of angled margins—typical for milling-implement production hammerstones at other quarries (Schneider 1993, 1995, 1996).

Figure 8. Overview of the southern slope of the eastern portion of Metate Hill. View to the north.

Figure 9. Circular production workshop, very characteristic of large milling-implement quarry and production sites elsewhere in the wider region. The circular area is relatively flat, contains production debris and hammerstones, and was likely reused on several occasions. View to the south-southeast. A small drainage runs at the base of the hill (in background of photograph). Cobbles within this drainage were used for hammerstones on the slopes of Metate Hill.

Figure 10. Scattered debitage and tested blocks on the southern slope of Metate Hill.

Figure 11. Abandoned metate preform at Metate Hill. Note the fresh surfaces exposed by the removal of flakes (see flake scars). Flakes removed from this preform lie about it.

Petroglyphs

Petroglyphs (motifs pecked or scratched into the rock-varnish coating of rocks) abound at Metate Hill and apparently are most abundant in the same areas as the debris of milling-implement production. The "rock writing," as the Hualapai People call it, is sacred, and the places where it appears are considered traditional cultural properties. According to Hualapai scholars, motifs may portray a history of events or maps to lead people to one place or another. For example, a particular motif may show where to find water in a region or where to find other natural resources such as plants or harvesting locations. In general, rock writing is sacred because it attests to the presence of the Hualapai in times past—and continuing into the present. In order to properly interpret the meanings of the rock writing, a Hualapai scholar indicates that one must spend two or three days to get the feeling and spirit of the place (Hualapai Tribe 1998).

a)

b)

Figure 12a,b. Typical hammerstones used for milling-implement production: (a) granite hammerstone with battering about the margins and at the apex. (b) quartzite hammerstone that has been split and continued to be used. Note the battering on the exposed margins (pencil for scale).

A recent comparative study of petroglyph motifs along the Gila and Colorado rivers provides some pertinent information concerning associations between certain motif styles and geographic, cultural, and linguistic groupings (Doolittle 2000). Great Basin Abstract Style of petroglyphs thought to originate in the Great Basin, first appears in the Southwest about 1000 B.C. and continues to about A.D. 1500 (Schaafsma 1980). This style is associated with the Archaic period in the Southwest and perhaps a portion of the Hohokam Pioneer period in southern Arizona (Doolittle 2000:87). Characteristic of this style are abstract, curvilinear, and linear designs. Many of the designs attributed to this style are evident on the outcrops at Metate Hill—concentric circles, wavy lines, meanders, grids, and "rakes" (see Figure 3).

The Gila Petroglyph Style is dated later in time: between A.D. 600 and A.D. 1450 (Doolittle 2000:87). The style is marked by the addition of stick-figure representational forms such as anthropomorphic and zoomorphic quadraped motifs. This style is apparently associated (but not uniquely) with Hohokam cultural groups and also may be represented at Metate Hill.

A third petroglyph style was defined by Doolittle (2000): the Patayan Petroglyph Style. This is tentatively dated to between A.D. 800 and A.D. 1900 and is, like the Gila Petroglyph Style, an outgrowth of the Great Basin Abstract Style, but is apparently associated with Yuman groups along the river systems in the Lower Colorado River region. Motifs are more detailed than simple stick figures (see Figure 4). For example, there are "handprints", human stick figures have fingers; and animal figures often have "D"-shaped bodies (Doolittle 2000:87).

Just how this information and the "styles" developed for what are complex and multifaceted symbolic phenomena fit into the story at Metate Hill remains unresolved. The preliminary work at the hill did not include a systematic recording of the petroglyph motifs there, but rather a superficial impression of abundance and variety. Further work is necessary to (1) fully record the petroglyphs at the hill and (2) determine if any correlations can be made between the rock writing there and that at other locations along the Colorado River or elsewhere in Arizona or beyond.

CULTURAL CONTINUITY

Undoubtedly, Metate Hill (which we have also called the Cerbat Quarry elsewhere) was a traditional place where metates were produced. Inquiry of elders of the Hualapai Tribe did not lead to any recollection of this activity within the orally transmitted culture. The hill, however, has been recognized by locals, having been given its local name of "Metate Hill." Blocks of stone were pried out of outcrops and shaped using large hammerstones (see Figures 7 and 12 a,b); naturally fallen blocks of stone were also worked (Figure 13).

In addition to the abundant petroglyphs and evidence of metate production, people evidently gathered and processed traditional subsistence items at Metate Hill. A number of used metates were found in and along the wash at the southern margin of the site (Figure 14). A number of native plant species that had important traditional uses grow in proximity to the hill. During our recent trip to the hill, Lucille Watahomingie, Hualapai elder, linguist, and ethnobotanist (Watahomigie et al. 1983, 1993), identified 11 different plant species that the Hualapai used for food, food preparation, or medicine (Table 1).

Figure 13. Worked blocks and percussion flakes found in the talus accumulation of naturally separated blocks on the southern slope of Metate Hill. View to the north.

Figure 14. Used metate in the wash. Note that the material is not from Metate Hill and that it has been used enough to require rejuvenation. Pencil for scale.

Table 1. Plants Identified by Lucille Watahomigie and their Traditional Uses

Hualapai Name	**Common Name**	**Use(s)**
Ivthi	Creosote Bush	Medicinal Gum is base for glue
Jimway	Indian Tea	General medicinal plant; e.g., to clean stomach out
Sete'	Stickleaf	Important traditional staple; seeds pounded into gruel
Hanyoq	Wild Onion	Edible root; used for seasoning
Impid	Wild Onion-Turnip	Edible root; smashed and made into cakes
Min'min ya'	Filaree (introduced)	Roots used to make tea for stomach disorders
Mil dad	Barrel Cactus	Used in earth oven pits for form of moisture in baking other plants and roots. Fruits are gathered, dried, and pounded
Daqui	Cholla	Fruits are baked or dried and pounded into cakes
Manad	Banana Yucca	Baked and dried, ground into meal. Stems used for ropes and sandals. Roots used for ceremonial washing
Na a'	Mesquite	Pods are pounded for cake or juice
Alav	Prickly Pear	Fruits dried and pounded for cakes or juice; pads also dried and bounded for food

Discussion

Schneider recognized a continuum of types of milling-implement quarrying and production, ranging from, at one end of the continuum, the casual choice of acceptably shaped stone that needed no further shaping to, at the other end of the continuum, people traveling or trading to get specific types of stone. The characteristics of those specific types of stone made tools of that stone especially valuable. Characteristics included abrasiveness, durability, and the capacity for use over long periods of time without the necessity of rejuvenating the working surface. That is, the milling surface remained abrasive even under heavy use (Schneider 1993, 1996, 2000).

The extensive metate quarry located on the bajadal slopes of the western side of the Black Mountains near Bullhead City, described by Huckell (1986), is here compared to the quarry at Metate Hill. The two quarries are different on a number of accounts. (1) The Bullhead City Quarry consists of a secondary deposit of isolated boulders that were worked down to make metates. At Metate Hill, quarrying and production is different from the Bullhead City quarrying and production in that it is more concentrated than at Bullhead City. (2) At Metate Hill, blocks were pried from bedrock exposures and used to make metates in addition to the use of blocks that had naturally separated from the outcrops, found on the talus slopes below. At the Bullhead City Quarry, the boulders that were exploited were widely spaced secondary deposits. The characteristics of the Metate Hill stone, however, are very similar to the boulders of the Bullhead City Quarry: the stone in both cases is a fine, homogeneous, vesicular basalt. Other *bedrock* quarries (like Metate Hill) that are known to exist in the region, are composed of arkosic sandstone or other varieties of volcanics such as andesite and rhyolite (Schneider 1993).

Motifs that are represented in the "rock writing" at Metate Hill may be connected with giving directions to resources, early agricultural plots, precipitation, ritual, or other activity (Lucille Watahomigie, personal communication 2001). Recent conclusions reached about the direct association of the abundant rock art and the intensive milling-implement production represented at Antelope Hill, an immense bedrock quarry, were that there was no direct functional association between the two activities, although, certainly there was indirect association (Schneider and Altschul 2000). We are uncertain if this is also the case at Metate Hill. Other milling-implement quarry sites also have rock art, but some do not.

Final Remarks

The traditional Coyote Tales told in the wintertime by Hualapai People are meant to instruct people in proper and improper ways to behave. In one story called "Eagle Man, Eagle Woman," Eagle Woman covets another man and his household belongings—in particular, the grinding stone. Eagle Woman tells her husband "Your grinding stone is not very good, it is old!" And since she was a woman, she would say, "They have a better grinding stone." Eventually Eagle Woman leaves her husband and commits adultery. From the adulterous union, venomous creatures were created. The lesson of this tale is that coveting another's possessions, whether it be a person or belongings, is improper behavior. So don't be going and lusting after better grinding stones! (Perhaps Eagle Woman wanted a metate from Metate Hill!)

In this traditional tale, the metate was considered an item of great value. The hill that is the subject of this paper was a quarry for metates, implements held in high esteem by the Hualapai People. Selection of the stone was precise: blocks were tested before attempts at production were made; the size of the block was chosen in order to get an adequately sized end-product; the homogeneous texture of the stone was also of great importance. Preliminary observations indicate that the success rate for metate production at Metate Hill was higher than at other quarries studied (Schneider and Altschul 2000). The proportion of broken and discarded preforms for metates left on the slope of the hill, in relation to the number of attempts at production, was less than was observed elsewhere. Because of the relatively small size of the quarry at Metate Hill in comparison to other bedrock quarries, it would be expected that most metates made there were for relatively local use rather than for long-distance transport.

The mutual enrichment of our knowledge was made possible through this cooperative project. Through archaeology, a traditional-use site of the Hualapai People was relocated, furthering our understanding of past lifeways. Plans are for further investigations of Metate Hill in a cooperative effort with the Hualapai.

Acknowledgements

John Rose of the Bureau of Land Management expedited communication between Schneider, Jackson, and the other Hualapai People who participated in this project. Schneider acknowledges the contributions of Professor R. G. Matson of the University of British Columbia to this report of Metate Hill. Lucille J. Watahomigie, Hualapai Elder, provided her learned insights. Neal Johnson and Philbert Watahomigie developed a videotape of a visit to Metate Hill, in which Schneider and Watahomigie discuss various components of the hill; the videotape is available from the Hualapai Cultural Resources Division and is used to teach Hualapai children about their heritage. Schneider appreciates the assistance of Matthew Des Lauriers on the first trip to the quarry. This paper was first presented in a symposium at the 66th annual meeting of the Society for American Archaeology in New Orleans. The symposium, the Archaeology and Ethnohistory of the Western Desert Region, Arizona, Nevada, and California, was organized by Deborah Dosh and John Rose.

REFERENCES

Casebier, Dennis G.
1980 *Camp Beale's Springs and the Hualapai Indians.* Tales of the Mojave Road Publishing Company, Norco, California.

Coues, Elliot
1900 *On the Trail of a Spanish Pioneer: the Diary and Itinerary of Francisco Garces (Missionary Priest) in His Travels Through Sonora, Arizona, and California, 1775-1776,* Volumes 1 and 2. Francis P. Harper, New York.

Dobyns, Henry F.
1954 Walapai Country. MS on file, Department of Anthropology, University of Arizona, Tucson.
1974 *Hualapai Indians, I: Prehistoric Indian Occupation within the Eastern Area of the Yuman Complex.* Garland Publishing, New York.

Dobyns, Henry F., and Robert Euler
1960 A Brief History of the Northeastern Pai. *Plateau* 32(3):49-57. Museum of Northern Arizona, Flagstaff.
1967 The Ghost Dance of 1889 Among the Pai Indians of Northwestern Arizona. Prescott College Studies in Anthropology 1. Prescott, AZ: Prescott College.
1976 *The Walapai People.* Indian Tribal Series, Phoenix.

Doolittle, Christopher J.
2000 Petroglyph Research at Antelope Hill. In *Of Stones and Spirits: Pursuing the Past of Antelope Hill,* edited by Joan S. Schneider and Jeffrey H. Altschul, pp. 85-11. Technical Series 76. Statistical Research, Inc., Tucson.

Euler, Robert Clark
1958 *Walapai Culture-History.* Ph.D. dissertation, University of New Mexico, Albuquerque.

Ewing, Henry P.
1903 The Origin of the Pai Tribes. Henry F. Dobyns and Robert C. Euler, eds. MS on file, Museum of Northern Arizona, Flagstaff.

Foreman, Grant, ed.
1941 *A Pathfinder in the Southwest.* University of Oklahoma Press, Norman.

Hualapai Cultural Resources Division
1993 Hualapai Tribe Ethnographic and Oral History Survey for Glen Canyon Environmental Studies and the Glen Canyon Dam Environmental Impact Statement. Report on file, Hualapai Office of Cultural Resources, Peach Springs, Arizona.

Huckell, Bruce B.
1986 *A Ground Stone Implement Quarry on the Lower Colorado River, Northwestern Arizona.* Cultural Resources Series Monograph No. 3. USDI Bureau of Land Management.

Kline, Dana, and Chris Coder
1993 Hualapai Tribal Map (with traditional Hualapai habitation sites and names). Map on file, Hualapai Office of Cultural Resources, Peach Springs, Arizona.

Kroeber, A. L., ed.
1935 Walapai Ethnography. Chapter authors: Fred Kniffen, Gordon MacGregor, Robert McKennan, Scudder Mekeel, and Maurice Mook. *Memoirs of the American Anthropological Association*, No. 42.

Malach, Roman
1975 *Peach Springs in Mohave Country.* Graphicopy, New York.

Manners, Robert A.
1957 Tribe and Tribal Boundaries: the Walapai. *Ethnohistory* 4(1).
1974 *Hualapai Indians, II: An ethnological Report on the Hualapai (Walapai) Indians of Arizona.* Garland Publishing, New York:

Matson, R. G.
1971 *Adaptation and Environment in the Cerbat Mountains, Arizona.* Ph.D. dissertation, University of California, Davis.

McGuire, Thomas R.
1983 Walapai. In *Handbook of North American Indians*, Vol. 10, Southwest, edited by Alfonso Ortiz, pp. 25-37. Smithsonian Institution, Washington, DC.

Mercer, Jean Ann
1995 Ethnohistory of Hackberry: a Hualapai Perspective. In *The Historic Hualapai Occupation at Hackberry, Mohave County, Arizona: Archival, Ethnohistorical, and Archaeological Investigations,* edited by David E. Purcell. Draft report 94-155 on file at SWCA, Inc., Flagstaff.

Schaarfsma, Polly
1980 *Indian Rock Art of the Southwest.* Scfhool of American Research, Santa Fe. University of New Mexico Press, Albuquerque.

Schneider, Joan S.
1993 *Milling-Implement Quarries in Eastern California and Western Arizona: a Behavioral Perspective.* Ph.D. dissertation, University of California, Riverside.

1995 Schneider, J. S., M. K. Lerch, and G. A. Smith
A Milling-Implement Quarry at Elephant Mountain, California. *Journal of California and Great Basin Anthropology* 17(2):191-221.

1996 Quarrying and the Production of Milling Implements at Antelope Hill, Arizona. *Journal of Field Archaeology* 23:299-311.

2000 *Bedrock Quarries for Milling Implements: Petrographic and Geological Characteristics. A Focus on the Lower Colorado - Lower Gila River Valleys. In , edited by John Herrmann and Norman Herz. Boston Museum of Fine Arts, Boston.* (in press)

Schneider, Joan S., and Jeffrey A. Altschul, eds.
2000 *Of Stones and Spirits: Pursuing the Past of Antelope Hill.* SRI Technical Series 76. Statistical Research, Inc, Tucson..

United States Congress, Senate
1936 Walapai Papers: Historical Reports, Documents and Extracts from Publications Relating to the Walapai Indians of Arizona. 74th Congress, 2nd Session, Senate Document 273. Washington, DC: U.S. Government Printing Office.

Watahomigie, Lucille J.
1981 Hualapai: Stories of the Land. *Plateau* 53(2):24-27. Museum of Northern Arizona, Flagstaff.

Watahomigie, Lucille J., Philbert Watahomigie, Sr., Malinda Powskey, Jorigine Bender, and Josie Uqualla
1983 *Ko* (Pinyon). Hualapai Bilingual Education Program, Peach Springs, Arizona.

Watahomigie, Lucille J., Malinda Powskey, Jorigine Bender, and Josie Uqualla
1993 Manap (Banana Yucca). Hualapai Bilingual Education Program, Peach Springs, Arizona.

DISCOVERING THE DESERT KAWAIISU

Jackson Underwood, Ph.D.
Senior Archaeologist/Ethnographer
RECON
1927 Fifth Avenue
San Diego, California 92101

Most of what has been published about the culture of the Kawaiisu derives from the research of Maurice L. Zigmond. He conducted ethnographic work in the Tehachapi and Paiute Mountains from 1936 through 1940 and from 1970 through 1974. According to Zigmond, traditional Kawaiisu territory included a large tract of desert lands east of the mountains that was used as foraging areas. His ethnographic descriptions focus exclusively on what I refer to here as the Mountain Kawaiisu. A re-examination of ethnohistoric sketches from 19th century Death Valley and the ethnographic research of Julian H. Steward and others suggests that the desert east of Tehachapi extending to Death Valley was permanently inhabited by people I suggest should be called the Desert Kawaiisu. With a dramatically different environment and mode of existence, it makes sense to think of the Desert Kawaiisu as a distinct cultural entity, separate from the Mountain Kawaiisu. An ethnographic sketch of the Desert Kawaiisu is provided.

Maurice L. Zigmond conducted ethnographic work in the Tehachapi and Paiute Mountains from 1936 through 1940 and from 1970 through 1974. His fieldwork work resulted in most of the published material on the Kawaiisu (Zigmond 1938, 1980, 1981, 1886). He suggested that while traditional Kawaiisu territory included a large tract of desert lands east of the mountains, "...just how far Kawaiisu territory stretched eastward into the desert it is impossible to say..." (Zigmond 1938:635). However, in later publications he did produce maps to delineate Kawaiisu desert territory (e.g., 1981; 1986). He always argued that these desert areas were unoccupied resource procurement areas. The "...desert was conceived of as an area for occasional excursions, never permanent residence..." (Zigmond 1938:637). The core area of the Kawaiisu was in the Tehachapi and Paiute Mountains of Kern County (Zigmond (1986:398-399; 1981:2).

The Kawaiisu traditional territory, according to Zigmond, encompassed the area south of a line extending east from Walker's Pass in the Paiute Mountains through the south part of the Coso and Argus Ranges, the southern part of the Panamint Valley, the southern part of the Panamint Range and ending in the southern half of Death Valley. The southern boundary extended southeast from Oak Creek in the southern Tehachapi Mountains. It passed northeast of Rosamond Lake and south of Rogers Lake and on southeast to the vicinity of Victorville. Zigmond's version of the boundary then crossed the Mojave River then swung east, then northeast, to re-cross the Mojave River in the vicinity of Midway (Figure 1). It continued north through the Avawatz Mountains before entering the

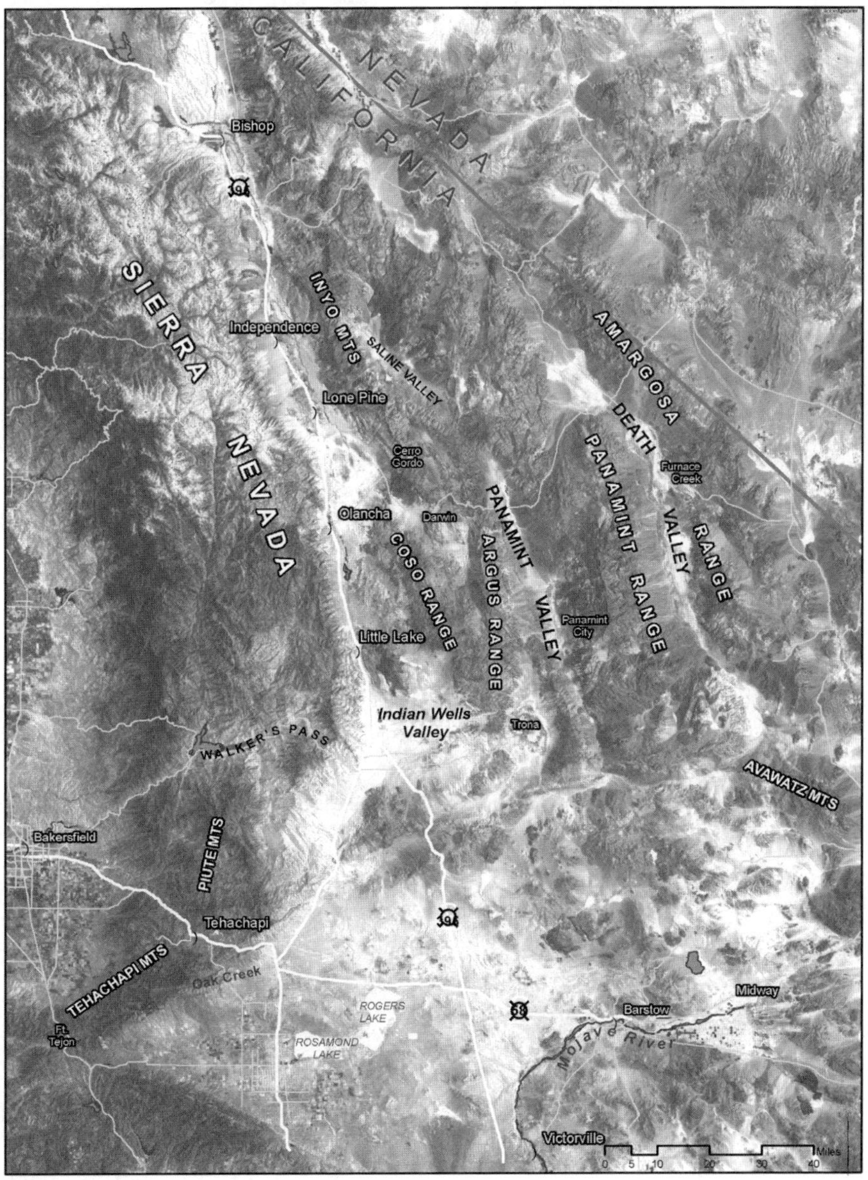

Figure 1. Regional map.

southwestern corner of Death Valley. (Zigmond (1978:398-399). Thus the desert portion of Kawaiisu territory according to Zigmond encompassed what are now the communities of Ridgecrest, Trona, Victorville, and Barstow, and the northeastern part of Edwards Air Force Base; the southern part of the Naval Air Weapons Station, China Lake; most of Fort Irwin and the southwest quarter of Death Valley National Park.

Kroeber provided a vague, small-scale map which suggested a similar northern boundary, but drew the southern boundary much farther north, well above the Mojave River (Kroeber 1925:Inside Front Cover). Kroeber (1925:601), viewed the boundaries of Kawaiisu lands in the desert as somewhat conjectural. He also mentioned interacting with a small group of Indians who were born in the Tehachapi area but were living in Victorville some years prior to 1925. He said that they may have been Chemehuevi, but since they were from Tehachapi, the Kawaiisu homeland, it seems reasonable to suggest that they may have been Kawaiisu. These people suggested Victorville was within Kawaiisu territory (Kroeber 1925:602) . Kroeber did not reflect this information about the Kawaiisu southern boundary in his 1925 map, however. On the other hand, the map of Zigmond (1978:398-399) may have been based to some degree on this or similar information.

Steward (1938:59) is the only researcher to depict the northern Desert Kawaiisu boundary in any detail, viz., the boundary between the Kawaiisu and the Coso Shoshone. He showed it passing south of Indian Wells Valley (well south of Little Lake and the Coso Range), then north along the crest of the Argus Range to include Maturango Peak, then east across Panamint Valley at about the latitude of Telescope Peak. It crossed northeast across the Panamint Range just south of Wildrose Canyon, but north of Telescope Peak and then east across Death Valley at the latitude of Furnace Creek. This accords well with the work of Mark Kerr, who collected ethnographic information from the Coso Shoshone of southern Inyo County in 1936 (Irwin 1980). Kerr also suggested that Little Lake was in Shoshone territory (Irwin 1980:x).

The neighbors of the Kawaiisu to the east were the Las Vegas Band of Southern Paiute (Kelly and Fowler1986:Figure 1). Steward (1938:59) places the boundary between the Kawaiisu and the Southern Paiute on the east side of Death Valley along the foot of the Black Mountains. East of Kawaiisu territory, the boundary between the Coso Shoshone to the north and the Southern Paiute in the south continues east at about the same latitude as Furnace Creek through Ash Meadows. Heizer and Whipple (1971:Map 1), like Kroeber, produced a vague, small scale map, but it appears to reflect the boundaries posited by Steward (1938). Kelly and Fowler (1986:Figure 1) provide a map indicating that the Southern Paiute east of Death Valley were the Las Vegas Band, while the Chemehuevi were found perhaps 100 miles farther south. Kelly and Fowler posit the boundary between them along a line just north of Bristol Lake and leading northeast to the area where the boundaries of California, Arizona and Nevada intersect. This intersection is some 11 miles north of Needles.

South of the Desert Kawaiisu were the Vanyume southwest in the southern Tehachapi Mountains were the Kitanemuk. To the west were the Mountain Kawaiisu and west of them in the San Joaquin Valley were the Yokuts. To the northwest along the south fork of the Kern River were the Tubatulabal. The Coso Shoshone (also known as the Panamint Shoshone or the Koso or Coso) held the northern half of Panamint Valley, the northern half of Death Valley and the Little Lake area, while the Desert Kawaiisu evidently held the southern portions of Panamint Valley and Death Valley and the Searles Lake area (Irwin 1980:x; Steward 1938:Figure 1; Zigmond 1986:398).

THE TERRITORY

As suggested above, nearly all that is published about the Kawaiisu comes from ethnographic research conducted in the Tehachapi and Paiute Mountains. A. L. Kroeber published some observations in 1907 and1925, evidently working among people in the Fort Tejon and Tehachapi areas (Kroeber 1907, 1925; Steward 1938:71). Harold E. Driver (1937:59) assembled a trait list, working with people from Kelso and Weldon in the foothills of the Paiute Mountains north of Tehachapi. However, the vast majority of ethnographic work among the Kawaiisu was conducted by Maurice L. Zigmond in two periods: from 1936 through1940 and from1970 through1974 (Zigmond 1986:411). Evidently, he worked exclusively with Kawaiisu consultants from the Tehachapi and Paiute Mountains. Because of the very different natural environments, much of the information gathered there is not applicable to the desert dwellers. As mentioned previously, Zigmond (1986:398) suggests that the people we would call the Mountain Kawaiisu, i.e., the people of the Tehachapi and the Paiute Mountains, made only collecting and hunting forays down into the desert from time to time, but did not live there. The presence of permanent Kawaiisu villages in southern Death Valley, Panamint Valley and Trona, as documented in ethnographic and ethnohistoric accounts cited below, suggests a separate division that we propose to call the Desert Kawaiisu.

Julian H. Steward (1938) is the only ethnographer to make a systematic attempt to gather information on the desert dwellers in the traditional territory of the people we are calling the Desert Kawaiisu. His fieldwork took place in 1935 and 1936, but he suggests the ethnographic memory of his consultants extended back to the 1870s (Steward 1938:48). It is important to underscore the fact that this was well after the area was significantly impacted by Whites.

In an account of the tragic Death Valley 49ers written by one of the survivors, there is a relevant, but brief passage about an Indian village in Panamint Valley. The author, J.W. Brier, was only 6 in 1849, but one of the few Death Valley 49ers who left an account.

Another rather thorough treatment is by William L. Manly, the leader of the party. Brier no doubt had the huge benefit of talking the material over with his mother, who was one of the heros of the disaster. He wrote that on the slow trudge out of Death Valley, they saw an Indian village, apparently in Panamint Valley, although his murky description makes it difficult to tell where he really was. According to Chalfant, the survivors probably went through Emigrant Pass, missed Wildrose Spring by going down Nemo Canyon, and came into Panamint Valley at the west end of Wildrose Pass. At some place south of Indian Ranch/Hall's Canyon in Panamint Valley, the party encountered an Indian village. This would be near the boundary between the Coso Shoshone and Desert Kawaiisu suggested by Steward (1938:59), so it is reasonable to suggest that the Indians in question may have belonged to either tribe or perhaps both, as was the case later.

> ...at the arrival at an Indian village among the mesquite groves, in which only one of the villagers remained, an aged squaw, who scolded us in a language we did not understand. Midway of the village [sic] was a vast heap of offal, and the bones of horses were scattered about freely, showing that the natives were accustomed to cross the great range of California for their regular supply of meat... (Chalfant 1936:29).

Raiding of Mission San Fernando and outlying ranchos of the Los Angeles region is well documented for the early and mid-19th century (e.g. Earle 1990:96). However, in this passage, we have specific evidence that the Indians of Panamint Valley were involved. This suggests that the native economy and trade patterns deep in the desert had been altered by the presence of Spanish/Mexican settlements near the coast. It is approximately 130 miles in a straight line from southern Panamint Valley to San Fernando Mission, so the passage also documents that these desert people were excellent, tough hikers. In their arid environment, it would have been rather uneconomical for them to maintain horses for transportation, as settlers, miners, and freighters soon discovered.

Surprisingly, prospecting in Panamint Valley began only a year after the disaster of the Death Valley Manly party. Dr. Erasmus Darwin French mounted an expedition into the area in 1850. There was a little prospecting in the Inyo-Coso region after that, and it was relatively quiet until 1860 when the Coso Mining District was formed. By June of that year, some 500 miners were working in the Coso Range a few miles southeast of Coso Peak. (Nadeau 1999:10-11). This was in Coso Shoshone territory a few miles north of Kawaiisu territory. During 1861 and 1862, prospectors were fanning out over the area between Little Lake and Death Valley and ranchers were beginning to enter the rich Owens Valley (DeDecker 1993:10), home of the Owens Valley Paiute, a Northern Paiute tribe. These incursions caused the inevitable friction with the Indians who already lived there. A series of skirmishes and atrocities against the Indians began in 1861 that were to last into the 1870s (Cragen 1975). Sometimes called an Indian war, the Native American resistance slowed the pace of exploration and development of mines and ranches in the area east of the Sierra Nevada.

In response to Indian attempts to defend themselves and their traditional territory, in 1862 the U.S. Army established Camp Independence on the southwest side of Owens Lake. This facilitated White takeover of the area, but probably saved some Indian lives as well (DeDecker 1999:10). A small town formed around Camp Independence to supply the soldiers. In 1863, Indians from around the area, including some from Panamint Valley, were tricked by the Army into coming to Camp Independence by a promise of peace, food, and clothing. Some 998 Indians were disarmed, penned up, then marched south to Walker Pass and across the San Joaquin Valley to the San Sebastian Reservation near Fort Tejon. This is in Grapevine Pass some 25 miles south of Bakersfield (Chalfont :1933:194; Irwin 1980:71). Prospecting soon resumed in the area. Several small gold discoveries were made in the Inyo Range on the east side of Owens Lake, and small towns sprung up, including Lone Pine northwest of Owens Lake. In 1863, there was a small silver strike north of where the town of Bishop is today, which established the town of Montgomery. Soon there was another small silver rush to the area that would become Benton (Nadeau 1999:66-67) north of Bishop some 35 miles.

By 1864, silver was discovered at Cerro Gordo (fat hill in Spanish) by Mexican miners in the Inyo Mountains northeast of Owens Lake, south of Saline Valley and west of Panamint Valley (Figure 2). The Mexican miners worked quietly on a small scale for a couple of years, but when the mine came to the attention of American miners, the word quickly spread about the silver deposits. In 1866, the Lone Pine Mining District was founded in Cerro Gordo. The district was named after the small town on the other side (northwest) of Owens Lake. Soon there was a silver rush, and Cerro Gordo became a large booming mining town. Also in 1866 Inyo County was created, with its county seat in the community of Independence. By the mid-1870s, two major mining towns were also booming in the heart of Desert Kawaiisu country. Panamint City was founded in 1873 (Nadeau 1999:134). A typical bustling, rowdy mining town, it was located in the Panamint Range up rugged Surprise Canyon on the east side of Panamint Valley (the west side of the Panamint Mountains). By 1874, there were 150 claims, a population of some 700 people and several saloons, stores, and shops (Nadeau 1999:134). Darwin, on the east central side of Panamint Valley, was also founded in 1874. By 1875, Darwin had seven saloons, three restaurants, a hotel, and several stores (Nadeau 1999:157)

It is reasonable to suggest that the mid-19th century was a time of massive cultural change for the Native people of the area. As miners and ranchers moved into their territory, they settled at the springs that the Indians needed for their mobile, seed-gathering economy. This had an immediate and disastrous effect on their ability to support themselves. At first they tried to accommodate the Whites, then in the early 1860s some turned to armed resistance. The Indians seem to have thought that since the cattle of the Whites ate their grass seeds and other plant foods, it was reasonable to harvest a cow now and then. Whites, on the other hand, tended to see this as a capital crime. Because of the racism, some American settlers came to view any Indian as guilty if some livestock turned up missing, for example, and attacked any Indian they could find rather that making an attempt to find the guilty party and assess guilt through the courts or military authorities. By the mid-1870s, their traditional economy and settlement system in tatters, many Indians along the eastern Sierra Nevada and the adjacent deserts tried to fit into the margins of the Anglo economy.

E. W. Nelson, writing in 1891, said of the Indians of Panamint and Saline Valleys, "They are very fair workers and find employment at times about the mining camps of Panamint, Darwin, and elsewhere" (1891:372) . It is fair to suggest that this situation probably prevailed by the mid-1870s when Panamint City and Darwin were booming and it may have began a decade or so earlier when Cerro Gordo boomed. Nelson may have been referring to all the tribes of the region; however, the groups closest to the mining camps he mentioned were the Desert Kawaiisu and the Coso Shoshone. Nelson's statement makes clear that the traditional socioeconomic system had largely broken down by the 1890s, forty years or two generations before the ethnographic research of Steward and Zigmond and others. Their pioneering ethnographic data need to be interpreted in light of Euro-American influences of the mid 19th century in order to arrive at reasonable reconstructions of pre-contact cultural systems and tribal territories.

Figure 2. Desert Kawaiisu environs.

Steward offers ethnographic testimony to document a permanent, albeit sparse, occupation of Kawaiisu in the southern portion of Death Valley, Panamint Valley, and Trona. John P. Harrington and Philip Cappannari left some brief ethnographic notes relating to the Desert Kawaiisu. The only other early published observations of the Indians of Panamint and Death Valleys consist of very brief observations of E. W. Nelson (1891), cited above; Frederick V. Coville (1892); and B. H. Dutcher (1893). Coville and Dutcher mention that they were on the Death Valley Expedition of 1891 to document the natural history of the area.

Steward (1938:71) suggests that the people who had occupied the southern Panamint Valley and southern Death Valley, viz., the people we are calling the Desert Kawaiisu, were sometimes known as the *Magunuwu*. Steward suggests that *Magu* means point and *nuwu* means people, so they would be "people of the point." The point referred to may be Telescope Peak, called *Mugudoya* (mugu = point; doyavi = mountain). At 11,048 feet, it is by far the tallest peak in the region. Steward suggests that *Mugu* may also refer to the Argus Mountains, located west of Panamint Valley. The Argus Mountains were called *Magu* or *Tinda'vy*.

Steward also notes that the Desert Kawaiisu were also called the Panumunt or Panamint Indians (1938:92). The term Panamint has been the source of some confusion because the Shoshone of northern Panamint Valley, northern Death Valley, Saline Valley, the Coso and Argus Mountains, and eastern Owens Valley were at times also called the Panamint Shoshone, sometimes shortened to Panamint (e.g., Miller 1986:99; Thomas et al. 1986:262; Zigmond 1986:398). These Indians of the Coso and Argus Mountains and eastern Owens Valley were also called the Koso or Coso Shoshone (Lamb 1958:98). The Panamint Shoshone and Coso Shoshone were treated as a single group by both Steward (1938) and Kroeber (1925). The mixed Shoshone village at

Furnace Creek in Death Valley today prefer to call themselves the *Timbisha* Shoshone. The term *Timbisha* derives from the word for rock (Steward 1938:92). Native people at this village at the time of Steward's research were a mixture of Coso Shoshone, Southern Paiute, and Desert Kawaiisu (Steward 1938:92).

As mentioned above, Harold E. Driver (1937) also conducted brief research among the Kawaiisu for his culture element distribution list of the area. One of his consultants, Bob Rabbit, was raised in Kelso among Mountain Kawaiisu, but his father was from the Panamint Mountains, implying the existence of a village there. Rabbit also confirmed that Kawaiisu territory extended east to the Panamint Mountains. Driver (1937:58) commented, "There must have been several local groups over an area so large, although informants knew no names for them and insisted that there had never been more than one chief for the area." Presumably by this Driver meant that the Desert Kawaiisu recognized one leader above and beyond various local village headmen. Driver continued (1937:58), "Steward, in conversation, July, 1935, reported Kawaiisu in the southern end of Death v. [sic] Informants knew the Chemehuevi by that name and were aware of their linguistic affinity to themselves." This passage is rather unclear, but it seems to say that the people in the southern end of Death Valley were also called Chemehuevi by his informants or that the Kawaiisu in general were also called Chemehuevi.

Steward used word lists from Kroeber's work among the Kawaiisu of the Fort Tejon and Tehachapi areas to confirm that the people of southern Panamint Valley and southern Death Valley were unquestionably Kawaiisu. He does point out a linguistic affiliation with the Chemehuevi, as one might expect given their geographic proximity and common linguistic heritage (Steward 1938:71). The cryptic comments of Driver are difficult to interpret except to underscore the probable existence of a permanent population of Kawaiisu in the desert east of the Tehachapi Mountains.

David D. Earle (2003:75) points out in his well-researched ethnographic overview of Fort Irwin, that John P. Harrington, Philip Cappannari, and Isabel Kelly also conducted research bearing on the Desert Kawaiisu question and left statements about the existence of Desert Kawaiisu in their unpublished notes. Harrington's main Serrano consultant, Santos Manuel, mentioned that two groups lived north of the Barstow area. In Serrano, these were known as the *Ak'utusam* and the *Jo'akayam*. While the *Jo'akayam* were southern Paiute, the *Ak'utusam*, according to Manuel, were people who were based in the Tehachapi area, viz., Mountain Kawaiisu. Earle (2003:75) suggests that the *Ak'utusam* were named for a ethnographically attested rancheria, *Akutushpea,* located in the Tehachapi Mountains and mentioned in the records at the Mission San Fernando. Manuel also suggested that the *Ak'utusam* people lived on the south side of the Panamint Mountains. These would be our Desert Kawaiisu. Another Harrington consultant, a Kitanemuk woman named Magdalena Olivas, also knew about the *Ak'utusam.* She said that these people lived in the desert and had a language very similar to Kawaiisu, but were not the same people (Earle 2003:76).

Philip Cappannari, who carried out Kawaiisu fieldwork from approximately 1946 to 1948, also noted that consultants told him that the Kawaiisu had permanent rancherias in Panamint Valley and that they also lived in the Panamint Mountains (Cappannari, n.d.). He suggested that, based on the information from his consultants, it was difficult to say if the Kawaiisu had rancherias in the southern part of Death Valley, as suggested by Steward (Earle 2003:76).

There is also some evidence from the Chemehuevi relating to the existence of the Desert Kawaiisu. Isabel Kelly conducted research among several Numic groups in 1933 and 1934. As Earle (2003:76) points out, one of Kelly's consultants, a Chemehuevi named Matavium, mentioned that Vanyume territory included Daggett (about 10 miles east of Barstow) and the Calico Mountains, while the *Panumint* lived to the north of the Calico Mountains. Matavium also suggested that the northern boundary of the *Panumint* was just south of a mountain called Kaiwi?it, which Earle suggests may be Telescope Peak. Mataviium's account accords well with the territory outlined by Steward (1938).

Judy Barras, a talented amateur anthropologist who lived in Tehachapi for many years, conducted extensive informal research from the late 1960s to the early 1980s. Baras (1984) wrote a popular-style ethnographic book about the Mountain Kawaiisu. Her consultants, among whom was Emma Williams, one of Zigmond's more important consultants, also suggested that the Kawaiisu lived in the Panamint Valley and in the desert between Panamint Valley and Tehachapi, e.g., Searles Lake (1984:19, 61).

It is clear that the ethnographic record for this region is rather thin. However, several sources reviewed here seem to document a mid-19th century permanent population of Kawaiisu people in the south part of Death Valley, the south part of Panamint Valley, and perhaps in the Searles Lake area. By the time the oldest consultants of Steward, Zigmond, and others were born, the traditional economy, settlement system, and trade system had been disrupted for some 50 or 60 years by the farreaching effects of Spanish/Mexican settlements near the Pacific Coast. The settlement system may have contracted by this time and some village consolidation may have taken place. We should keep this in mind when using these ethnographic observations as a starting point for archaeological reconstructions utilizing the direct historical approach (e.g., Steward 1977).

SETTLEMENT

The only published source for detailed settlement information is in the work of Steward. At the time of Steward's fieldwork, i.e., 1935 and 1936, the Indians in the region near Panamint and Death Valleys lived in small villages of mixed ethnic and linguistic backgrounds or remembered such villages from their youth. For example, the village at Ash Meadows, about 30 miles east of Death Valley's Furnace Creek, consisted of a mix of Southern Paiute and Coso Shoshone (Steward 1938:71). Steward mentioned that "north of Ballarat, Panamint Valley was predominantly Shoshone, with some admixture of Kawaiisu. South of Ballarat it was largely Kawaiisu" (1938:84). The Furnace Creek area was at the intersection of three traditional territories: the Coso Shoshone to the north and west, the Southern Paiute to the east, and the Kawaiisu to the south and west. The village population at Furnace Creek was a mixture of these three tribes and the people spoke all three languages (Steward 1938:92).

> ... Death Valley south of Furnace Creek was occupied by mixed groups, especially Kawaiisu, and it's [sic] political affiliations are undetermined. Furnace Creek was, from all accounts, independent of the villages in the upper part of the valley, though it no doubt associated with them at times(1938:76) [e.g., for communal rabbit drives and fall fiestas].

Although it is difficult to know for sure, it may be that this ethnic mixing at villages around Death Valley and Panamint Valley was a result of Euro-American influences. As suggested above, with the arrival of White prospectors in the 1850s and ranchers by the 1870s, there was a contracting Indian population in the area (Coville 1892:351-352). As populations became too low for a viable village, perhaps remnants of two or more villages may have moved in together to form a single, mixed village. It seems reasonable to suggest, as Driver seems to have done (1937:58), that at one time the Desert Kawaiisu must have had a relatively large number of small villages considering the size of their territory and the large number of springs.

At the time of his fieldwork in 1935 and 1936, Steward found no living full-blooded Desert Kawaiisu, but he did have an important consultant, TSP, who was one quarter Desert Kawaiisu. Steward suggested that the only living full-blooded Desert Kawaiisu at that time was a fellow named Long Jim living in Pahrump Valley, Nevada (Steward 1938:71). The fact that the Desert Kawaiisu were all but wiped out by the mid-1930s is not surprising considering Coville's 1892 comments:

> The Panamint Indians of Inyo county [sic], California, belonging to the widespread Shoshonean family. . . are now nearly exterminated and their customs and language little known, the writer desires to put on record the following observations, made among them while he was performing the duties of botanist of the Death Valley expedition in 1891 (Coville 1892:351-352).

Because of the previously mentioned confusion surrounding the term Panamint, it is not known with certainty if Coville was referring to the Desert Kawaiisu or to the Coso Shoshone. Evidently he did not make any distinction between them, so he may have been referring to both. Later in the same article he mentioned Indians at two villages at what are now known as Indian Ranch and Johnson's Ranch. These villages were both inhabited by a mixture of Coso Shoshone and Desert Kawaiisu, according to Steward. Since Coville didn't mention this ethnographic fact, he was apparently unaware of or unconcerned with the difference between them. However, it is also possible that these villages were not mixed in 1891. Indian Ranch may have been in Kawaiisu territory, while Johnson's Ranch definitely was (as discussed below).

Steward (1938:84) suggests that Panamint Valley was so low and arid that the population must have been quite sparse. The principal and possibly the only permanent village in the valley was at the mouth of what is now known as Hall's Canyon. This village area is also known as Indian Ranch. It is located about six miles north of the old mining town of Ballarat on the east side of the valley. Surprise Canyon is about three miles south. The site of the mining town of Panamint City is located at the top of Surprise Canyon. Indian Ranch is in Desert Kawaiisu territory according to Steward's map (1938: Figure 7), but elsewhere Steward (1938:84) says that "North of Ballarat, Panamint Valley was predominantly Shoshone with some admixture of Kawaiisu. South of Ballarat it was largely Kawaiisu." Indian Ranch was evidently on the traditional territorial boundary, and the inhabitants were a mixture of Coso Shoshone and Desert Kawaiisu.

According to Steward's Indian consultants, there was also a small village consisting of a few families at Wildrose Springs. This would have been on a prehistoric travel route between Panamint Valley and Death Valley. There is now a rough road through Wildrose Canyon, passing by the spring, and over Emigrant Pass linking Panamint Valley with Death Valley. Wildrose Springs was in Coso Shoshone territory according to Steward, but probably had a mixed Shoshone/Kawaiisu population similar to other villages in the area.

Likewise, there were a few families at a place that Steward (1938:93) refers to as Hungry Bill's Ranch, located on the west side of Death Valley at an elevation of approximately 5,000 feet up Johnson Canyon. This is also known as Johnson's Ranch (Nelson 1891:372). Coville also mentions this village as being located in the mountains above Johnson Canyon (1892:352). This would be upslope and southwest of Bennet's Well and the Eagle Borax Works, wellknown landmarks on the west side of Death Valley. The inhabitants of this Johnson Canyon village were apparently a mixture of Kawaiisu and Coso Shoshone although it was clearly within Kawaiisu territory (Steward 1938:93). The Trona/Searles Lake area was in Kawaiisu territory but may have also been inhabited by a mixture of Shoshone and Kawaiisu during the early or mid 19th century. Steward states that these Trona people were called *Owa'dzi*. This implies that there had been a village in the area because people were known to one another by the village to which they belonged (Steward 1983:71). Steward does not mention any other Desert Kawaiisu villages or named groups, so the large area south of Death Valley to the Mojave River, including the Avawatz and Granite Mountains (now Fort Irwin), remains unaccounted for. There are numerous springs and other resources in these mountains, so there is some possibility that Desert Kawaiisu may have once had some rancherias in the Late Prehistoric.

The very hot, arid conditions in the region made a dependable water source critical for the placement of villages. The settlement system must have consisted of a regional variant of what Walter Taylor (1961) called tethered nomadism. That is, because of the arid landscape, people had to live near springs and forage primarily in the surrounding area. Forays for plant food must have consisted primarily of day trips and overnight trips of short duration (Steward 1938:84), that is, human populations were essentially tethered to springs and tinajas. Under prehistoric, pedestrian conditions, it does not seem physically possible to dry camp in the desert for more than a day or two, even in cool weather, considering the difficulty in carrying water.

The Coso Shoshone and Desert Kawaiisu apparently migrated from "permanent" winter villages or rancherias at springs in foothills or at the edges of valleys to higher-elevation springs in the spring, summer, and fall in order to live in a cooler climate and avail themselves of the abundant mountain food resources. Families moved camp independently of one another rather than as a community. Typically families made multiple moves throughout the year to be near particular plant resources when they became available, but they would return to the winter villages or similar locales after the fall piñon harvest (Steward 1938:84).

For the Desert Kawaiisu living in the southern Panamint Valley, these spring, summer, and fall camps would have been primarily in the Panamint and Argus Ranges. Those living in the southern part of Death Valley would have focused on the Panamint Range. The Panamint Range has yearround springs in the higher elevations on both the Panamint and Death Valley sides. The range extends south to Wingate Pass, and springs are found throughout. Most springs are above 3,000 feet in elevation, while Wingate Pass is at an elevation of about 2,300 feet. South of Wingate Pass springs become scarce.

Other than the Panamint Range, the mountains in the Panamint and Death Valley area are too low to support piñon pine trees, a staple prehistoric food resource. This would include the Slate Range west of the southern part of Panamint Valley, the Owlshead and Avawatz Mountains southeast of Wingate Pass, and the Granite Mountains south of Wingate Pass. As mentioned above, there are some springs in these mountains. Grasses and other seed plants also grow there, and these probably would have been attractive subsistence resources for Desert Kawaiisu.

SUBSISTENCE

As mentioned above, most food resources were gathered within a relatively short distance of villages and camps (Steward 1938:84). The most important plant food for people living in traditional Desert Kawaiisu territory was pine nuts. In a good year, enough could be gathered to last most of the winter. In this part of the desert, piñon pines grow above about 6,000 feet in elevation. The most substantial area for piñon is the Panamint Range, but they also grow in the Argus and Coso Ranges to the north and along the eastern edge of the Sierra Nevada. Desert Kawaiisu from the southern Panamint Valley would have gathered piñon primarily in the Panamint Range, but probably also in the Argus and Coso Ranges. In 1891 B.H. Dutcher stayed for two nights with a party of "Panamint Indians" who were collecting piñon in the Hunter Mountain area, between northern Panamint Valley and southern Saline Valley (Dutcher 1893). As suggested previously, these "Panamint Indians" may have been Coso Shoshone or a combination of Coso Shoshone and Desert Kawaiisu.

> . . . before sunrise a meager breakfast had been eaten. . . Immediately after the meal several of the women equipped themselves with large, conical pack-baskets and beating sticks and sallied forth to gather the cones from the trees in the vicinity. The baskets were made of light wicker-work. . . about two to two and one-half feet high and nearly as broad. . . A leather thong was fastened into the side just below the rim, passed around the forehead, and similarly inserted into the wicker-work on the other side. . . . thus serving to bind the basket to the carrier. The beating or pulling sticks were straight rods, about an inch in diameter and five or six feet long, stripped of bark and with all the branches removed, save one at the outer extremity, which was cut off about six inches from its union with the main staff. . . . Thus equipped with basket and stick, a squaw would advance to some untouched tree and proceed to beat and pick the cones from the limbs until her receptacle was full, when she would return to camp, empty the load onto the ground, and start off again to refill the basket or remain to open cones, as necessity dictated (Dutcher 1893:378).

Piñon produces prodigious amounts of seeds, but each grove does not produce well every year. Consequently, people would travel substantial distances to access groves that were productive in a particular season. The Western Shoshone did not typically assert grove ownership or ownership over other plant resource patches or springs (Steward 1938:106; Thomas et al. 1986:266), but Southern Paiute sometimes did (Kelly and Fowler 1986:380; Steward 1938:73)). As suggested above, the Desert Kawaiisu shared many cultural traits and their language with the Southern Paiute and had shared residence with the Shoshone in a number of villages. Undoubtedly the Desert Kawaiisu were influenced by both groups, but it is not known if they asserted spring or grove ownership. Zigmond (1986) does not mention grove ownership among the Mountain Kawaiisu. When a particular grove was productive, it produced considerably more pine nuts than small, mobile groups could process, so evidently there was not much economic incentive for territoriality (Steward 1938: 78).

Steward (1938:27) estimated that a person could harvest about 50 pounds of piñon per day, and a family of four could gather as much as 1,200 pounds in the month or so when piñon nuts were available for harvesting. Steward's figures suggest that either not all family members worked or that the family worked only one day per week during the month they were harvesting piñon, i.e., 4 people x 50 pounds = 200 pounds per day. 1200 pounds/200 pounds per day = 6 days. If only the woman of the family worked the harvest, as the quote from Dutcher (1893) implies, then 1200 pounds/50 pounds per day = 24 days. It seems reasonable to assume he was correct about the 1200 pounds and less so about the rate of harvest, the level of effort, and the division of labor. This 1200 pounds of pine nuts might last the family the four months of winter when few other foods were available. After the harvest, people would often cache some pine nuts in the mountains and take some back to the winter village, located near a spring near the valley floor (Steward 1938:73). The Indians were able to save impressive amounts of seeds for the lean times of winter. Chalfant (1933:190) reported that in May 1863, during the height of the so-called Owens Valley Indian War, four soldiers from Camp Independence destroyed some 300 bushels of stored seeds in the Bishop Creek area. The population size of this Indian village is not mentioned, but it may have been considerably larger than villages in the desert. By way of comparison, a bushel of clover, wheat, or alfalfa weighs about 60 pounds, while timothy grass is about 45 pounds and brome is about 14 pounds (Rowlett 2005).

After piñon, next in importance according to Steward's Native consultants (1938:72) were the seeds of mentzelia (*Mentzelia* sp.), a low-growing yellow flower found in the foothills and mountains. The seeds of sage (*Salvia* sp.) especially chia (*Salvia columbariae*), were also important. Sage also grows in the foothills and mountains. A number of other seeds and berries were also eaten. Coville (1892:352) observing "Panamint Indians," emphasized the importance of Indian rice grass (*Oryzopsis hymenoides*) a plant found on the desert floor and in the foothills. Coville described its harvest method (1892:353):

> In gathering it, the squaw carries in one hand a small basket and in the other a paddle made of wicker-work, resembling a tennis racket, but of only about two-thirds the size. With this she beats the grass panicles over the basket.

Coville also mentioned the seeds of the cottontop cactus (*Echinocactus polychephalus*) and Mormon tea (*Ephedra nevadensis*) and the berries of boxthorn (*Lycium andersonii*). Seeds were often parched in a shallow basketry tray and/or ground with a mano and metate. Among the Mountain Kawaiisu, the portable metate was most often used to grind seeds, along with a mano. Bedrock metates were less important (Zigmond 1986:400). This may hold true for the Desert Kawaiisu as well; there is very little bedrock in the alluvial basins which form a large part of their territory. Young buds of Joshua tree (*Yucca brevifolia*) were also eaten, as Coville described it (1892:353):

> The four or five uppermost mature leaves are drawn together over the apex of the bud and grasped by the hand; then, by a twist and a quick, sidewise bend, the head is broken off. This is rendered possible by the skill of the worker and by the fact that the stem, although in age exceedingly tough, in its young stage is quite brittle. The mature leaves and the green tips are discarded, leaving an egg-shaped, solid, cream-like, juicy mass. This is roasted on hot coals and eaten at once or kept until cold. The tissues of these young buds are full of sugar and other formative carbohydrates and are undoubtedly very sweet and nutritious.

A variety of greens were eaten in the spring, the most important of which included prince's plume (*Stanleya pinnata* and *S. elata)* and wild cabbage (*Caulanthus crassicaulis)*(Coville 1892:354). These are members of the mustard family. These greens had to be boiled then rinsed five or six times in order to remove bitterness (Balls 1962:27; Zigmond 1981:65). Balls (1962:27) suggests that the seeds of these plants were also harvested, ground into meal, and used as pinole. Balls (1962:6) described pinole and atole:

> The two staple forms of food, atole and pinole, were used generally throughout most of California. Atole was the thick, souplike food made from ground, leached acorns or any of a number of other nutlike seeds. Pinole was the fine flour made by grinding seeds of Tansy-Mustard, Chia, or many of the grasses and flowering annuals. This was eaten either dry, pinch by pinch, or made into a mush with water.

The floor of the Panamint Valley has a number of mesquite bosques associated with stabilized dune formations around the two major playas that cover much of the valley floor. Mesquite provided substantial amounts of food for Native inhabitants (Driver 1937:64; Thomas et al. 1986:267). The Furnace Creek area of Death Valley also had a great deal of mesquite in the past. and there are still stands in the Mesquite Flat area north of Stove Pipe Wells. Much of the mesquite has been cut down in the Furnace Creek Resort area or has been crowded out by invasive, non-native tamarisk.

Mesquite was gathered in the late spring and early summer. Mesquite pods are similar in appearance to long, thin green beans. The pods were dried, then pounded in a large wooden mortar with either a wooden or stone pestle. These pestles can be rather long, 40-60 cm. The pods themselves were made into a flour. The seeds, with an appearance reminiscent of bird shot and nearly as hard, were discarded. The mesquite flour was pressed into cakes for storage and transport (Driver 1937:68-69; Thomas et al. 1986:267). The flour was also mixed with water and drunk as pinole (Driver 1937:64).

A wide variety of animals were eaten, although the scarcity of game meant that meat was not a major item of the diet. Mountain sheep and deer were hunted in the Panamint Range. Coville (1892: 352) sums it up:

> Their animal food consists principally of jack-rabbits, cotton-tail rabbits, and quails, occasionally mountain sheep or deer, and sometimes wood rats, kangaroo rats, white-footed mice and a large lizard known as the chuckawara (chuk'-a wa-ra).

The primary sources of protein for desert dwellers were jackrabbits and rodents. Jackrabbits were typically taken in communal rabbit drives. Chalfant (1936:20) suggested that the Death Valley 49ers encountered a narrow valley just east of Death Valley where jackrabbits were particularly numerous and some 500 were killed. While he doesn't mention how they were taken, evidently there were very large rabbit populations in some desert places and that drives could be very productive.

> In the drives they used one or two nets, each about 2 feet high and 100 or more feet long, propped at intervals with sticks. Eight or ten men beat the brush, driving the rabbits into the nets, while the owners remained behind their nets to dispatch the ensnared animals with clubs (Steward 1983:82-83).

Families might travel substantial distances to gathering places or to participate in communal rabbit hunts and fall fiestas. Some of the main locations for rabbit drives were Darwin Wash, Little Lake, and Olancha. The longest practical distance to travel for one of these events, according to Steward (1938:82), might be about 50 miles. People from Panamint Valley, for example would not go as far as Olancha for a rabbit drive, but people from Saline Valley would. Panamint Valley people might go to Darwin Wash, Death Valley, or some other place closer to home (Steward 1938: 82). On the other hand, evidence from the Death Valley 49ers previously mentioned suggests that horse rustlers might travel some 130 miles to ply their trade. Fiestas and cries (mourning gatherings) were occasionally held in the Stove Pipe Wells area (Steward 1938: 80). Steward (1938:85) continues:

> It is improbable that any important communal activities were held within Panamint Valley. Saline and Death Valleys were within convenient distance for festivals. Panamint Tom, however, seems, at least in post-Caucasian times, to have gained some prominence in connection with hostilities against the White man. Whether he was a Kawaiisu is uncertain. TST thought that he had been chief of the Death Valley Kawaiisu.

Coso Shoshone also hunted deer and mountain sheep in the Inyo Mountains (Steward 1938:78), and the Desert Kawaiisu may have hunted there on occasion, given the fact that they and the Coso Shoshone shared villages in Panamint Valley and Death Valley. However, it is likely that the main big game hunting area of the Desert Kawaiisu was probably the Panamint Range.

Apparently the Indians of the region were able to store considerable amounts of meat for the winter by drying it. In 1862 during the Owens Valley Indian War, Whites destroyed a rancheria in the Alabama Hills west of Lone Pine that contained about a ton of dried meat (Chalfant 1933:153). Big game was taken with bow and arrow. Among the Mountain Kawaiisu, the bow was typically made of juniper and backed with sinew (Zigmond 1986:401). The bow string was made of twisted sinew. A compound arrow was used. The main shaft was carrizo grass (*Phragmites australis*) into which was inserted a foreshaft of some hard wood, e.g., *Prunis* sp. For big game, a flaked stone projectile point was affixed to the foreshaft with sinew. For small game the hardwood foreshaft was simply sharpened. For birds, two twigs were tied to the foreshaft in the form of an x (Zigmond 1986:401). Since the Desert Kawaiisu had easy access to these materials (i.e., juniper, carrizo, and prunis), it might be reasonable to surmise that their bows and arrows were similarly constructed.

A number of basketry items were used in gathering seeds: the large conical burden basket, winnowing and parching trays, and storage containers. These working baskets were twined and typically made of willow (Zigmond 1986:399). In addition to these items, canteens were also made of twined basketry made waterproof by a coating of pitch. The use of coiled basketry in the cooking process of the Panamint Indians was described by Coville (1892:357):

> The name pot-basket is used to designate wicker-work receptacles that have the form of a pot or bowl. The one in most common use has the shape of a rather deep bowl with curved sides and a flat bottom, and has a capacity of about three pints. The squaw uses it as a general measure, as a bowl for dry food and for soup, and often, when in the sunshine, as a hat. Its texture is so close that, when well made, it holds water. It is probable that before the advent of the iron kettle, the Panamints used these wicker-work pots for the purpose of boiling and stewing food, applying the heat by the ordinary process of dropping hot stones into the liquid. Most of their starchy food, however, is roasted dry by mixing seeds, before they are ground, with hot coals and stirring them in a basket.

The Mountain Kawaiisu also made a number of very fine decorative coiled baskets. A number of different plant materials were used for different colored patterns over a deer grass foundation (Zigmond 1978). While the Mountain Kawaiisu are renowned for their fine basketry, it is not known if the Desert Kawaiisu had such a well-developed basketry tradition beyond utilitarian items.

In the late 19th century, E. W. Nelson (1891:372) and Frederick V. Coville (1892:352) noted that the Indians at Hungry Bill's or Johnson Ranch on the Death Valley side of the Panamint Mountains and at Indian Ranch in east-central Panamint Valley both practiced agriculture. As mentioned above, these settlements were both inhabited by mixtures of Coso Shoshone and Desert Kawaiisu. While there is evidence of very late, pre-contact agriculture among the Southern Paiute (e.g., Stoffle and Zedeño 2001), the Western Shoshone, Coso Shoshone, and the Mountain and Desert Kawaiisu were evidently not agriculturalists prior to the mid- to late-19th century (Steward 1938:72-73; Thomas et al. 1986:266; Zigmond 1986:399).

Structures

The only description of a Panamint structure in the published literature is provided by Dutcher. It is a windbreak constructed at a piñon camp in the Hunter Mountain area, between northern Panamint Valley and southern Saline Valley. This windbreak structure would leave little more than a cleared circle in the archaeological record. Again, these "Panamint" Indians may have been Coso Shoshone or a combination of Coso Shoshone and Desert Kawaiisu (Dutcher 1893:378).

> In the shadow of a small group of piñon trees a number of small circles or "corrals" had been built. Perhaps there were some five or six of these in all, each seeming to accommodate one family or that fraction of a family that was present in the camp. In diameter they measured eight or ten feet, and their walls consisted merely of the broken piñon branches and of the small bushes that grew around, piled up into a loose row two or three feet thick and about as many high. The circle was broken or imperfect where entrance or exit was needed, and where two of these circles became tangent a passage was generally made from one to the other. Their uses seemed to be few—to secure a little privacy for the occupants; to serve as a slight wind-break during the night, when the family slept inside, and during the day to serve as a rack in holding out of the dirt the blankets, extra clothes, cooking utensils, and other paraphernalia of the household. In the center of each circle was a small area where the fire was kindled, around which all gathered for a short time during the chill of the early morning and in the evening before retiring. The floor was smooth, clear of stones and weeds, and carpeted by a thick layer of fine gray dust (Dutcher 1893:378).

Given their propinquity to the Southern Paiute and Coso Shoshone, the typical Desert Kawaiisu winter house was probably similar to the conical or domed structures covered with brush reported for these groups. Thomas et al. (1986: 268) provide a description of a Shoshone house that might be equally applicable for Southern Paiute and Desert Kawaiisu houses in the Death Valley region if one substitutes locally available brush for slabs of bark :

> The typical winter house was a conical hut, housing a family of about six. The light frame was covered with slabs of bark, sometimes surrounded with a single tier of stones to keep the supports firmly planted... Since few Western Shoshone structures involved subterranean construction, the only vestiges of such structures are often stone circles...

Driver (1937:66) reported semi-subterranean round, conical houses for the Coso Shoshone and the Mountain Kawaiisu. These were excavated down approximately two and three feet respectively. Zigmond (1986:401), on the other hand, suggests the Mountain Kawaiisu had a ground-level, circular house built on a vertical, forked-pole frame. The house described by Zigmond for the Mountain Kawaiisu was covered with brush. Horizontal battens were used inside and outside the structure to secure the brush to the frame. Bark and tule mats were also used. Zigmond also mentions use of ramadas, or rectangular flat-roofed shades, and the circular brush windbreaks that Dutcher reported for the "Panamint" Indians, recounted above. Earth-covered sweathouses were also used among the Mountain Kawaiisu. One can reasonably assume these structure types, adapted to the building materials available in the desert, were probably used by the Desert Kawaiisu. A conical house is described for the Southern Paiute of northern Arizona and southern Utah by Stewart (1942). Kelly (1964), working in the same area, describes both dome-shaped wikiups and conical houses, some with ridgepoles. The Shoshone winter house was also typically conical (Thomas et al. 1986:268). Zigmond reported a conical house with two upright posts and a gable for Death Valley (reported in Thomas et al. 1986:268).

Eric Ritter (1980) reported a standing conical house in Pleasant Canyon just east of Ballarat in the heart of Desert Kawaiisu territory. It was a conical structure approximately 4 meters in diameter and 2 meters tall, constructed of piñon and juniper logs leaned against two central posts with short beams in between to form a short gable, similar to the type described by Zigmond for Death Valley. It was built on the ground surface, had cobbles as chinking; and may have once been covered or partially covered with dirt. It appeared to date to about 1870.

Bedrock milling slicks, a metate, petroglyphs, incised green slate, steel ax-cut wood, wire, and cans at the site suggest it was occupied by people practicing modified traditional economy. The site inhabitants may have worked for the nearby Pleasant Canyon Mine in the pattern reported by Nelson (1891) and mentioned previously.

The domed or conical houses with a circular plan view described by Dutcher (1893), Thomas et al. (1986:268), and Zigmond (1986:401) leave a cleared circle, or a cleared circle with a rock ring around the perimeter, in the archaeological record. These are relatively common in traditional Desert Kawaiisu territory, while semi-subterranean house remains are less well known (e.g., Deis and Underwood 2004).

Political Organization

As suggested previously, the villages of the Desert Kawaiisu, at least at the time of the visits by Coville (1891) and Steward (1935-1936), were quite small. Steward does not offer many observations on Desert Kawaiisu political organization, but it is reasonable to infer that it was somewhat similar to the Coso Shoshone in Panamint and Death Valley since they lived in some of the same villages, at least in historic times (Steward 1938:84).

Individual families decided their own movements throughout the spring, summer, and fall gathering season. While people were known by the village or group of villages to which they belonged, village membership was rather fluid. Families could move independently out of one winter village and into another, and they could make independent movement decisions at other times of the year. Since families chose individually where to travel to harvest foods, village composition often depended upon which resource patches particular families traveled to.

This familial particularism and the strength of kinship bonds and controls made further political organization somewhat superfluous. Beyond the family, cooperation and friendship extended to other members of the winter village. Village residents most frequently cooperated in games, dances, and hunting. Opportunities for meeting and visiting with people in other villages were apparently limited to fall fiestas, cries, and communal hunting activities. Villages were named, often after a prominent geographic feature, as suggested previously.

Steward was unable to obtain information about village headmen in the Death Valley and Panamint Valley areas, but he suggests that, "There is little doubt that, like the headmen in Nevada Shoshone villages, their functions did not extend beyond keeping informed on the few matters of village interest, such as the ripening of pine nuts" (1938:75). At least in the 1860s and 1870s, prestige and positions of leadership were evidently acquired by demonstrating resistance to the encroachments of Whites. Oratorical ability as also prized and rewarded (Steward 1938:75, 85). Steward suggested that there was little evidence that they had anything but family-band-level political organization (e.g., Steward 1938:75). However, Carobeth Laird suggested (1976:24) that the Chemihuevi had pre-contact regional chiefs. Whether this came as a result of close Chemehuevi contact with the Mojave, who did have chiefs and/or interactions with Whites in the 19th century is an open question. Kelly and Fowler (1986:380) suggest the Southern Paiute did not have regional chiefs, with the possible exception of some "appointed" by Whites. Thomas et al. (1986:276), referring to the Western Shoshone (including the Coso Shoshone), suggested that no supra-village organization existed except perhaps for initiating and managing communal rabbit or antelope drives or fall fiestas. However, as mentioned previously, Driver (1937:58) commented that his consultants said that the Desert Kawaiisu had a single chief. In light of the other evidence, it does not seem likely that there was any formal regional leadership among the Desert Kawaiisu, except perhaps in organizing multi-village rabbit drives, fiestas, and cries as was the case among the Coso Shoshone (Thomas et al. 1986:276).

Conclusion

The ethnographic work of Zigmond focused on the people we call here the Mountain Kawaiisu. While this was a rich contribution, it largely ignored the vast desert area to the east of the Tehachapi Mountains that all authorities attest was traditional Kawaiisu territory. The ethnographic work of Steward, the brief ethnographic sketches of the late 19th century (Coville, Dutcher, and Nelson), and some other brief passages in publications and unpublished notes reviewed here suggest that there was a permanent Kawaiisu population living in the south half of Panamint and Death Valley and in the Searles Lake area. We suggest that the term Desert Kawaiisu be used to refer to these people. They lived in a very different environment from the Mountain Kawaiisu and had a very different economy, settlement system, and possibly other sociocultural institutions. Based on these lines of evidence, it seems reasonable to acknowledge the Desert Kawaiisu as an ethnic entity separate from the Mountain Kawaiisu. There is ample precedent for such a distinction in southern California. For example, the Kumeyaay of the San Diego coast and mountains have long been considered distinct from the Kamia of Imperial Valley despite sharing lineages and a common language because of their distinct environments and lifeways (e.g., Kroeber 1925). Likewise, the Mountain, Pass, and Desert Cahuilla of San Bernardino County are considered separate entities (e.g., Kroeber 1925). We have attempted to provide an outline of the Desert Kawaiisu culture based on the very thin ethnographic record and inferences drawn from what is known about their close neighbors. Further archaeological research will surely shed some additional light on their lifeways and the locations of their former villages, but unfortunately, we will probably never know much about these desert people.

Acknowledgements

Many thanks to Richard Deis, Russell Kaldenberg, Judyth Reed, Dr. James H. Cleland, and Dr. Alan Gold (Garfinkle) for reading and commenting on various versions of this paper. David D. Earle and I had many productive and stimulating conversations about the Desert Kawaiisu in the mid-1990s when we both work at Edwards Air Force Base. Recently Earle (2003) brought to light several valuable unpublished research sources relating to the Desert Kawaiisu that were previously unavailable. Muchas gracias, David. Of course, any oversights or shortcomings in the paper are my responsibility.

References

Balls, Edward K.
1962 Early Uses of California Plants. University of California Press, Berkeley.

Barras, Judy
1984 Their Places Shall Know Them no More. Published by Judy and Bud Barras, Tehachapi, California

Cappannari, Philip
n.d. Ethnographic notes on the Kawaiisu, from approximately 1946 to 1948. Notes on file, Department of Linguistics, University of California, Los Angeles

Chalfant, W. A.
1936 Death Valley, The Facts. Stanford University Press, Stanford, California.

1933 The Story of Inyo. W. A. Chalfant, Bishop, California.

Coville, Federick V.
1892 The Panamint Indians of California. American Anthropologist, 5: 351-361.

Cragen, Dorothy C.
1975 Boys in Their Sky-Blue Pants: The Men and Events at Camp Independence and Forts of Eastern California, Nevada, and Utah. Limrose Publishing, Fresno.

Deis, Richard and Jackson Underwood
2004 Cultural Resources Evaluations and Survey, South Range, Naval Air Weapons Station, China Lake, California. Document on file with the Base Historic Preservation office Naval Air Weapons Station, China Lake, California.

Driver, Harold E.
1937 Culture Element Distributions: VI Southern Sierra Nevada. Anthropological Records 1(2):53-154. University of California Press, Berkeley.

Dutcher, B. H.
1893 Piñon Gathering Among the Panamint Indians. American Anthropologist, vol 4: 377-380.

Earle, David D.
1990 New Evidence on the Political Geography of the Antelope Valley and Western Mojave Desert at Spanish Contact. In Archaeology and Ethnohistory of Antelope Valley and Vicinity, edited by Bruce Love and William H. De Witt. Antelope Valley Archaeological Society Occasional Papers 2, Lancaster, California.

2003 Ethnohistorical and Ethnographic Overview and Cultural Affiliation Study of the Fort Irwin Region and the Central Mojave Desert. Document on file with the historic preservation officer, Fort Irwin, San Bernardino County, California.

Heizer, R. F. And M. A. Whipple (editors)
1971 The California Indians: a Source Book. University of California Press, Berkeley.

Irwin, Charles N. (editor)
1980 The Shoshoni Indians of Inyo County, California: The Kerr Manuscript. Ballena Press Publications in Archaeology, Ethnology, and History No. 15. Co published by the Eastern California Museum, Independence, California.

Kelly, Isabel T.
1953 Notebooks of Las Vegas Band, Southern Paiute and Chemehuevi. University of California, Archives No. 138.1m Anthropology Documents 17-18

Kelly, Isabel T. And Catherine S. Fowler
1986 Southern Paiute. In *Great Basin*, edited by Warren L. D'Azevedo, pp. 368-397. Handbook of North American Indians, vol. 11,
W. C. Sturtevant, general editor, Smithsonian Institution, Washington, D.C.

Kroeber, A. L.
1925 Handbook of the Indians of California. Dover Publications, Inc. New York.

Nadeau, Remi
1999 The Silver Seekers: They Tamed California's Last Frontier. Crest Publishers, Santa Barbara, California.

Nelson, E. W.
1891 The Panamint and Saline Valley Indians. American Anthropologist vol 4: 371-372.

Ritter, Eric W.
1980 A Historic Aboriginal Structure and Its Associations, Panamint Mountains, California. Journal of California and Great Basin Anthropology 2(1):97-113.

Rowlett, Russ
2005 U.S. Commercial Bushel Sizes, Grains. Electronic document: www.unc.edu/~rowlett/units/scales/bushels.html Accessed 8/26/2005

Spier, Leslie
1923 Southern Diegueño Customs. University of California Publications in American Archaeology and Ethnology 20:297-358.

Steward, Julian H.
1938 Basin-Plateau Aboriginal Sociopolitical Groups. University of Utah Press, Salt Lake City.

1977 The Direct Historical Approach to Archeology. In Evolution and Ecology: Essays on Social Transformation by Julian H. Steward, edited by Jane C. Steward and Robert F. Murphy. University of Illinois Press, Urbana.

Stoffle, Richard W. and Maria Nieves Zedeño
2001 Historical Memory and Ethnographic Perspectives on the Southern Paiute Homeland. Journal of California and Great Basin Anthropology 23(2): 229-248.

Thomas, David H., Lorann S.A. Pendleton, and Stephen C. Cappannari
1986 Western Shoshone. In Great Basin, edited by Warren L. D'Azevedo, pp. 262-283. Handbook of North American Indians, vol. 11, W. C. Sturtevant, general editor, Smithsonian Institution, Washington, D.C.

Thomas, David H.
1973 An Empirical Test for Steward's Model of Great Basin Settlement Patterns. American Antiquity 38(2):155-176
1974 An Archaeological Perspective on Shoshonean Bands. American Anthropologist 76(1):11-23.

Zigmond, Maurice L.
1938 Kawaiisu Territory. In Tribal Distribution in the Great Basin, edited by Willard Z. Park. American Anthropologist 40(4):634-638.
1986 Kawaiisu. In Great Basin, edited by Warren L. D'Azevedo, pp. 398-411. Handbook of North American Indians, vol. 11, W. C. Sturtevant, general editor, Smithsonian Institution, Washington, D.C.
1981 Kawaiisu Ethnobotany. University of Utah Press, Salt Lake City.
1978 Kawaiisu Basketry. Journal of California Anthropology 5(2): 199-215.

WHATCHAMACALLITS, DOO-DADS, AND THINGAMAJIGS: ENIGMATIC ROCK FEATURES IN THE MOJAVE DESERT

Michael R. Walsh, Ph.D., and C. W. Clewlow, Jr., Ph.D.
Ancient Enterprises, Inc.
P.O. Box 5138
Santa Monica, California 90409

ABSTRACT

If we have learned anything from the career of Jay von Werlhof, it is that some times the darndest things turn into truly exciting and important elements of the prehistoric landscape. We just have to look at enough of them long enough and in the right light. How else to bring about the discovery of fish weirs in the middle of the desert or of geoglyphs discernable only from the air? This lesson is applied to a number of enigmatic rock cairns recently discovered near Searles Dry Lake in the northwestern Mojave Desert. We have no idea what these features represent, and at present simply detail our attempts to make some sense of them.

BACKGROUND

In the course of a 2000-acre surface survey of the Christmas Canyon region, adjacent to and south-southwest of Searles Dry Lake in the extreme northwestern Mojave Desert (Figure 1), we encountered and recorded 45 enigmatic rock piles or cairns (Walsh and Clewlow 2003; Wells 2003). We were at first absolutely willing to attribute these cairns to mining activities (see below) coincidental and unrelated to prehistoric materials that were often observed surrounding the features. But it quickly became clear that so mundane an explanation for these rock features was simply not adequate, at least not for all of them. The 45 recorded cairns are, moreover, only a fraction of the numbers of cairns in the larger region. We know of many more cairns in the immediate vicinity, in fact, based on our own informal observations made en route to and from our survey areas and on anecdotal evidence provided by parties knowledgeable about the region, including other archaeologists. In this paper, we detail a few of our efforts to narrow the field of possible alternative explanations for the functions of the features. But, in the end, we are most convincing and successful in our efforts to determine what these features are *not*.

We will not describe these physically in any great detail here. They are small piles of 10 or more cobble-to-boulder-sized local igneous and metavolcanic rocks lying atop desert pavement (Figure 2). Cairns usually occur singly, but in one case (CA-SBR-10836) some 23 cairns are noted within an area under 10,000 m^2. In this case, we are unable to discern an overarching pattern to the cairn distribution, which extends down a ridgeline fronting an easily accessible prehistoric rockshelter site. In another case, four cairns form a rough quadrilateral some 3 m on a side (CA-SBR-10917). Desert pavement outlined within the quadrilateral is undisturbed, suggesting something other than a shelter foundation (Figure 3). The four component cairns are otherwise indistinguishable from cairns found singly in the region.

Individual cairns typically cover a diameter of from 1 to 2.5 m and range from 0.5 to over 1 m tall. Rocks are generally haphazardly piled (not coursed); most rocks are irregularly shaped, and none appears intentionally modified. Many of the cairns appear deflated or collapsed but, overall, all seem fairly well intact. Several show topmost rocks that are precariously balanced. It is not clear whether this suggests a fairly recent vintage—else these wobbly stones would surely have fallen by now—or a greater vintage, where they perhaps succumbed through time to their current unstable condition. There is generally little or no obvious accretion of soils on the up-slope sides of cairns; but neither is there dramatic accretion at the up-slope margins of local in situ bedrock daylighting through the desert pavement.

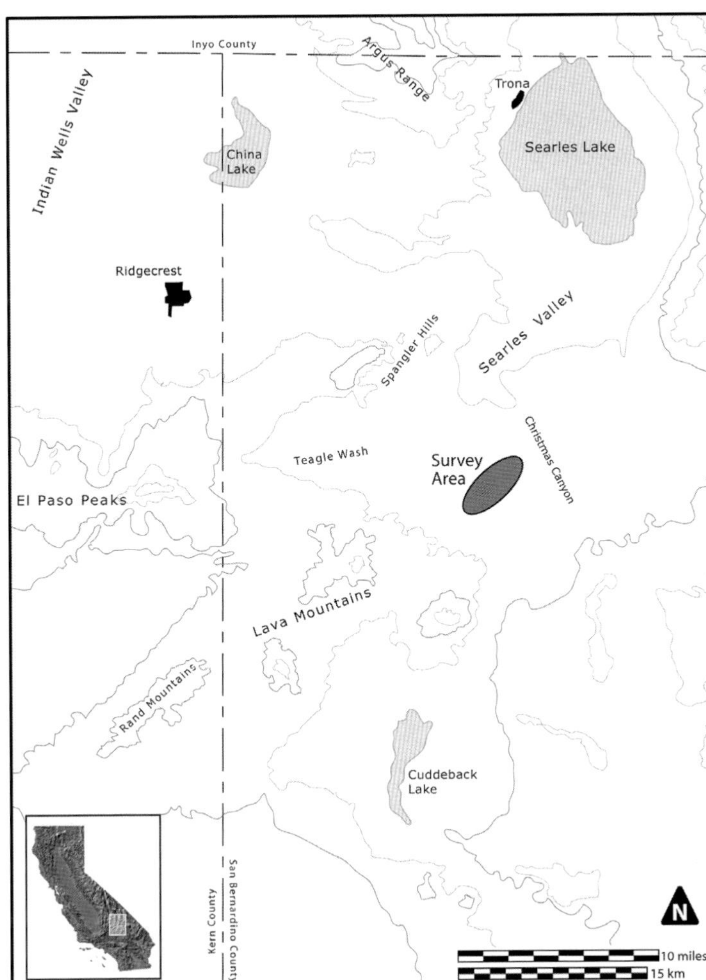

Figure 1. Regional map.

Figure 2. Lone cairn.

Figure 3. Four cairns forming a quadrilateral.

Of the total 45 cairns, some 42 were found on or immediately adjacent to a prehistoric site. Significantly, we were careful to draw site boundaries based on surface artifacts only. That is, parsimony (and initial skepticism) discouraged us from extending the boundaries of archaeological sites simply in order to include a cairn. The strong correspondence of cairns with prehistoric sites thus appears to be genuine, despite the fact that in no case do cairns appear to be the spatial focal point of an associated artifact distribution. Of 46 sites recorded during our survey, 13 were accompanied by one or more cairns.

An additional two "cairns" were also noted, but these are definitively historic in origin: each is accompanied by a heavily weathered, embedded vertical post of 4-x-4-inch pine. The posts are separated in space by some 40 m, within 300 m of the *only* mining prospect recorded for the survey area. In truth, these features are not exactly "cairns," because each post is surrounded by a single course of support boulders (the rocks are not piled), and each rock distribution covers an unusually small diameter (<0.45 m). These post-cairns are not further considered below; but in fairness to the supporters of an historic origin for all (or many) of the cairns, it may be countered that, in the presence of posts, larger cairn size is unwarranted. *All* cairns may thus be historic, one could argue, and the larger, taller ones are simply those without posts.

In this last sentence alone it is possible to see our dilemma with many of these features: there is a "yeah, but…" lurking behind practically every deduction. And what *really* confounds the issue is this observation: the historic post-cairns are found virtually at the center of a prehistoric lithic site (CA-SBR-10850)! What does this say about any argument concerning cairns and "associated" prehistoric artifacts? These are the sorts of problems we seek to resolve.

The functions and the ages of the cairns are essentially reciprocal: knowing one will hint strongly at the other or at least narrow the field considerably. Unfortunately there are very real difficulties in establishing dates for the cairns, even in relative terms. Besides inconclusive (or ambiguous) geophysical indicators such as precarious placement or accretion of alluvium, no chronologically diagnostic historic or prehistoric artifacts (or obsidian, etc.) were recovered in unequivocal spatial association, within the cairn matrices. (It should be noted that the entire survey yielded very few chronological diagnostics overall.) Indeed, no artifacts of any sort were noted *within cairn matrices* and, recalling the post-cairns above, we obviously should be cautious about our conclusions drawn from ill defined spatial associations.

It may be possible to date the cairns through cation ratio assessment of rock varnish, and indeed studies are underway to examine the feasibility of the method for the Christmas Canyon area (Dorn and Cerveny 2002). Serious problems continue to plague the technique, however (Harry 1997), with two important problems from the outset with particular respect to these cairns. First, sigma values (plus and minus figures) for dates obtained tend to be very large (see Wells 2003:35:Table 2). This is worrisome enough, but it additionally will make it difficult to distinguish cairns piled in the later prehistoric era from those made in the historic era—a potentially momentous distinction in terms of cairn function. Second, many dates for rock varnish are untenably old, and call into question not the cation ratio assessment of varnish age per se, but rather whether the dated sample refers to varnish that began to accumulate before or after the rocks were arranged by humans (a syndrome familiar to students of the Calico claims). These rock varnish dating problems are *not* irresolvable, however, and we applaud and will support continued efforts to refine the method. For the moment, however, the method does not solve our problem.

But even chronologically neutral functional assessment is beset with a dilemma, that of "equifinality." The problem is this: in certain cases, mutually exclusive behavioral or functional interpretations may be reasonably drawn from one and the same appearance of the archaeological record. For an easily accessible hypothetical example of the equifinality issue, consider that *both* technological change and social migration may be reasonably deduced from the single observation, that of change in projectile point style (see Rondeau [1996] for an actual example and discussion of this hypothetically stated problem). In the case of the cairns of Christmas Canyon, virtually every suite of expectations for any given function may be argued to be equally (or at least potentially) suggestive of another.

The "simple" solution to the equifinality dilemma is, of course, multiple and exclusive explanatory tests. We should simply not rely on single tests for verification of our propositions. Unfortunately, for piles of rocks of unknown age and function, several of the explanatory propositions put forth by us and others essentially rest on the expectation that we find *nothing* in association! In other cases, possible functions are not well enough understood in material terms (as yet) to suggest multiple tests. In any event, the truth is, our present archaeological observations may point in widely differing directions.

For this reason, the focus of this exercise is really on *eliminating* several proposed functional explanations. In fact, some alternatives we have discussed among ourselves are not considered below simply because we haven't yet figured out tests useful to eliminating them from consideration! We do not expect to arrive at functional explanation by way of default, however, or by waiting for the dust to clear to see which alternative explanation still stands. But by narrowing the field of potential alternative functional explanations, we may begin to consider the practical aspects of focused testing, including additional surface survey and selective excavation of a sample of these enigmatic cairns.

What follows is series of potential alternative explanations for the functions of these cairns. Certain of these are "regionally specific" and may not have validity as potential explanations in other regions where cairns occur. That is, we have discovered cairns in a region that, while perhaps not singular for this trait, has seen a tremendous amount of human activity, both prehistoric and historic. As we shall see, many of these activities have included the extreme likelihood of cairn-making. These propositions are in no particular order of favor, disfavor, importance, or likelihood of acceptance or rejection.

Propositions

"It's a Horse, not a Zebra"

Probably the simplest proposition for the functions of the cairns is that they are boundary markers for historic mining claims. This is the proposition most often suggested by archaeologists working the region (ourselves included), particularly at first glance. Without a doubt many—even most—of the cairns *look* like mining claims that dot the Mojave Desert, and we have by no means been able to entirely rule the proposition out. We're just not so sure now that we've seen the sheer numbers of cairns in dubious geological and topographic contexts (from a miner's perspective), and the repeated co-occurrence of cairns and prehistoric artifacts. More particularly, we discount the proposition as universal based on the absence of other independent evidence for mining activities *in the immediate area*.

The "smoking gun" for the "It's a Horse, Not a Zebra" Proposition is evidence for mine-claim paperwork that officially stakes a mining claim. As anyone who has surveyed in the Mojave Desert knows, this paperwork (a copy of the official claim stake that is filed with the local Division of Mines) is often stashed within a cairn, typically in a tobacco tin or other sealed container of metal or glass. While the paperwork itself has often decayed, the tobacco can survives as a "smoking gun."

In this regard, in the greater Spangler Hills area, numerous cairns have been recorded with telltale tin cans, many with intact paperwork, spanning at the least the 1930s-1970s. But cairns in the Christmas Canyon ACEC proper have yet to reveal even one smoking gun. The "It's a Horse, Not a Zebra" Proposition is effectively ruled-out *within the Christmas Canyon ACEC*. For now.

"The New Agers Did It"

Blaming "New Agers" for many things we can't explain in the archaeological landscape has become something of a universal remedy for us. Perhaps this has become a too-convenient catch-all but, unfortunately, the proposition is often valid. Whether it is leaving quartz crystals at sacred hot springs in Northern California, converting rockshelters into sweathouses, or scattering the ashes of the dearly departed into kivas at Chaco Canyon in New Mexico, a few misguided New Agers leave a lot of red herrings in the archaeological record, and it can be difficult to test for and eliminate many of their activities. However, it is our experience that ersatz (and incongruous) medicine circles are the features of choice in the Mojave Desert. In view of this, we note no firm evidence for New Age activities in the Christmas Canyon ACEC. Bear in mind, however, that for many of us, the activities of New Agers are as unfathomable as prehistoric ritual. We cannot rule the proposition out in absolute terms, and we are particularly aware of the implications this proposition has for the correlation of cairns with archaeological sites, which may be specifically targeted for New Age activities.

"Blame It on the Bikers"

This leads us to the "Blame It on the Bikers" Proposition. We know from recent BLM-sanctioned motorcycle racing in the more general Spangler Hills area that racecourses are painstakingly and clearly delineated prior to a race. Pink flagging tape or the like is used for this purpose today, affixed to bushes or to embedded wooden slats. It is reasonable to suggest use of cairns to mark trails, particularly where there are few bushes for affixing flagging and where flimsy slats are impossible to drive (as on desert pavement). In some sense, this is one of the more "testable" propositions: we expect such cairns to associate directly with motorcycle trails.

In a few cases, cairns do indeed appear near recent motorcycle trails; but far more often, they do not. There is obviously a second test for this proposition: evidence for flagging (slats, etc.) near the cairns. We do note a few cairns loosely associated with a wooden slat or two; but in no case is there direct association of a cairn and a slat and, given the wide distribution of both cairns and slats, it is perhaps inevitable that one would be found near the other at some location. Finally, we could simply ask some bikers. Organized motorcycle racing clubs using the area today deny that they are responsible for erecting any cairns, in the strictest adherence to the BLM prohibition of the practice. If you ask a biker or two, though, you'll also find that they never drink on weekdays and don't cheat on their spouses. Without meaning offense, we shouldn't rule out the "Blame It on the Bikers" Proposition based entirely on an interview or two. The proposition may have some validity, for some cairns at least.

"Walk This Way"

Motorcyclists aren't the only ones who mark their trails, of course, and it is equally plausible that campers or hikers have marked their way points. The "Walk This Way" Proposition is perhaps not so easily tested, however, because ephemeral foot traffic alone is not likely to leave much of an associated trail. Also, it is difficult to generate a convincing set of expectations for artifact or other material associations, so supporting tests are lacking. It is well worth noting, though, that many of the cairns are not highly visible until you are virtually on top of them, even for a slow-moving hiker, and would thus provide poor way markers. Significantly, moreover, many cairns are not located at logical places for hiking trails. A motorcycle racecourse may not purposefully seek the line of least resistance (over a ridge, for instance), but we may expect hikers to.

"*Walked* This Way"

There is a past-tense to the previous proposition: the "*Walked* This Way" Proposition. It remains an interesting possibility that some of the cairns are prehistoric trail markers. Because aboriginal trails—in use for a generation or more—may leave a tangible scar in the landscape, the "Walked This Way" Proposition is abundantly testable *and through multiple lines of testing*. That is, aboriginal trailsides are known to have been the focus of ritualistic offerings (Davis 1961:7; Heizer 1978:692) as well as other activities embedded in ritualistic journeys, including raw-material assay and lithic reduction (Cleland 2004). Indeed, in the Colorado Desert, such locations were often accompanied by rock features, including cairns (James Cleland 2004: personal communication). In testing this proposition, we might expect a visible trail and an accumulation of token materials (whatever they may be).

Indeed, we have firm suggestion of an aboriginal trail within the Christmas Canyon ACEC. But perhaps significantly, we have yet to discover a cairn along its route. We *may* have some additional evidence for prehistoric trails within the Christmas Canyon ACEC, some near cairns, others not. But teasing these trails out from motorcycle trails, other (recent) hiking trails, game trails, and so on presents *exactly* the same dilemma we have in explaining the cairns in the first place: trails are largely intractable to conventional dating and are difficult to associate with a predictable artifact assemblage. As for tokens, we note only one potential example: in one case (CA-SBR-10841) we found a single cairn in spatial association with five nondiagnostic bifaces (all within 7 m of the cairn). The overall site (a 2,500-m^2 low-density lithic scatter) revealed no other bifaces. A putative motorcycle trail is found at a distance of some 30 m from the cairn.

Despite this single provocative example, however, the "Walked This Way" Proposition is not well-supported in either of two critical tests: we know of at least one trail with no cairns, and we know of many cairns with no trails. Still, the proposition is testable and, more significantly, cannot be ruled out at this time.

"The Devil's Advocate"

When we first encountered the cairns and tried to make some sense of them, we noted that most were spatially associated with demonstrable archaeological sites. This led us to the "Devil's Advocate" Proposition: rather than allow that the cairns were indeed prehistoric features, extreme parsimony led us to wonder whether these may instead purposefully identify the locations of archaeological sites, but as modern phenomena. The idea is perhaps over-cautious, but not that crazy. We know that the area was previously surveyed by archaeologists Sylvia Winslow and Emma Lou Davis in the 1960s. We also understand that they considered dabbling in aerial photography of the general vicinity (Russell Kaldenberg 2002: personal communication). We wondered (and still do) whether some of these cairns might serve as cadastrals, identifying the locations of archaeological sites for future recording, from the air or otherwise. A search of files at the Maturango Museum, Ridgecrest, a repository for Winslow's notes, failed to shed any light on this "Devil's Advocate" Proposition.[1] Note once again, however, that the cairns by themselves are fairly inconspicuous and provide poor visibility from any vantage—let alone from the air.

---

[1] It may be significant that neither Winslow nor Davis make any mention of cairns. Unfortunately, we cannot tell whether this represents a simple oversight, dismissal of the cairns as mere mining claims or other historic phenomena, or even a post-Winslow/Davis vintage for the cairns (1960s). It should be noted that explicit survey focus of Winslow and Davis was on locating stone tools presumed to be associated with Pleistocene Lake Searles.

We also wondered whether site vandals may have placed the cairns for similar purposes. However, we have not observed any evidence for pothunting or other unauthorized collection of archaeological materials.[2] For that matter, most or all of the low-density, debitage-rich sites in the region have considerable professional interest, but little presumed value for private collectors (there are few whole or identifiable tools). In any event, we are struck with the observation that nearly all of the cairns (n = 42 of 45) are spatially associated with surface lithics.

"Grave Markers"

A recent discovery during earth-moving activities in nearby Owens Valley suggests historic and/or protohistoric use of cairns to mark Native graves. Adjacent to the town of Keeler, on the eastern shore of Owens Lake, human remains appear to be in association with *stacked* (not piled) rocks. We await formal reporting of the salvage operation, but for the moment we suggest that any analogies to the Christmas Canyon cairns are thin, and we reject the "Grave Marker" Proposition for now. First, in no case do we observe evidence for prior digging at or near any Christmas Canyon cairn. It may be countered that the cairns themselves may mask such evidence but, judging from usual cairn size alone, any Christmas Canyon inhumation confined to only the cairn footprint would have to be a quite small package—such as a near-total cremation.

It may be countered that, over a long time span, pavements may have reformed, and past digging may simply be obscured. But pavement reformation requires a (perhaps implausibly) long time and, in any case archaeological evidence for the use of cairn markers in the prehistoric era is lacking for the greater Mojave. Finally, many or most cairns are found in locations of fairly shallow soils, which are unlikely to have been favored burial locales.

The beauty of the "Grave Marker" Proposition is that it is abundantly testable, and can be either demonstrated or refuted with precision, simply by digging a one-by-one. Note, however, that demonstration or refutation of the proposition obviously requires total destruction of the cairn and is, in any case, effective only on a cairn-by-cairn basis (refutation or demonstration at one cairn does not perforce apply to the next cairn). As we will note in a minute or two, one irritating thing about the search for cairn functions is that any or all propositions may be simultaneously true—on a cairn-by-cairn basis.

"All Roads Lead to Coso"

A provocative explanation has been offered by David Whitley (personal communication 2004). He notes that ethnographic, shamanic, and other vision quests are often accompanied by exhaustive labors, which may in this case have taken the form of cairn construction. He finds it to be no coincidence that massive fields of Coso rock art are located at and surrounding Little Petroglyph Canyon, about 70 kilometers to the northwest of Christmas Canyon. He suggests that the cairns represent a stage in a complex of ritual behavior, which culminated eventually in the formation of Coso-style rock art. This is the "All Roads Lead to Coso" Proposition.

Unfortunately, the proposition is not easily testable. Whitley has cast his net far to the north for ethnographic notice of cairns as components of vision quests (Whitley n.d.). While a perfectly valid anthropological methodology, this does not lend itself to suggesting specific additional characteristics and associations we might expect to accompany a vision quest in Christmas Canyon. Next, and perhaps significantly, the cairns do not at all appear to represent an exhaustive undertaking, unless the quester was so stupefied by drugs that breathing itself would have been laborious. We can't exclude this possibility, of course; but neither can we test it. Finally, the proposition essentially requires that we find *nothing* in association with the cairns, which is the precise test for many other propositions we put forth in this exercise.[3] How do we sort-out which cairns are a part of a vision quest and which are not?

"The Weasel"

The worst case scenario, and probably closest to the truth, is the "Weasel" Proposition, which states that all or at least some of the above are true in some combination, on a cairn-by-cairn basis. This proposition may explain our unsuccessful efforts to discern some sort of spatial patterning to their distribution. The proposition derives from a sound assumption: that there are only so many ways to pile rocks. In short, even highly varied functions of individual cairns should not necessarily be expected to discriminate according to regularized sets of differing physical characteristics. Whether a cairn was piled by a vision-questing shaman, a hopeful miner, a lost hiker, or a bored New Ager, a pile of rocks is still a pile of rocks.

---

[2] Rockhounds have apparently gleaned some chert from a quarry site (CA-SBR-10835) but there is no indication that they were aware of its prehistoric significance. Their objective appears to have been not artifacts per se, but colorful red and yellow chert, mistakenly believed to be jasper.

[3] Whitley et al. (1999) have found a significant correlation between quartz (hammerstones, etc.) and rock art and, by extension, with other elements of shamanic activities. This important finding presents a potential "smoking gun" for the actions of questers at or near the Christmas Canyon cairns. Quartz was not found in association with the present cairns, however.

Discussion

The myriad reasons for cairn construction in prehistory alone (Begole 1974; Von Werlhof 1987) mitigate against simple solutions; adding historic activities to the mix only heightens the problem. We have mentioned some of the uncertainties inherent to dating the cairns; but even in the unlikely event that *all* of them are demonstrably quite old, we still are no closer to providing functional explanation for individual cairns, at least with the current data. Any functional "explanation" is presently pure speculation and, we submit, is likely to remain so for the foreseeable future. Excavation of a selected sample of cairns will almost certainly shed considerable light *but on the excavated cairns alone*. Then again, given the issue of equifinality alluded to above, maybe not.

There is an awful lot of work still to be done, and we fear that this will include a lot of awful work. We will continue to see speculation without reasonable documentation and, perhaps worse, wholesale disregard of the features. Much of this disregard will stem from the assumption that the cairns represent mundane historic features. But let's remember this bit of cynicism: If we had only begun to notice a few fish dams in the middle of the Colorado Desert *today* we would almost certainly attribute them to some sort of New Age monkey business. If we had only begun to notice geoglyphs *today*, we would almost certainly attribute these to some sort of crop-circle prank. Instead, like von Werlhof before us, we should try to remain unjaded, and just keep at it. Our lack of concrete answers today does not mean that our questions are bad or unworthy. In fact, the opposite is true.

For now, we rely on von Werlhof's Law: Keep recording the trees, and the forest will appear.

Acknowledgements

The authors wish to acknowledge the pioneering work of Emma Lou Davis and Sylvia Winslow in the China Lake and Searles Lake basins. If anyone knows where their artifacts and notes are stored they could help solve some of the mysteries of these features if they would share the information. CWC,Jr.

References

Begole, Robert S.
1974 Archaeological Phenomena in the California Deserts. *Pacific Coast Archaeological Society Quarterly* 10(2):51-70.

Cleland, James H.
2004 The Sacred and the Mundane: Cultural Landscape Concepts and Archaeological Interpretation in the Colorado Desert. Paper presented at the 38th Annual Meeting of the Society for California Archaeology, Riverside.

Davis, James T.
1961 Trade Routes and Economic Exchange among the Indians of California. *Reports of the University of California Archaeological Survey* No. 54, Berkeley.

Dorn, Robert and N. Cerveny
2002 Final Report, BLM Order No. BAP010177, August 1, 2002. Submitted to Bureau of Land Management, Ridgecrest District, Ridgecrest, California.

Harry, Karen G.
1995 Cation-Ratio Dating of Varnished Artifacts: Testing the Assumptions. *American Antiquity* 60(1):118-130.

Heizer, Robert F.
1978 Trade and Trails. In, *Handbook of North American Indians Volume 8: California,* ed. by R. F. Heizer, pp. 690-693. Smithsonian Institution, Washington, D. C.

Rondeau, Michael F.
1996 When is an Elko? In, *Stone Tools: Theoretical Insights into Human Prehistory*, ed. by G. Odell, pp. 229-243. Plenum, New York.

von Werlhof, Jay
1987 *Spirits of the Earth,* Volume I. Imperial Valley College Museum Society, El Centro, California.

Walsh, Michael R., and C. W. Clewlow, Jr.
2003 Intensive Archaeological Inventory of an Additional 1000 Acres Adjacent to Ancient Lake Searles in the Christmas Canyon ACEC, San Bernardino County, California. Report Submitted to Bureau of Land Management, California State Office, Sacramento, California. Order No. BAP020093, Modification No. 2.

Wells, Helen
2003 Intensive Archaeological Inventory of 1000 Acres Adjacent to Ancient Lake Searles in the Christmas Canyon ACEC, San Bernardino County, California. Report Submitted to Bureau of Land Management, California State Office, Sacramento, California. Order No. BAP020093.

Whitley, David S.
n.d. Rock Art and Rites of Passage in Far Western North America. In, *Talking to the Past: The Ethnography of World Rock Art*, ed. by J. Keyser and G. Poetschat. University of Utah, Salt Lake City (in press).

Whitley, D. S., R. I. Dorn, J. M. Simon, R. Rechtman, and T. K. Whitley
1999 Sally's Rockshelter and the Archaeology of the Vision Quest. *Cambridge Archaeological Journal* 9(2):221-247.

THE CARRIZO COLLAPSE: ART AND POLITICS IN THE PAST

David S. Whitley, Ph.D. and Joseph M. Simon
W&S Consultants
447 Third Street
Fillmore, California 93015

and

Johannes H.N. Loubser, Ph.D.
New South Associates
Atlanta, Georgia 30506

ABSTRACT

The Carrizo Plain National Monument is renowned for its pictographs. It also has a significant concentration of prehistoric villages, recently identified during studies conducted for the Bureau of Land Management. These habitation deposits cover 10% of the ground surface in the Caliente Range foothills: 98% of this coverage dates to the Middle Period; 2% is Late Horizon. This suggest a tenfold population decrease over time: from about 300 to 30 people. A three-phase pictograph chronology (based on superpositioning) charts development of a distinctive Carrizo tradition, illustrating the increasing prominence of Middle Period shaman-headmen and their disappearance, with final-phase art consisting of occasional "intrusive" Yokuts-style motifs. The combination of the pictograph and settlement data has implications for two ongoing California archaeological debates: the importance of the Medieval climatic anomaly and the rise of the Chumash chiefdom.

Jay von Werlhof is well known for his many contributions to California desert archaeology, a task that has occupied much of the last few decades. Perhaps less well known today are Jay's contributions at the start of his professional career. One of these was his recording of Tulare County pictograph sites during the 1950s. His half-century-old site records still constitute the primary data set on this important corpus of south-central California rock art, and they have proven invaluable to subsequent researchers in that region (e.g., Heizer and Clewlow 1973; Whitley 1992, 2000a). They also reflect Jay's fundamental recognition of the need for a holistic understanding of the prehistoric past; one that is not just based on technology and subsistence, but also considers the artistic and spiritual heritage of our archaeological forebears.

We have recently conducted a series of studies and projects in the Carrizo Plain National Monument (CPNM), in southeasternmost San Luis Obispo County, California, for the Bureau of Land Management (W&S Consultants 2003a, 2004). The Carrizo Plain is of course world-renowned for its concentration of Chumash pictograph sites (e.g., Grant 1965; see Figure 1). In many respects, these can be argued to provide a parallel to the Yokuts pictographs that Jay recorded on the opposite side of the San Joaquin Valley a half-century ago. Regardless of the possible parallels between these two corpora of art, Jay's earlier rock art research inspired and motivated our work on the Carrizo. This work emphasized the remainder of the CPNM archaeological record, that is, the archaeology beyond the rock art, which had been barely explored yet which was obviously existed in the international shadow of the pictographs. Our BLM-sponsored projects were, in part, an effort to develop sufficient information about the prehistoric settlement pattern as a whole to facilitate the development of a management plan for this newly created monument. This gave us an opportunity to integrate the well-known rock art sites with the "dirt" archaeological record and along the way to discover some interesting information about the prehistoric past in this part of interior south-central California.

Figure 1. Carrizo pictograph (digitally enhanced) Washburn Ranch (CA-SLO-100).

THE CARRIZO PLAIN

The Carrizo Plain is located along the boundary of Kern and San Luis Obispo Counties, California, roughly midway between Bakersfield in the southern San Joaquin Valley and the Pacific coast. The plain is in fact a broad inland valley that (physiographically) forms a southern part of the larger South Coast Ranges. It is bordered on the west by the Caliente Range and on the east by the San Andreas Rift Zone and the Temblor Range. Although once entirely internally draining, the northern end of the plain has been hydrographically captured by the Salinas River system whereas the majority of the southern portions of the plain, including the CPNM, continues to drain into Soda Lake. Soda Lake is currently a large, usually dry mud playa. Elevation at its bottom is approximately 1900 feet above mean sea level. The high stand for this lake, presumably achieved at the end of the Late Pleistocene, is only 1940 feet in elevation, indicating that this was once a broad but relatively shallow body of water.

The Caliente Range, running SE-NW, is a relatively low and narrow series of ridges separating the Cuyama Valley to the southwest from the Carrizo. The Temblors and the San Andreas Rift Zone, running parallel to the Calientes immediately northeast of the plain, separate the Carrizo from the southern San Joaquin Valley further to the northeast. One aspect of local geology is relevant to the archaeological record on the plain and thus warrants mention. This is the Vaqueros Sandstone formation, exposures of which are common on the northeastern front of the Caliente Range, that is, the front abutting the plain. This formation is slightly more indurated than other sedimentary units in the region, one result of which is the existence of vertical cliff faces, caves, caveates and rockshelters, and monolithic outcrops. The Vaqueros provides the context for the numerous and well-known pictograph sites in the CPNM, which together constitute the Carrizo Plain Rock Art Discontiguous National Register of Historic Places (NRHP) District. Furthermore, the formation supports more gradual foothills and slopes above the Carrizo Plain bottom, thereby creating a broader, wetter (due to its predominantly northern aspect), and more ecologically diverse transition between the open flats and the mountain ridges. The transitional foothill zone of the Calientes apparently was very favorable for prehistoric human occupation and, because of this fact, contains by far the highest density of archaeological sites in the CPNM.

Prior to Euro-American impacts and changes, vegetation within this region varied depending upon topography, aspect, slope, and available water. Prehistorically much of this region was covered by chaparral and scrub, especially the more mountainous zones, with foothill areas also containing grasslands and juniper and oak woodlands, whereas the Soda Lake margin as well as springs along the foothills would have supported riparian environments. Much of the plain itself, however, was covered by a Valley Grassland ecosystem dominated by perennial bunch grasses such as purple needlegrass (*Stipa pulchra*) along with annual grasses and herbs (Schoenherr 1992).

During the contact period the Carrizo Plain was occupied by a small Chumash-speaking population, which formed the northeasternmost extension of this large ethnolinguistic group. As we shall see, our archaeological evidence tends to supports continuity in Chumash occupation, within the CPNM, at least to the beginning of the Middle Period which, for various reasons, we set at approximately 4000 YBP[1].

Archaeology of the Carrizo Front of the Caliente Range

Our fieldwork within the CPNM involved both Class I (purposive) and Class II (sample) surveys (W&S Consultants 2003a, 2004, respectively), the latter of which consisted of a stratified random study intended to relate site densities and types to environmental zones. Total survey coverage for the two studies was about 13,000 acres, or roughly 6% of the CPNM as a whole. In broadest terms the CPNM can be characterized as having very low archaeological site density, with one exception. This is the Carrizo front of the Caliente Range, where the vast majority of the prehistoric sites have been recorded. Notably, this is the area that contains the well-known Carrizo pictograph sites, including Painted Rock, Saucito and Washburn Ranch (Sulphur Springs), among others. But for sake of computational clarity and analytical precision in the following discussion, we consider only those sites which fell within our survey units (which generally, though not invariably, were drawn to exclude the previously recorded pictograph sites).

We were able to date 67% of the sites found in our survey transects using two chronological indicators, the first of which is temporally diagnostic artifacts. These were primarily limited to two classes: projectile point types, with well-known chronological implications; and mortars/pestles, which we took to indicate Middle Period or later, age. Second, we paid careful attention to soil conditions, with evidence for artifacts eroding out of oxidized paleosols in particular as (conservatively) suggesting pre-Late Horizon ages. (Concomitantly, the presence of middens still exhibiting organic enrichment was considered an indication of relative site youthfulness inasmuch as, over time, these organics are leached out of soils.) Although mortars/pestles were relatively common, the more temporally sensitive projectile points were somewhat rare, and unfortunately other temporal diagnostics that are common in the Chumash region, such as shell beads, were not found on any of our sites during the survey. Two qualifying points, nonetheless, need be noted. It is possible that the ages of some of our earlier sites or site components have been underestimated and that they may be older than we have inferred, but it seems less likely that sites are younger than our estimates. Perhaps more importantly, the temporal patterning that we observed is so robust that it is unlikely that changes in the age assignments to a few of our sites would appreciably effect our conclusions.

When all of our sites are considered, including those that we could not date, 56% were assigned to the Middle Period (~4000-800 YBP) and only 11% to the Late Horizon (for our purposes, post-800 YBP). Habitation sites—that is, villages and camps—comprised 53% of our total; 74% of these are assigned to the Middle Period and only 18% to the Late Horizon. As is immediately clear, these data suggest a significant disparity between Middle Period and Late Horizon occupation of the Carrizo Plain; indeed, they suggest a dramatic reduction in occupation over time.

But counts of clerically defined sites can be misleading, especially for any kind of population estimate. Following accepted archaeological practice (cf. Yellen 1977; Hassan 1981), we calculated the acreage covered by these sites as a more useful approximation in this regard and found that Middle Period habitations cover about 205 acres, whereas Late Horizon habitations account for only 5 acres. This is a 98% versus 2% difference in site size for these two time periods. As is then clear, our data suggest a dramatic change in population from the Middle Period to the Late Horizon. Moreover, it is worth noting what these acreage figures represent: they constitute roughly 10% of the total area of the Carrizo front of the Calientes. Given that much of this foothill zone is slope and ridgeline, it is likely that these archaeological deposits cover somewhere in the vicinity of a fifth to a third of the flat ground in this zone. Simply put, this is the largest and densest concentration of habitation remains that we are aware of, away from the coast and islands, in south-central California.

It is also worth noting that the pattern evident in the CPNM data was not entirely unexpected. Instead it verified what we have seen, over the last decade in our archaeological work in the interior and inland portions of this region, albeit in much more piecemeal fashion. In fact, we had previously observed and argued for a dramatic increase in Middle Period population in the inland zone, followed by a site abandonment at the Middle Period to Late Horizon transition (e.g., W&S Consultants 1994a). But we had never heretofore had a data set of sufficient size to adequately verify that this site abandonment and population collapse were anything more than localized phenomenon.

Comparative data illustrating that the pattern seen on the Carrizo Plain was widespread is provided by Phase I survey and Phase II test excavation data from the Hidden Creek Ranch, a 4200-acres property in Moorpark, inland Ventura County, about 20 miles from the Pacific Coast (W&S Consultants 1994b, 1995, 1997). Note in this regard that our data for the Hidden Creek Ranch include test excavation results and they are for this reason more fine-grained than the Carrizo data. At Hidden Creek Ranch we found that 82% of our habitation sites/site components date to the Middle Period and only 18% to the Late Horizon. These differences are exacerbated when acreage figures are calculated: Middle Period habitation deposits cover about 64 acres whereas Late Horizon deposits include only 3.5 acres. This represents a 95% versus 5% difference, which are figures that are remarkably close to our calculations for the Carrizo.

Clearly then, major changes occurred in inland and interior populations sometime during the transition from the Middle Period to the Late Horizon in south-central California. We can estimate these changes, in absolute population sizes, from our site deposit acreage figures. The result is surrogate or ballpark estimates that, though far from precise, are useful as relative measures of population change over time. Deriving these surrogate figures requires, first, normalizing the acreage estimates in terms of the relative lengths of the two time periods. A computation made in this fashion suggests that the Middle Period in the interior Carrizo Plain had roughly ten times the population of the Late Horizon, whereas the Middle Period in the inland Moorpark area had about four times the population of its Late Horizon. In both cases the reductions were dramatic.

Absolute estimates for the change in population on the Carrizo can be computed, second, using historical mission records as an indicator for the Late Horizon population and then working backwards. There are 14 individuals identified in the mission records from the Carrizo, with population for the CPNM area estimated at 14 (most conservatively but least likely) to about 32 people (more likely; see Johnson 1988, 1999). This suggests that the Middle Period population in the CPNM area was roughly 140 to 320 peoples. But regardless of the exact figures, one conclusion is clear. The Middle Period to Late Horizon transition in inland and interior south-central California was marked by a widespread population collapse.

The Medieval Climatic Anomaly and the Carrizo Collapse

Early reconstructions of Chumash prehistory tended to envision the Holocene archaeological record in terms of a slow but steady increase in population which peaked at about the time of European contact, accompanied by parallel upward progressions in adaptive and technological sophistication and socio-political complexity. More recently, researchers have recognized that socio-cultural, adaptive, and technological changes in the past were somewhat more complicated than initially recognized. Particularly important in this regard is the potential effects of the Medieval Climatic Anomaly, a period stretching from about A.D. 800 to 1350 that was marked by catastrophic droughts. Whereas neo-Darwinian adaptationists have tended to dismiss the effects of this climatic instability on culture change (e.g., Bettinger 1999; Basgall 1999), others have cited it as an important factor during the Middle to Late transition (e.g., Jones et al. 1999; Arnold 2001; Kennett and Conlee 2002; and many others), including the role that environmental stress played in the appearance of the Late Horizon Chumash chiefdom of *Lulapin*.

Three points can be made with respect to this topic and our Carrizo data. First, an analysis of cores from Soda Lake provides a study of Holocene climate change in the Carrizo Plain that is near-decadal in its detail (Eigenbrode 1999; Eigenbrode et al. 1997, 1999), and that correlates very well with similar records from Walker and Great Salt Lakes. The general paleoclimatic trends revealed by this study are as follows:

- ~4000 - 3400 YBP–Water level in the lake rose
- ~2900 - 2600 YBP–High lake stand achieved with 3 - 4 m of water
- ~2240 YBP–Second high lake level
- ~2100 - 1500 YBP–Extreme desiccation
- ~1350 YBP–Moister conditions
- ~1200 - 800 YBP–Lowest lake levels, drought

The effects of the Medieval Climatic Anomaly were locally manifested, on the Carrizo Plain by the desiccation of Soda Lake between 800 and 1200 YBP. Since this is a relatively small, internally draining basin, the implications are straightforward: available moisture was substantially reduced throughout the Carrizo during this period. This fact can be reasonably assumed to have had adverse impacts on the biotic carrying-capacity of the region as a whole.

Second, there is an observable pattern in site distributions that appears to correlate with these changing environmental conditions (Figure 2). Large Middle Period village complexes are typically located on open ridges/terraces above stream drainages in areas that in the modern environment lie below the limits of normal surface water flow. The much smaller and rarer Late Horizon villages, in contrast, are invariably associated with contemporary flowing springs. This suggests that settlement locations followed available

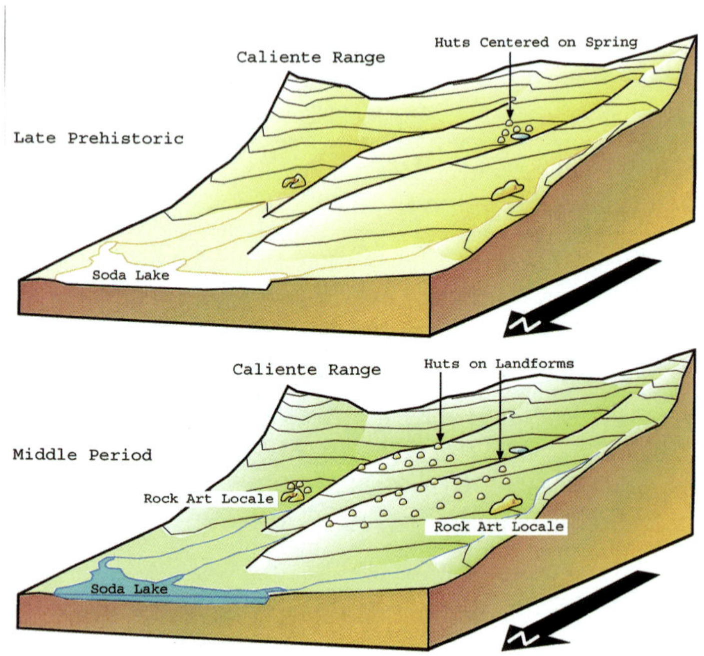

Figure 2. Schematic illustrating the change in the settlement pattern from the wet Middle (lower) to the dry Late (upper) periods.

water, responding to drought with a move to the most reliable water sources—those that are still flowing today. It is worth noting that we have observed this same associational pattern on numerous other occasions in the inland/interior region and it does not appear to be restricted to the Carrizo Plain.

This last conclusion is amplified by another observation about our surveyed sites. We noted a significant number of tarring pebbles on the ground surfaces of the Late Horizon villages but none on the Middle Period deposits. Assuming that tarring pebbles were used to waterproof baskets, this suggests that water management, transportation and storage were important concerns during the Late Horizon. Again, this observed Carrizo pattern appears to be widespread: Wallace (1955) much earlier noted that the Late Horizon was marked by extensive asphaltum use, and at least some of this almost certainly involved waterproofing baskets.

Third, although survey level data are inherently coarsely grained (especially in an area like the Carrizo Plain, where temporally diagnostic artifacts are rare), there nonetheless is a plausible basis for linking the Carrizo collapse to the effects of the Medieval Climatic Anomaly. Basgall (1999) apparently would disagree inasmuch as he has argued that much more direct empirical linkages are required to make this kind of an argument. Gross correlations, in his terms, such as we have offered here, require too many inferential steps for any kind of interpretive confidence.

While Basgall's argument sounds reasonable at first glance, it overlooks the key analytical point. This concerns the scale of the putative causal process under study and the resulting scale of the data needed to evaluate the proposed cause and effect relationship. Paleoclimatic change is obviously a very widespread process, often manifested at a global scale. For this very reason its potential effects can be evaluated with broad-sweep data, contrary to Basgall's assertion, which implies that all analysis must be occur at the most particularistic of levels (and which implies, e.g., that all analysis must be equivalent in scale to micro-economics, which would certainly surprise the many global and macro-economists who are employed in university economics departments today). In this particular case we know as a well-established fact that the Medieval Climatic Anomaly and the subsequent Little Ice Age correlate chronologically with population collapse and settlement abandonment in the Puebloan world. They also correlate with the demise of the Toltec state in northern Mesoamerica and the retraction of nuclear Mesoamerican civilization southwards to the Basin of Mexico during the so-called Chichimec Interregnum, as well as the Classic Maya Collapse (Gill 2000).

Localized historical and other factors obviously play a role in any specific response to climatic catastrophes, and not all societies respond to paleoclimatic change in an identical fashion. But it is also clear that, when considering natural phenomena that affected major regions across North America, wide-ranging survey-level data certainly have their place in the development of reasonable inferences. To argue that the interior/inland Chumash collapse was not then related to the effects of the Medieval Climatic Anomaly, in light of the comparative data from other regions in western North America, is to argue against the obvious; that is, at least until a plausible alternative hypothesis is proposed explaining why these widespread population disruptions occurred at the same time, during a period of drought and environmental instability, yet in which this environmental catastrophe played no part and the parallel timing of which was somehow coincidental.

Dating the Carrizo Pictographs

The obvious question, in light of the renown of the Carrizo pictographs, is the part that the rock art plays in regional prehistory. A primary issue here involves chronology. Previously we (along with many other archaeologists) argued that the Carrizo pictographs were mostly Late Horizon in age (e.g., Whitley 1996), based largely on impressionistic beliefs about pictograph preservation and longevity. But the empirical evidence that we have obtained during our recent projects suggests otherwise. We believe there is reasonable evidence supporting a Middle Period origin for many, but not all, of the Carrizo paintings. This includes site associations, motif depictions, condition, comparative chronometric data, and the implications of an internal (relative) chronology.

In the first instance the pictograph sites are typically found in locational association with Middle Period village deposits rather than Late Horizon sites, and this seems more than merely coincidental. The motifs, next, include a significant number of depictions of the Pacific Pond turtle (*Clemmys marmorata*; see Figure 3). This is the only indigenous water turtle in California yet the Carrizo Plain, currently, is beyond its range, because of the absence of perennial steams. Although alternative explanations could be offered, the emphasis on this aquatic species in the art intuitively supports its creation during wetter (i.e., Middle Period) times, when this chelonian likely inhabited the area. Rock art condition also supports a generally older, rather than younger, age: many of the sites and site panels are heavily weathered; essentially all are coated with mineral skins. While the speed of natural weathering and degradation and the growth of mineral skins can obviously vary substantially, these consistent conditions again support relative antiquity rather than youthfulness. (Put another way, the absence of these conditions would support a Late Horizon age.) Moreover, one direct AMS radiocarbon date has been obtained from a Chumash pictograph site in this general region, Carneros Rocks, and this chronometric age (~1500 YBP; see Scott and Hyder n.d.) is a Middle Period date. Obviously this radiocarbon assay does not date the Carrizo pictographs. But it does indicate that paintings in this region can survive since the Middle Period. And this disproves the preservational logic upon which the shorter chronology for the rock art was originally based.

We have also developed a provisional internal chronology based on an analysis of motif superimpositions at Painted Rock, Saucito and Washburn Ranch. This identifies three phases (Figure 4):

Early: Small, dispersed, inconspicuous Chumash-style motifs
Middle: Large, conspicuous nucleated Chumash-style motifs
Late: Medium, conspicuous, dispersed Yokuts-style motifs

Figure 3. Aquatic turtle (*Clemmys marmorata*) motif from the Saucito Rocks site (CA-SLO-103). Such motifs are relatively common at the Carrizo Plain sites. That the plain is currently beyond the range of water turtles supports the interpretation that much of the art dates to the wetter Middle period. This interpretation of the age of the art is also supported by the generally heavily weathered condition of many sites and motifs, along with their association with Middle period villages.

A few points are important to emphasize concerning this sequence. All of the art, from Early to Late, conforms to neuropsychological expectations linking its production to shamanistic visionary experiences (cf. Blackburn 1977; Lewis-Williams and Dowson 1988; Whitley 1992, 1994a, 2000a). We have every reason to infer, in other words, that the Early through Late phases all involved shamanistic rock art production.

Note further that the transition from the Early to Middle pictograph phases involves a change in motif emphasis and location, but not a change in the motifs used or the style employed, and thus not a cultural change in the sense of an ethnolinguistic group replacement. The transition from the Middle to the Late pictograph phases, in contrast, involved a shift from Chumash to Yokuts-style motifs.[2] Importantly, the later are commonly found at Late Horizon and ethnographic Yokuts villages (Whitley 2000b) and therefore appear to largely post-date the Middle Period. The Carrizo pictograph sequence, in other words, is "capped" by intrusive, Late Horizon motifs, further supporting the plausibility that the two earlier pictograph phases date to the Middle Period. Yet the presence of this late Yokuts style rock art must be matched against another fact: the continued Chumash occupation of the Carrizo Plain into the Mission Period.

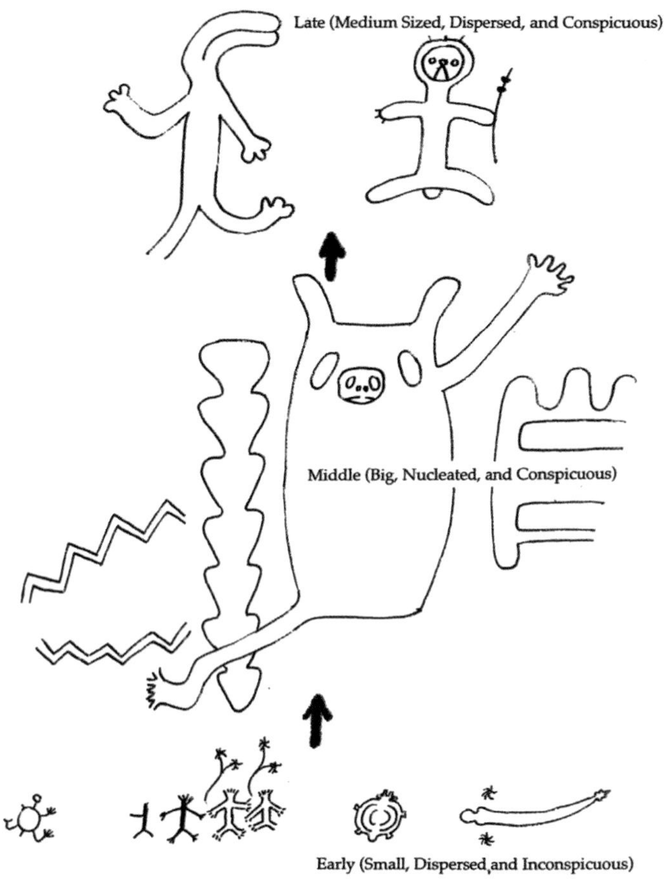

Figure 4. Preliminary rock art chronology, based on the superimpositional sequence at Painted Rock (CA-SLO-79), Washburn Ranch (CA-SLO-100) and Saucito (CA-SLO-103).

Art and Politics

Placing these various kinds of data and inferences into a larger prehistoric interpretation requires some speculative leaps. We think this worthwhile at this time, if only to provide hypotheses to guide additional research in the region. Our preliminary reconstruction of Carrizo Plain prehistory, then, is as follows:

(a) Substantial evidence now demonstrates that the Middle Period in the inland/interior Chumash region experienced dramatic population growth. In the Carrizo Plain specifically, the population within a relatively circumscribed area ultimately reached the size of ethnographic tribelet, that is, the basic autonomous land-owning political unit in Native California.

(b) The Middle Period population increase was accompanied by changing circumstances for the shaman-artists who created rock art. This involved a kind of ritual intensification that is unparalleled elsewhere in Chumash territory. It resulted in increasingly more prominent social roles for the shaman-artists, culminating in their creation of large, public compositions (such as the largest panel at Painted Rock), which visibly broadcast their powers and importance. Logically, shamans were the headmen of the Middle Period Carrizo polity[3] which, for sake of argument, we label the Carrizo tribelet. This further implies the importance of religious, rather than secular, authority.

(c) Approximately at the Medieval Climatic Anomaly—the Middle Period to Late Horizon transition—population throughout the inland/interior region collapsed. The Carrizo Plain was almost abandoned, and there was a complete or near-complete cessation in Chumash rock art production. We take this to imply the disappearance/dissolution of the previously existing tribelet, followed by minor Chumash occupation that was most likely linked to political systems centered outside of the Carrizo Plain. Moreover, this suggests an end to the religious authority on the plain.

(d) During the Late Horizon the larger Carrizo rock art sites were sporadically used by Yokuts shamans,[4] who probably recognized the previous art as signs of the supernatural importance of these locations. Regardless of how the Yokuts shamans viewed the sites, "foreign" use of them further implies the absence of local religious authority during the Late Horizon.

Our reconstruction, in other words, emphasizes two general points. The first involves the importance of changing population sizes in the interior/inland region during the Middle Period to Late Horizon transition, a circumstance we take to be empirically well verified by comparative data throughout this region. More inferentially, we believe that some form of autonomous political unit, perhaps similar to a tribelet, developed during the Middle Period in the Carrizo Plain and then collapsed, most likely due to the environmental disruptions caused by the Medieval Climatic Anomaly.

The second point concerns the evidence for ritual intensification, seen in the near unique concentration of Chumash rock art on the Carrizo Plain and its disappearance prior to the Late Horizon. We chart the rise and fall of the Carrizo tribelet as a parallel rise and fall in local religious authority, expressed in the changing intensity and nature of rock art production. This last inference is important to emphasize, as this change likewise appears to have been generalized across the Chumash region. Martz (1984), in particular, documented this process independently in her analysis of mortuary data from the Santa Monica Mountains region. She found that, in a cemetery dating from AD 300-700, 78% of the burials with "ritual status artifacts" also had "political status artifacts", this mixture dropped to about 40% in later cemeteries, indicating increasingly secular authority divorced from any form of religious trappings. Martz's data document "a rise in the power of the new lineage dominated political groups at the expense of the traditional religious leaders" (1984:487), especially after about AD 1100.

SOME WIDER IMPLICATIONS

We emphasize again that our current reconstruction is speculative, especially in some respects, and in this sense should probably be treated as a working hypothesis. That said, the Carrizo data and reconstruction have some valuable implications that even at this stage are worthwhile to note. The first is the simple fact that California archaeology has a historical bias towards coastal/island, as opposed to inland/interior, research. This is understandable on one level, but it has its downside, as the Carrizo population data illustrate. Native Californian societies certainly may have been focused on the coast and islands during the Late Horizon and ethnographic period. But they were still cultural *systems*, and these systems included inland populations, resources and events.

Moreover, the late socio-cultural focus on the coast and islands itself may have changed over time, as the dramatic evidence for Middle Period ritual intensification on the Carrizo Plain, and what this implies, suggests. Certainly there is no coastal or island evidence (that we know of) which comes close to exhibiting the kind of ritual intensification seen in the rock art on the Carrizo Plain. But equally important are the implications of the population collapse at the Middle to Late transition. Johnson (1998) estimates Chumash population at contact at about 20,000 people, approximately two-thirds of whom he places on the islands and coastal strip, leaving roughly 6000 in the interior/inland zone. Our surrogate population estimates for the Middle to Late demographic collapse from the Carrizo Plain and Hidden Creek Ranch suggest that, during the Middle Period, the inland/interior Chumash zone may have contained somewhere between 24,000 and 60,000 people—which is to say more than the coast/islands at their peak. Even if these estimates are off by 50%, this is still a massive inland/interior population that heretofore has been largely ignored.

Where then was the focus of Chumash culture during the Middle Period? And if the focus of society changed from the inland/interior to the coast, or even if it shifted from a kind of Middle Period parity to a coastal/islands emphasis during the Late Horizon, what does this imply about the data needed for confident reconstructions and explanations?

The implication here is simple: reconstructions and explanations based on coastal/island data ultimately may not prove *wrong*. Indeed, as we have emphasized a few times above, certain of the Carrizo Plain processes and events are also evident on the coast/islands, including in some cases settlement abandonment during the Middle to Late transition. Overall, in fact, there seems to be substantial complementarity between the regions, perhaps with the exception of ritual intensification. But this last issue proves the point: exclusively coastal/island explanations will always be *incomplete* and sometimes in ways that, without the inland/interior data, we cannot fully judge.

This last point is important to emphasize in slightly more theoretical terms, because it plays to the issue of the rise of socio-political hierarchies, especially chiefdoms, in coastal California, about which much has been written (e.g., Arnold 2001; Erlandson and Jones 2002) and substantial constructive debate has recently occurred. In slightly simple terms, explanations for the appearance of hierarchical societies tend to emphasize intrinsic population growth accommodated by resource abundance, or intensification resulting from resource scarcity, often combined with human agency as a central casual factor. Some, but by no means all, archaeologists argue that changing climate also plays a central role; others, that it is irrelevant. It is this last issue—climate change and its relevance—that we address here.

Our point is a simple one, and it partly turns on our earlier comment that hierarchical coastal/island societies were necessarily cultural systems with numerous links to inland/interior groups. As we have emphasized above, we believe that the disruptions caused by the Medieval Climatic Anomaly played a central role in the Carrizo collapse. Put another way, the Carrizo tribelet's reactions to the climatic and environmental disruptions caused by this event were inadequate to sustain this polity, whatever its precise form. Meanwhile, as Arnold (2001) and others (e.g., Kennett and Conlee 2002) have convincingly argued, this same period of environmental instability contributed to the development of the Chumash chiefdom, due to the stress that this created and the way that this stress was accommodated and manipulated by a developing elite to their own ends.

Our Carrizo data emphasizes two related points. The first is that, *even within one cultural system*, response to environmental stress can differ dramatically. The Medieval Climatic Anomaly in this sense appears to have created conditions that at once set the stage for the development of a hierarchical society on the coast, meanwhile destroying at least one other political unit that was potentially poised to make this same transition prior to environmental catastrophe. The appearance of the Chumash chiefdom was in this sense truly a transformation (not an evolution) of previously existing conditions, one aspect of which involved the consolidation of disparate, perhaps even warring, elements into a whole. And while one element in this mix did emerge at the end of the process as the victor in this transformational process, this almost certainly was matched by many more elements, like the Carrizo Plain tribelet, that devolved from possible autonomous polities to peripheral backwaters.

Likewise, and perhaps more importantly, ritual intensification cannot be easily correlated with a single kind of circumstance or attributed to one overriding process. As one of us has argued earlier (Whitley 1994b), the period of environmental stress associated with the Medieval Climatic Anomaly resulted in a marked ritual intensification in the western Great Basin, illustrated by the florescence of Coso petroglyphs emphasizing rain shamanism and the appearance of Numic bands and headmen. But the preceding

Holocene Climatic Optimum appears to have led to ritual intensification on the Carrizo Plain and probably the development of a tribelet and headmen, whereas the Medieval Climatic Anomaly contributed to its demise.

The ultimate conclusion that one can draw here is the fundamental complexity of the human condition and career. Life, for all of us, is a complex welter of choice and circumstance mediated by processes that too often are out of our control. And it would be rebarbative in the extreme, to assume that somehow the prehistoric past was any different in this regard. But this does not mean that the past is therefore intrinsically unknowable or that there is no value in plumbing its secrets. Indeed—as we suspect Jay von Werlhof would be quick to agree—it is that very process that brings meaning to life, and builds meaning for a past that otherwise would be unknown, if not unknowable.

Acknowledgments

Our work in the Carrizo Plain National Monument has benefitted greatly from the assistance, cooperation, and enthusiasm of the BLM and monument staff. We are particularly grateful to Duane Christian, Marilyn Braun, Johna Hurl, Pat Apply, and Cindy McWhinney. We also thank Mike Hakim, Michael Kuhn, Rob Snibbley, and Tamy Whitley, who have assisted with various aspects of our studies; John Johnson, who is always generous with his ethnohistorical information and insights; and Erica Seibert at the office of the Keeper of the NRHP, who has been particularly helpful concerning our many inquiries about NRHP and NHL nominations. We thank Russ Kaldenberg for organizing the SCA session for Jay, from which this paper originates, and editing the resulting volume, and Marcia Sandusky for her editing and production efforts. But mostly we would like to thank Jay himself, who has always served as a central humanizing beacon in this science that we call archaeology.

References

Arnold, Jeanne E., editor
2001 *The Origins of a Pacific Coast Chiefdom: The Chumash of the Channel Islands*. Salt Lake City: University of Utah Press.

Basgall, Mark
1999 Comment on "Environmental Imperatives Reconsidered: Demographic Crisis in Western North America during the Medieval Climatic Anomaly," *Current Anthropology* 40:157-158.

Bettinger, Robert
1999 Comment on "Environmental Imperatives Reconsidered: Demographic Crisis in Western North America during the Medieval Climatic Anomaly," *Current Anthropology* 40:158-159.

Blackburn, Thomas
1977 Biopsychological Aspects of Chumash Rock Art. *Journal of California Anthropology* 4:88-94.

Eigenbrode, J.L.
1999 *Sedimentological, Carbon-Isotopic and Molecular Records of Late Holocene Climate in the Sediments of Soda Lake, Carrizo Plain, California*. Unpublished M.S. Thesis, Dept. of Geological Sciences, Indiana University.

Eigenbrode, J.L., C. Davis, L.M. Pratt, J.R. Arrowsmith, D.D. Rhodes and L. Rossbacher
1997 Recent Climate Change Recorded by Sedimentary Organic Matter in Soda Lake, Southern California. *Geological Society of America Abstracts with Programs* 29(6):A155.

Eigenbrode, J.L., L.M. Pratt, D.D. Rhodes, J.R. Arrowsmith and G. Mora
1999 Sedimentological and Carbon-Isotopic Records of Late Holocene Climate in the Sediments of Soda Lake, Carrizo Plain, California. Geological Society of America Abstracts with Programs, v. 31. *Annali di Geofisica* 39:1301-1307.

Erlandson, Jon M. and T.L. Jones, editors
2002 *Catalysts to Complexity: Late Holocene Societies of the California Coast*. Perspectives in California Archaeology, 6. Cotsen Institute of Archaeology, UCLA.

Gill, Richardson B.
2000 *The Great Maya Droughts: Water, Life and Death*. Albuquerque: University of New Mexico Press.

Grant, Campbell
1965 *The Rock Paintings of the Chumash*. Berkeley: University of California Press.

Hassan, Fekri A.
1981 *Demographic Archaeology*. Academic Press, New York.

Heizer, R.F. and C. W. Clewlow, Jr.
1973 *Prehistoric Rock Art of California*. Ramona: Ballena Press.

Hyder, W., G. Lee and M. Oliver
1986 Culture, Style and Chronology: The Rock Art of the Carrizo Plain. *American Indian Rock Art* 11: 43-48.

Johnson, J.R.
1988 *Chumash Social Organization: An Ethnohistoric Perspective*. Ph.D. dissertation, UCSB.
1999 Chumash Population History. In S. McLendon and J.R. Johnson, editors, *Cultural Affiliation and Lineal Descent of Chumash Peoples in the Channel Islands and the Santa Monica Mountains*, pp. 93-130. Santa Barbara Museum of Natural History.

Jones, T.L., G.M. Brown, L.M. Raab, J.L. McVickar, W.G. Spalding, D.J. Kennett, A. York and P.L. Walker
1999 Environmental Imperatives Reconsidered: Demographic Crisis in Western North America during the Medieval Climatic Anomaly. *Current Anthropology* 40:137-170.

Kennett, Douglas J. and Christina A. Conlee
2002 Emergence of Late Holocene Sociopolitical Complexity on Santa Rosa and San Miguel Islands. In J.M. Erlandson and T.L. Jones, editors, pp. 147-165, *Catalysts to Complexity: Late Holocene Societies of the California Coast*. Perspectives in California Archaeology, 6. Cotsen Institute of Archaeology, UCLA.

Lee, G. and W. Hyder
1991 Prehistoric Rock Art as an Indicator of Cultural Interaction and Tribal Boundaries in South-central California. *Journal of California and Great Basin Anthropology* 13:15-28.

Lewis-Williams, J.D. and T.A. Dowson
1988 The signs of all times: entoptic phenomena in Upper Palaeolithic art, *Current Anthropology* 29:201–245.

Martz, P.C.
1984 *Social Dimensions of Chumash Mortuary Populations in the Santa Monica Mountains Region*. Ph.D. dissertation, University of California, Riverside.

Perry, C.A. and K.J. Hsu
2000 Geophysical, archaeological, and historical evidence support a solar-output model for climate change. *Proceedings of the National Academy of Science* 97(23):12433-12438.

Schoenherr, A.A.
1992 *A Natural History of California*. Berkeley: University of California Press.

Scott, David A. and William D. Hyder
n.d. A Study of Some Californian Indian Rock Art Pigments. Unpublished manuscript, Getty Conservation Institute.

W&S Consultants
1994a Phase II Test Excavations and Determinations of Significance at CA-LAN-2133, -2233, -2234, -2235, -2236, -2240, -2241 and -2242, Los Angeles County, California. Manuscript on file, CSUF AIC.
1994b Phase I Archaeological Survey and Cultural Resources Assessment for the Moorpark Specific Plan Area #8, Ventura County, California. Manuscript on file, City of Moorpark.
1995 Phase II Test Excavation and Determinations of Significance at CA-VEN-508, -1132, -1133, -1134, -1135, and -1136, Moorpark, Ventura County, California. Manuscript on file, City of Moorpark.
1997 Phase II Test Excavations and Determinations of Significance at CA-VEN-1130, -1137, -1138, -1139, -1140, -1141, -1142, -1143, -1144 and -1147, Moorpark, Ventura County, California. Manuscript on file, City of Moorpark.
2003a Reconnaissance Level Archaeological Survey of Portions of the Carrizo Plain National Monument, San Luis Obispo County, California. Report on file, BLM Bakersfield office.
2003b Documentation and Analysis of the Rock Art at CA-ORA-269, Irvine, Orange County, California. Manuscript on file, LSA Associates, Inc.
2004 Class II Inventory of the Carrizo Plain National Monument, San Luis Obispo County, California. Report on file, BLM Bakersfield office.

Wallace, W.
1955 A Suggested Chronology for Southern California Coastal Archaeology. *Southwestern Journal of Anthropology* 11 (3):214-230.

Waugh, Georgie
1999 A Study in the Prehistory of the Santa Clara River Valley: Archaeological Data Recovery at CA-LAN-2233, Los Angeles County, California. Cultural Studies Office, CALTRANS, Sacramento.

Whitley, David S.
1992 Shamanism and Rock Art in Far Western North America. *Cambridge Archaeological Journal* 2:89-113.
1994a Shamanism, Natural Modeling and the Rock Art of Far Western North America. In *Shamanism and Rock Art in North America*, edited by S. Turpin, pp. 1-43, Special Publication 1, Rock Art Foundation, Inc., San Antonio.
1994b By the Hunter, For the Gatherer: Art, Social Relations and Subsistence Change in the Prehistoric Great Basin. *World Archaeology* 25: 356-377.
1996 *Guide to Rock Art Sites: Southern California and Southern Nevada*. Missoula, MT: Mountain Press Publishing.
2000a *The Art of the Shaman: The Rock Art of Native California*. Salt Lake City: University of Utah Press.
2000b Use and Abuse of the Ethnohistorical Record. *American Indian Rock Art* 26:1-12.

2004 Hawai'ian Cupules And Numic Vulva-Forms: Two Case Studies In Gender And Rock Art. Paper presented at the "International Conference on Gender and Rock Art: from Cave Art to Contact," Foix, France.

Yellen, John
1977 *Archaeological Approaches to the Present: Models for Reconstructing the Past.* New York: Academic Press.

END NOTES:

1. Our reasons for setting the beginning of the Middle Period in the inland/interior Chumash region at about 4000 YBP are archaeological, in the larger sense, and paleoclimatic and require some explanation. First, we do not use the current coastal chronology (particularly the dating of the Early to Middle transition) because it is based on typological distinctions in artifacts that, in over 25 years of field experience, we have very rarely if ever encountered in the inland/interior region. There is in our minds no value to imposing a coastal chronology on the inland/interior region when the coastal chronology is effectively unrecognizable in the archaeological record. Wallace's (1955) classic chronology, in this sense, much better fits the inland data and sites, and it is his chronology that we modify slightly here. Second, research has demonstrated the existence of a number of inland/interior sites dating to what traditionally was referred to as the terminal Early Horizon and early Middle Period transition (e.g., Waugh 1999). This suggests that a cultural break exists between the terminal portions of the Early Horizon (or Early Millingstone) Period and earlier times, with cultural continuity in the inland/interior region then existing between the Terminal Early Horizon phase and the Middle Period. The implication is that the meaningful cultural-historical distinction is the break between middle and late Early Horizon versus terminal Early Horizon phases, when intensive inland/interior occupation first occurred, and not Early Horizon versus Middle Period, per se, as the existing chronologies emphasize. This change can be set at roughly 4000 BC. Third, the Holocene Climatic Optimum, which has been described as the optimal climatic conditions for the Holocene, starts at about 3800 YBP (Perry and Hsu 2000). It is impossible for us not to link the explosion of inland/interior population at essentially this time with this dramatically improved environment.

2. The presence of both Chumash and Yokuts-style motifs in the Carrizo Plain has been noted by a number of researchers (e.g., Hyder, Lee and Oliver 1986; Lee and Hyder 1991). Heretofore this has been seen as a simple mixing of motifs, reflecting more or less simultaneous use of the sites by different ethnolinguistic groups. Our relative chronology suggests instead that these stylistic distinctions also reflect temporal trends in site use.

3. Justification for this inference is more than simply intuitive, and we have discussed it at some length elsewhere (W&S Consultants 2003b, 2004; Whitley 2004). In very quick summary, it involves four issues. First, religion and politics are commonly homologous in traditional, small-scale societies and overtly religious rather than purely secular authority and for this reason, to be expected (cf. Whitley 1994b). Second, the rock art itself needs be understood not as some socially ambiguous aesthetic expression but instead as a manifestation of the manipulation of supernatural power and thus an illustration of control of what was thought by Native Californians as the most powerful force and principal causative agent in the universe. The creation of rock art illustrated an individual's access to power, and this power was thought much more than simply a spiritual conceit. Third, Chumash rock art sites are not evenly distributed on the landscape but in fact are quite sporadically distributed, including occasional dense concentrations, as in the Carrizo case. It follows that the distribution of sites illustrates the distribution of religious power and authority across the landscape. Fourth, as discussed above, the Chumash transition from religious to increasingly secular authority has been independently documented in mortuary analyses (Martz 1984).

4. We have observed Yokuts-style pictographs at Painted Rock and Saucito, the two largest sites on the Carrizo. Washburn Ranch, which is much smaller and less accessible, appears stylistically to be exclusively Chumash.

HIGH-STAND SHORELINE SURVEY IN THE SEARLES LAKE AREA, SAN BERNARDINO COUNTY, CALIFORNIA

David S. Whitley, Ph.D., and Joseph M. Simon, Ph.D.
W&S Consultants
447 Third Street
Fillmore, California 93015

Niccole Cerveny, Ph.D.
Mesa Community College
1833 W. Southern Avenue
Mesa, Arizona 85202

Ronald I. Dorn, Ph.D.
Arizona State University
P.O. Box 870104
Tempe, Arizona 85287

The Late/Terminal Pleistocene shorelines of the California deserts have fascinated archaeologists at least since the 1930s, when the Campbells conducted their pioneering studies in the Lake Mohave area (Campbell et al. 1937). Their interest in the relationship of human occupation to ancient lake margins was extended into the northern Mojave Desert by Emma Lou Davis (1978) who, with a series of colleagues, spent a number of years working in the Ridgecrest area, mostly at Pleistocene Lake China and the Panamint Valley. Though Davis' contributions to the archaeological literature are well-known, she also had a significant local impact, especially among those who volunteered for her projects. One of these was the artist Sylvia Winslow, whose personal interests in local archaeology were fanned by Davis. As a result of their working relationship and friendship, Winslow scoured much of the region looking for potential early sites. One of the areas that looked particularly promising to her was the Christmas Canyon sub-basin of Searles Valley, located south of the town of Trona. The Christmas Canyon Area of Critical Environmental Concern (ACEC) was created by the Bureau of Land Management (BLM) at her suggestion in order to preserve what was clearly a dense concentration of sites on the south side of this geomorphological feature.

The Christmas Canyon area is a remnant of the Pleistocene system of lakes that covered the Great Basin until the end of the Pleistocene (Figure 1). Between roughly 30,000 and 15,000 years before present (YBP) (Bishoff and Cummings 2001) and again about 11,500 YBP (Smith et al. 1983, 1993), when these lakes were at their high stand, paleohydrological and geomorphological conditions resulted in the creation of a beach ridge where Christmas Canyon enters Searles Valley. This large beach ridge separated this small embayment from Searles Lake (and Valley) as a whole. This had two consequences. First, it contributed to the preservation of an intact Late Pleistocene and Early Holocene landscape, one that has not suffered from the subsequent erosion and degradation experienced by the vast majority of the California Desert. Second, and largely due to this first fact, this intact ancient landscape contains a dense, very well-preserved and extremely important concentration of Late Pleistocene and Early Holocene archaeological sites; that is, sites dating between 7,000 and 11,000 YBP, if not potentially earlier.

From 2002 to 2004, we conducted a series of projects within the Christmas Canyon sub-basin for the BLM. These included site micro-mapping within the ACEC itself, reconnaissance-level and intensive site surveys more widely across the sub-basin, and dating research. While there is still considerable work to be done within this area, the following is a preliminary report on our studies, intended to highlight some of the key findings so far. One of these involves rock cairns and geoglyphs, which are unusually common in this area.

Rock cairns and geoglyphs are effectly ignored by many archaeologists, but they were of course particularly interesting to Jay von Werlhof (1987, 1995, 2004). We hope for this reason that the comments we offer below on their age, distribution and possible functions will contribute to the research on this topic that Jay pioneered.

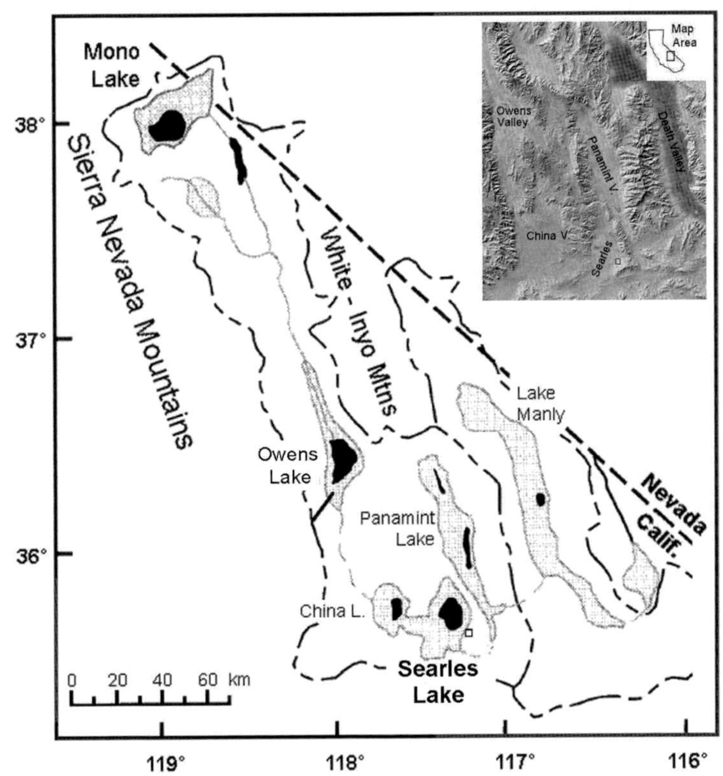

Figure 1. Series of closed hydrologic basins created by the tectonics of eastern California. During the wetter late Pleistocene, the surface hydrology of eastern California (adapted from Smith et al. 1983) connected a series of paleolakes that terminated at Death Valley about 140,000 YBP and Panamint Valley about 30,000 to 15,000 YBP. The timing of these lakes corresponds with glacial advances in the Sierra Nevada (Bishoff and Cummings, 2001). The small box on the southeast side of Searles Valley indicates the location of the Christmas Canyon field site.

Surface Exposure Dating

The first of our projects provided the initial evidence pointing to the great age and therefore importance of the sites in this area (Cerveny et al. 2006). Using a variety of chronometric techniques, 30 chronometric ages have been obtained from ACEC sites. These derive from surface stone tools, cairns, rock rings, rock quarry surfaces, and geoglyphs. Note that all of these types of archaeological remains are found exposed on the groundsurface rather than buried in the soil and the dating techniques employed specifically address surface exposure dating. They include varnish microlamination (VML) dating; cation-ratio (CR) dating, lead-profile dating, and AMS ^{14}C dating of calcium carbonate coatings (see Dorn 2001 for a summary of these approaches). These are typically combined for any given specimen.

While all of these approaches represent experimental techniques, the combination of multiple independent techniques on individual specimens promotes confidence in the results. Whereas other methods may take an abundance of time (e.g. varnish microlaminations) or funds (e.g. accelerator radiocarbon dating), CR dating provides researchers a chance to obtain a snapshot — a quick and relatively inexpensive "overview" of the chronometry.

The CR ages range from 500 ± 200 YBP to 14,800 ± 1500 YBP (Table 1); that is, from the recent prehistoric past to the Pre-Clovis Period. (Note that the error factors given for these dates represent 2 standard deviations or 98% confidence interval). Fully one-third of these dates are greater than 7000 YBP (at 2 standard deviations) and thus pertain to the periods of greatest archaeological interest and significance, in the California deserts. Of particular note in this regard are ^{14}C ages of 7260 ± 40 YBP and 8440 ± 60 YBP on two rock rings, and ages of 14,800 ± 1500 and 8800 ± 2000 YBP on two different worked exposures of a jasper quarry. Put another way, ages were obtained from 13 different sites, and six of these sites—or 46% of them—had ages greater than 7000 YBP.

The general strategy of much of the dating work, particularly concerning rock features, involved identifying human alterations to natural landforms, settings, and contexts. Consider a boulder in a desert pavement. A key characteristic of the general Christmas Canyon geomorphological context is long-term stability. In the case of the boulder, this means that it has a particular orientation and position relative to the air, the groundsurface, and the subsoil, and this orientation has been stable over time (Figure 2A). Note in this regard four different kinds of the rock coatings that will form over time on our boulder: black rock varnish, in the area exposed to the air; a centimeter-wide black line of very shiny varnish that occurs as a kind of ground-band; an orange iron film that occurs on the undersides of boulders; and a white pedogenic carbonate coating that will develop, given enough time and given that the boulder is emplaced deep enough (e.g. 40 cm) in the subsoil.

Human modification, such as re-positioning the boulder, can change this orientation and thereby change the kind of rock coating that any single portion of the boulder develops over time. Because each of the original kinds of rock coatings signaled a particular position in the ground, changes in these coatings reveal both that the boulder has been moved, and that the move results in a kind of palimpsest of different rock coatings that can be used to constrain the age of this positional modification; that is, the human action that moved the boulder.

Table 1. Cation-Ratio Ages on Materials in the Christmas Canyon Area, Using the Cation-Leaching Curve Developed for the Coso Range Petroglyphs (cf. Dorn 1998).

Site	Sample	Cation Ratio Age
Site #1		
	01-BLM-XMAS-1c Chert Core	2100±500
	01-BLM-XMAS-1a Rhyolitic Flake	2300±600
	01-BLM-XMAS-2c Chert Core	8600±800
	01-BLM-XMAS-2a Quartz Flake	8000±900
Site #13		
	01-BLM-XMAS-3a Quartz Flake Scar from a scatter	7100±100
	01-BLM-XMAS-4a Quartz Flake Scar from a scatter	6700±120
	01-BLM-XMAS-5a Quartz Flake Scar from a scatter	6000±100
Site #11		
	01-BLM-XMAS-27a refitted flake	800±250
	01-BLM-XMAS- 28a refitted flake	950±250
Site #2		
	01-BLM-XMAS-6a Quartz Flake Scar from a scatter	2100±600
	01-BLM-XMAS-7a Quartz Flake Scar from a scatter	1600±300
	01-BLM-XMAS-8a Quartz Flake Scar from a scatter	1800±600
	01-BLM-XMAS-9a Quartz Flake Scar from a scatter	1500±200
	01-BLM-XMAS-10a Quartz Flake Scar from a scatter	2000±600
Site #3		
	01-BLM-XMAS-11a refitted quartzite cobble	600±200
	01-BLM-XMAS-12a refitted quartzite cobble	500±200
	01-BLM-XMAS-13a refitted quartzite cobble	600±200
Site #15		
	01-BLM-XMAS-23a chert flake	1800±400
	01-BLM-XMAS-24a chert flake	1500±500
	01-BLM-XMAS-25a chert flake	1500±400
	01-BLM-XMAS-26a chert flake	1900±700
	01-BLM-XMAS-3c jasper corer	1800±400
Quarry Area		
	Davey Flake	6600±130
	Quarry Face 1	14,800±15
	Quarry Face 2	8800±200
	Quarry Face 3	800±200
	"Running Man" Geoglyph Chip	900±350
	Snake Geoglyph Chip (on flipped orange bottom varnish)	<300

If this boulder were rotated (Figure 2B), for example, in a way that exposed the former groundline/subsurface orange varnish to the air, this rotation can be readily identified. (Confidence that this change was due to human modification occurs if boulders in the immediate surrounding area lack any evidence of this repositioning, indicating that it did not occur for example due to general downslope movement). The varnish microlamination method (Liu 2003) can be used in such cases. In the case of our boulder from the Christmas Canyon ACEC, the method yielded evidence of a terminal Pleistocene age for the creation of this cairn (Figure 2C). that carbonate coatings themselves have a kind of micro-stratigraphy, and we felt that by AMS dating only the lowest carbonate layers that are partly silicified, we would have a better chance of obtaining an inferentially useful AMS age. Using specimens graciously provided by Luz Ramirez de Bryson, we applied this technique to two samples from a site called Lagunitas, located near the playa bottom in the ACEC, where she was working.

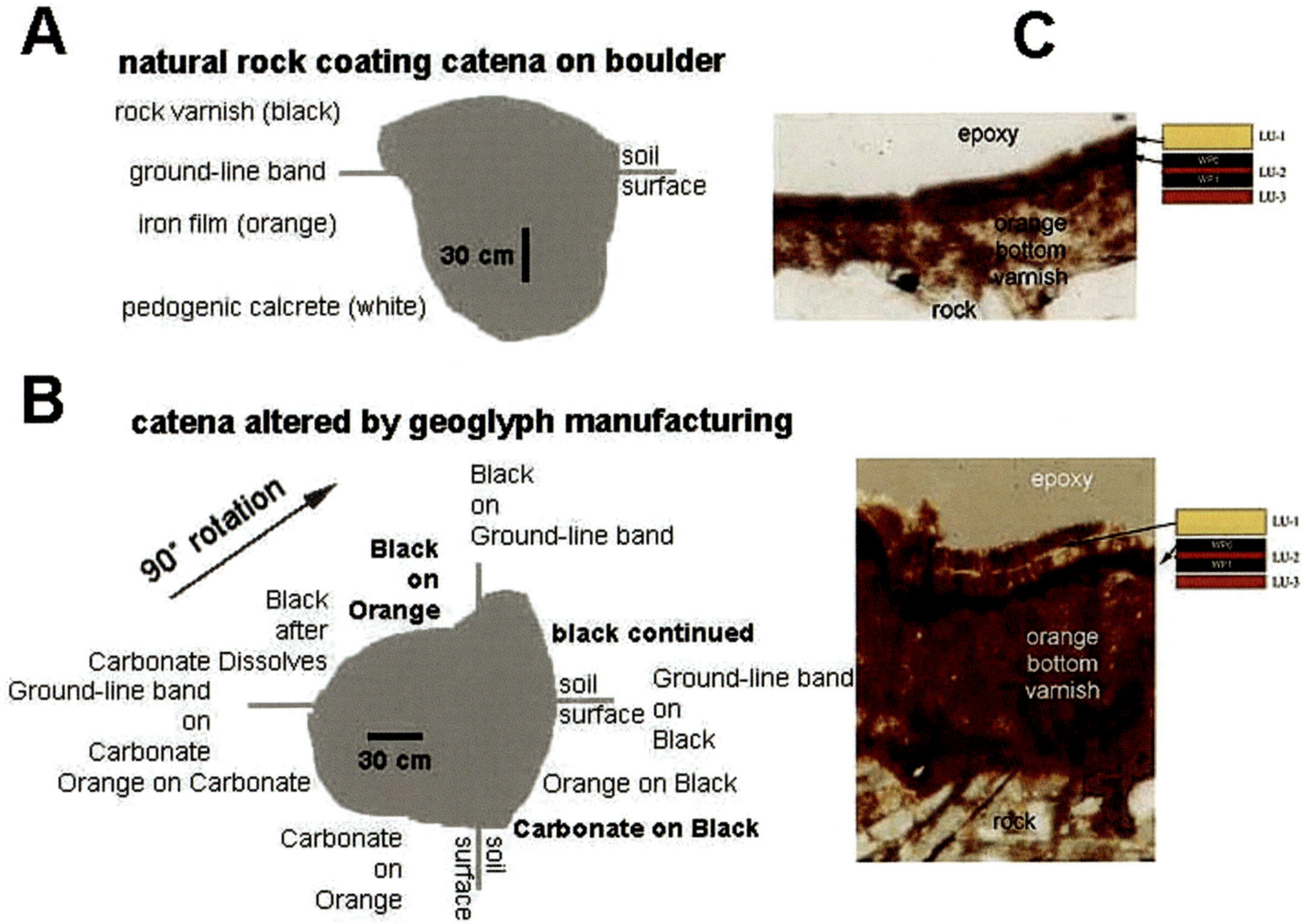

Figure 2. Changes to a boulder's position alter the orientation of rock varnish at a particular point in time. In this case, a cairn boulder was moved by humans sometime after the 14,000 YBP black layer was deposited (i.e. not seen in the microstratigraphy) and sometime before the 11,000 YBP black layer was laid down (i.e. seen in the microstratigraphy).

One experimental area of dating research that we investigated involved the AMS dating of the carbonate rinds that coat buried portions of rocks (whether artifactual or natural). Carbonate dating is known to be problematic because carbonate continues to form while the coated specimen is buried and because it can be an open system where newer carbon comes and goes. But we recognized

The innermost carbonate rinds surrounding a littoral cobble and a possible artifact both had a platey texture, probably as a result of partial silica replacement of the carbonate. The outer carbonate was powdery and after mechanical removal was not included in further sample processing. The innermost plates were mechanically separated from the host rock and cleaned of rock detritus. The platey-textured material surrounding the artifact yielded a radiocarbon age of 10,200±50 YBP (Beta-175484), correcting for stable isotopes. The platey-textured material surrounding natural cobbles in the same deposit yielded a radiocarbon age of 11,160±50 YBP (Beta-175485).

Note that these two chronometric ages do not overlap, even at 2 standard deviations. This may be due to a failure in our experimental technique, or due to simple ("true") differences in the specimen ages, resulting from the fact that the sampling process was not precise enough or that they came from different depths in a stratigraphic unit and therefore are not quite identical in age. Still, they fall within the same temporal ballpark and, for this reason alone, they can be provisionally accepted. Moreover, note that the error margins on AMS ages such as these represent instrumental errors; that is, errors in the measuring precision of the instrument. While AMS dating is very *precise*, whether a given analytical age is *accurate* or not is another question entirely. (This is a fundamental distinction that unfortunately too few archaeologists adequately apprehend.) If we could calculate the accuracy of AMS ages rather than the precision of the instruments used to calculate the ages, we might find that the true uncertainties of these two ages overlap.

An inferentially stronger indication of the age of at least one of the ACEC sites was obtained through the chronometric analysis of a projectile point base collected from a desert pavement (Figure 3). This large-stemmed base and the site where it was recovered are discussed in more detail below. Suffice it here to note that the base was dated using the VML technique, which yields a minimum age for the exposure of the underlying artifact surface. Using Liu's (2003) calibration verified in a blind test (Marston 2003), the VML sequence indicates that the artifact is likely between 14,000 and 11,000 ^{14}C years old. The dating work establishes that a large and significant concentration of early sites has been preserved in the Christmas Canyon sub-basin, including sites that might contribute to our understanding of the first peopling of the Americas. Moreover, the early use of the ACEC suggested by these dates is supported by independent evidence in the form of two additional projectile point discoveries. In addition to the VML-dated large-stemmed base, these include an unfluted Paleoindian point base and a Lake Mohave point base. These two artifacts and their archaeological and geomorphic context minimally confirm Terminal Pleistocene/Early Holocene use of the sub-basin, as suggested by our dating project.

Micro-Mapping and Site Reconnaissance

A second project involved micro-mapping 16 sites at the northern end of the ACEC, and reconnaissance level surveying in the high-stand lakeshore area within the ACEC. This work resulted in a fairly detailed characterization of the archaeological remains within the ACEC specifically (although the density and nature of archaeological sites in other parts of the sub-basin were then of

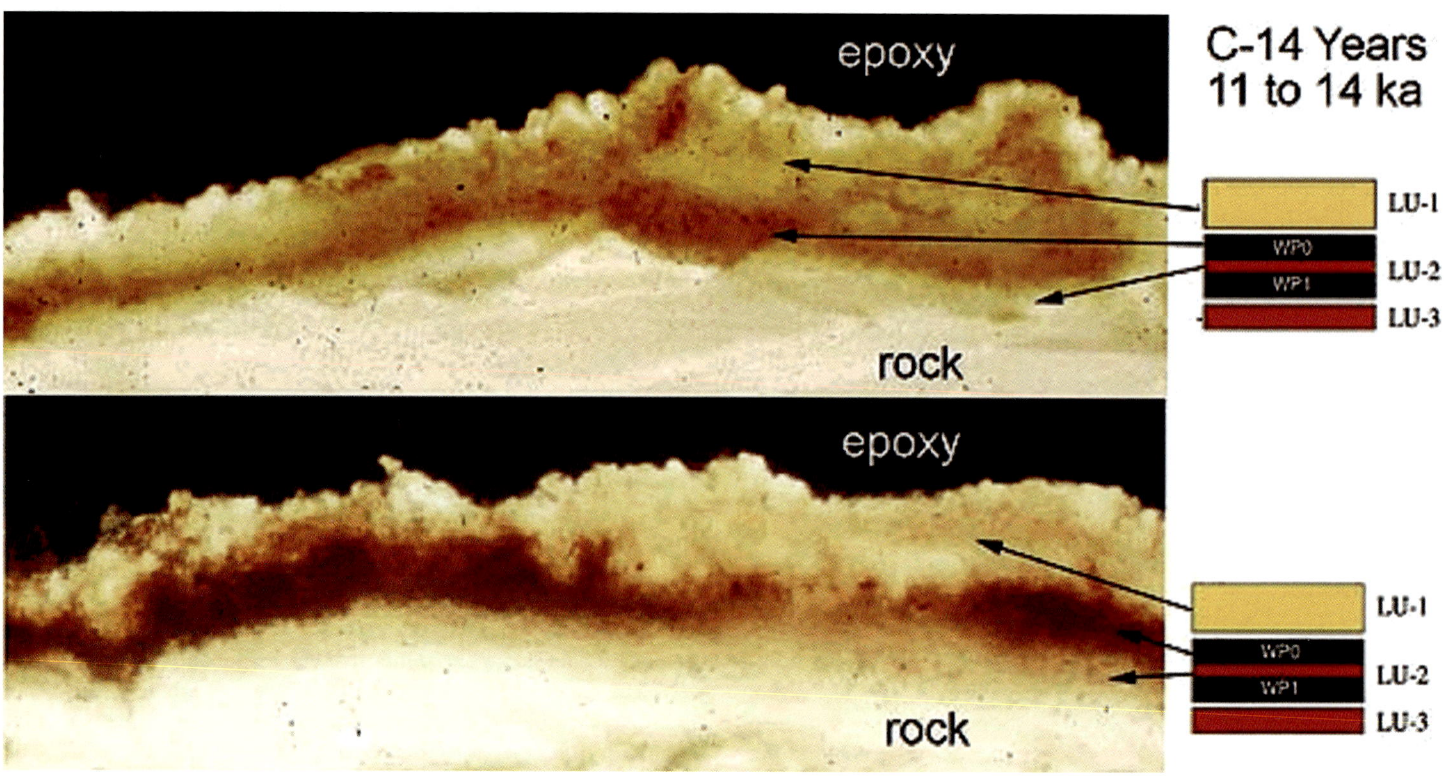

Figure 3. A large-stemmed point base (see Figure 6) found at the Christmas Ridge site shows a microstratigraphic sequence consistent with a terminal Pleistocene age.

course unknown). The groundsurface at these 16 sites, and much of the existing ACEC, is covered by an intact desert pavement. Included in this pavement are numerous pieces of "float" jasper, a crypto-crystallate that was commonly used for the manufacture of stone tools. Because this lithic resource is dispersed across the desert pavement in large cobbles mixed in with other kinds of rock (such as basalt), prehistoric exploitation of it was a widespread and somewhat opportunistic activity (that is, in comparison to the one known jasper quarry within the ACEC, where exploitation was intensive). The result, however, was the creation of a large number of discrete and intact "chipping stations": locations where prehistoric peoples attempted to transform natural jasper cobbles into stone tools. These chipping stations are found across the areas of the ACEC covered by desert pavement, and from an archaeological perspective, they are extremely useful for interpreting the past because they represent a single behavioral event. How common these were in other portions of the sub-basin was initially unknown.

An example of the density of these types of sites within the ACEC, nonetheless, is provided by a survey transect that was approximately one kilometer long by 150 meters wide. This resulted in the recording of nine separate sites—or almost one site every 100 meters, suggesting that there are probably thousands of these sites in the ACEC as a whole. The chronometric dates on these chipping stations range from 500 $\pm$ 200 to 8600 $\pm$ 800 YBP; that is (at 2 standard errors), fully into the Paleoindian Period.

A second characteristic of the ACEC—and, as we ultimately discovered, of the sub-basin as a whole—is an unusually large number of rock features. These range from small piles of cobbles to larger, almost sentinel-like cairn structures that are visible on the horizon, to small and also large rock rings, to concentrations of cobbles embedded in the groundsurface. The functions of these features are not yet known, although it is likely that different ones were made for different purposes (see below). Blair and Fuller-Murillo (1997), for example, documented over a dozen kinds of rock circles in the ethnographic record. These ranged from hearths, to sleeping circles, hunting blinds, pot rests, caches, and storage areas of various kinds and ritual constructions.

Certain of the Christmas Canyon rock rings appear to be hearths, while others are likely "habitation" structures of some kind: perhaps foundations for brush shelters, perhaps "spirit circles" or vision quest structures. Some of the rock piles may also be remnants of deadfall traps. But a concentration of over a dozen small rock cairns at a site on a low fan in the southern extreme of the ACEC belies any argument that all of these were made for "economic" or subsistence purposes. Almost certainly, this concentration of cairns was ceremonial in origin, signaling the diversity of kinds of archaeological topics and questions raised by the sites in the district. Indeed, the importance of the sites in the Christmas Canyon sub-basin with respect to questions of prehistoric religion (cf. Whitley n.d.) is further seen in the large numbers of geoglyphs ("earth figures", formed by low rock alignments) that are also present. Historically at least, these were created as part of ritual activities conducting during vision quests and it is likely that a similar purpose existed in the prehistoric past (Whitley et al. 1999a; 2004).

Again, chronometric ages obtained on a few of these rock features emphasize their importance: dates of 7260 and 8440 YBP place two of them in the Early Archaic Period. Features associated with older sites suggest that some of them may be even older (see Figure 2).

One overriding goal of our studies was to determine whether the archaeological conditions identified in the ACEC were characteristic of the Christmas Canyon sub-basin as a whole. Of obvious and particular concern was the presence of intact Late/Terminal Pleistocene and Early Holocene landforms, with associated sites.

Because the Christmas Canyon sub-basin was part of an extensive Pleistocene lake system (which included the Owens, China Lake, Searles, and Panamint basins), and because this lake system was influenced by widespread paleoclimatic trends, the implications of well-established ancient climatic events can be used to guide preliminary archaeological research and interpretation, especially concerning Late/Terminal Pleistocene and Early Holocene prehistory. For example, the sub-basin was filled with water, to an elevation of about 2263.8 ft or 690 meters (above modern sea level) from approximately 30k to 15k YBP (Bischoff and Cummings 2001), creating a lake that was roughly 300 feet deep. The lake waters receded dramatically and quickly thereafter, due to drying conditions after 14,000 YBP, probably creating marshlike conditions on the very bottom of the basin during Paleoindian times. A resurgence in lake waters occurred circa 11k YBP, probably reaching the overflow point (Smith et al. 1983) before the lake finally dried out entirely after 10k YBP.

The general hydrological pattern is shown in Figure 4, which summarizes the circumstances for the Owens River system that connected the Sierra Nevada with Searles Valley (see Figure 1). Some initial observations are relevant with regard to this reconstruction and its implications for the Searles area. First, the Late Pleistocene Tioga (post-16,500 YBP) high-stand is easily identified geomorphologically and is known to fall at about 690 m elevation (above modern sea level). Its precise age is yet to be determined, but roughly 14-15k YBP is a reasonable approximation for the high-stand at Searles Lake. Confidence in this high-stand is provided by the correspondence of available ages (cf. Smith et al. 1983) with the timing of overflow of Owens Lake (Bischoff and Cummings 2001).

Second, the circa-11,000-YBP wet pulse and partial basin refilling is controversial and still ambiguous, but based on decades of study by George Smith (Smith et al. 1983). While there is general agreement that such a wet event occurred throughout the southwestern Great Basin, there is no concurrence on its magnitude. This issue is particularly problematic with regards to Searles Valley and the Christmas Canyon sub-basin, which contain numerous minor transgressive and regressive lake terraces and beach strand-lines. Despite considerable effort, geomorphologists have been unable to conclusively identify the elevation of the Terminal Pleistocene high-stand in the Searles region.

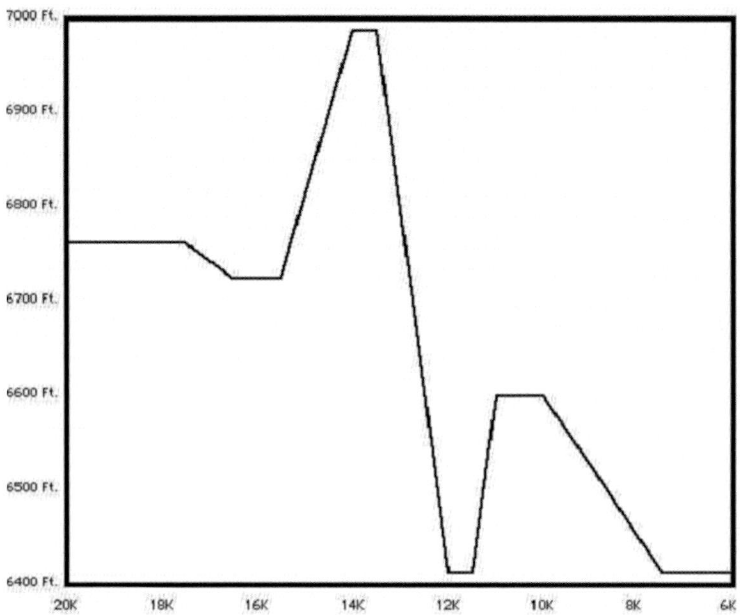

Figure 4. General levels of Pleistocene Lake Russell (after Lajoie 1968; Smith and Street-Perrott 1983).

Third, archaeologists commonly perceive of lake in-filling as a time-progressive process involving the filling of the lake at the top of the series, its overflow into the next lake, and so on down the line. The implication here is chronological, with the last lake filling after all the others. In fact, much of the movement of water down a drainage system is subsurficial, with lakes filling due to hydrostatic charge, not solely due to a spill-over effect when upstream waters reached a particular elevation. Although certain of the Pleistocene lakes certainly did coalesce, overflow alone was not responsible for the creation of the lake system, and the complete lake system existed before the overflows occurred.

The implications of these general conditions are relatively straightforward. Any archaeological sites in the sub-basin that pre-date about 30k YBP—if such exist—are most likely buried in lake deposits. Preserved sites that date to the earliest part of Pre-Clovis Period, before 14-15k YBP, necessarily must be located above the high-stand lake level, at 690 m elevation. Pre-Clovis sites that are between 14-15 ka and 11 ka as well as Paleoindian-aged sites that are 10k YBP or less, on the other hand, potentially may occur anywhere above the playa bottom. Given that the lake edge would provide a good habitat for hunting and gathering during both of these time periods, however, we can predict that Pre-Clovis sites would be located immediately above the Tioga high-stand whereas Paleoindian sites would be most common immediately alongside the (then-marshy) playa bottom, roughly 275 feet lower. Sites during the low stand between 14 ka and 11 ka could have experienced terminal Pleistocene reworking as Searles Lake rose and then fell.

Moreover, and perhaps most importantly, the water that filled Searles and the sub-basin was saline, hence the massive deposits of mineral salts that have accumulated in these playa bottoms and are now mined in nearby Trona. Habitation sites logically would be located along the lake shore or marsh edge, providing access to the rich plant, animal, and fish resources present in these locations. But they would also be tied to fresh water sources: streams that entered the lake or marsh, providing potable water.

This "coastal geomorphology" model for Late/Terminal Pleistocene settlement location was developed from our archaeological work along the southern California coast. The model was tested in the Christmas Canyon sub-basin during our reconnaissance-level survey of the ACEC high-stand shoreline area. Based on a dry-lab map exercise, we predicted the existence of possible Late/Terminal Pleistocene settlements on a high-stand terrace area immediately west of the largest drainage. This terrace area, called "Christmas Ridge," is unusual in that it incorporates a portion of the Garlock Fault scarp. This scarp created a kind of "peninsula" and (slightly lower) "reef" that jutted out into the lake during the high-stand, thereby forming a protected embayment. We suspect this embayment may have been an important environment inasmuch as, during the high-stand the south side of the sub-basin was largely a steep beachface yielding an unprotected reflective shoreline and would have suffered from wave action.

Figure 5 illustrates the site complex that proved to be present in this area, containing 12 sites and an aboriginal trail. We have not yet had an opportunity to test these sites and their ages are for this reason still unknown. But two circumstances support the possibility that some or all of them may be Late/Terminal Pleistocene in age. First, they are consistently associated with a heavily oxidized full-glacial soil which necessarily formed before roughly 10k YBP. If these sites contain sub-surface components, these necessarily will be of equivalent age. Second, a projectile point was found on the surface of one of these sites. This is a large basalt-stemmed base, and it is typologically unlike existing point types for the Far West (Figure 6). VML dating, discussed above, suggests that this artifact is older than 11k years (Cerveny et al. 2006:Figure 3).

Micro-mapping within the ACEC, in other words, demonstrated the existence of a dense concentration of sites, particularly small chipping stations and rock features. Varnish dating of a sample of the sites and features suggested that some might be Late/Terminal Pleistocene in age. Reconnaissance-level surveying provided additional evidence potentially supporting our provisional interpretations; that is, that the Christmas Canyon sub-basin contains early sites associated with preserved shoreline features.

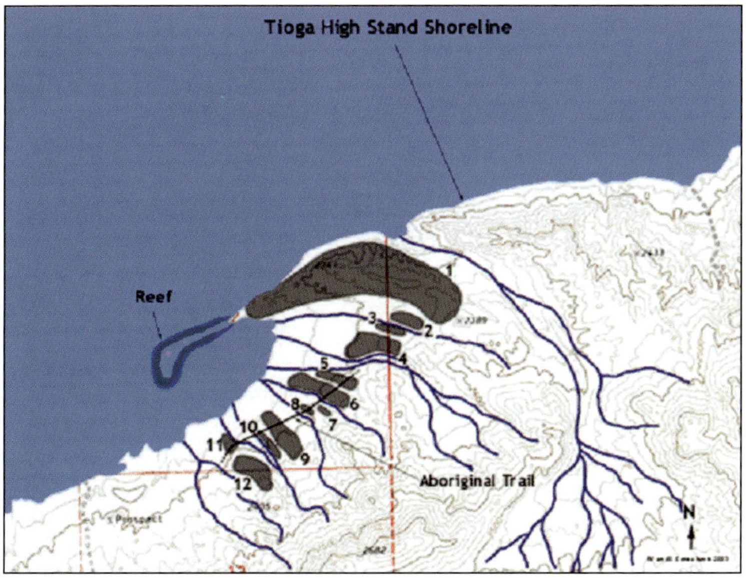

Figure 5. Recorded archaeological sites within the ACEC, located above the Tioga high-stand shoreline.

Intensive Shoreline Survey

The final project we conducted was part of a much larger survey of the Spangler Open Area, which was jointly undertaken by ASM Affiliates, Inc., Ancient Enterprises, Inc., and W&S Consultants. Our segment of the larger survey project was partly motivated by the archaeological conditions that had been previously identified in the nearby Christmas Canyon ACEC. Specifically, our contribution was designed to determine whether the conditions in the ACEC, including the presence of possible early sites, were duplicated elsewhere in the sub-basin. It was also aimed at testing our model for a Pleistocene village site location within the Christmas Canyon sub-basin as a whole.

Fieldwork for our portion of the project involved intensively surveying a transect that followed the Tioga high-stand shoreline around the Christmas Canyon sub-basin. This involved a corridor that is 17 by 1/4 miles in size, centered on the 690-m contour elevation. It also included a 2434-acre study area referred to as Block 1. This consisted of a 1-km-wide by approximately 9.5-km-long transect that crossed the sub-basin N-S, near the mouth of Teagle Wash on the sub-basin's west side.

Figure 6. Christmas Ridge site projectile point base.

A total of 39 cultural resources were recorded. Twenty-nine of these are archaeological sites, per se, while the remaining 10 are isolated artifacts. Habitations were defined as sites with a diversity of artifacts and features, including especially the presence of tertiary flakes and/or evidence for hearths. (Although midden soil in some areas is also characteristic of villages, none was noted during the project.) At least some evidence for the existence of a subsurface deposit is also characteristic. Five of these types of sites were recorded. These sites are assumed to have served, probably repeatedly, as seasonal living areas. Their size and artifact/feature density varied, presumably as a function of intensity and longevity of seasonal use. Thus this type includes what are commonly referred to as villages as well as smaller camps.

Lithic scatters are sites predominated by (if not exclusively containing) debitage. Eight such sites were recorded. Lithic scatters are generally (but not invariably) associated with the presence of raw lithic materials. They varied significantly in size, probably partly as a function of immediately available raw lithic materials. Some of the larger lithic scatters may in fact have been camps or villages, with this fact unrecognized during the survey based on a paucity of surface evidence.

Rock cairns/structures are piles, concentrations, and/or arrangements of natural rock cobbles and small boulders. In some cases these features were present on other of kinds sites. Sites were recorded generically as cairn sites when archaeological remains were exclusively or nearly exclusively limited to these kinds of features.

All but one of the cairn sites are assumed to be prehistoric rather than historic. This attribution were based on a variety of factors and circumstances, including the presence of associated artifacts and evidence for antiquity in the features (e.g., surficial revarnishing over crack varnish). Negative historical evidence (e.g., lack of any evidence for mining, inappropriateness of geological setting for mining, morphology of the cairn) also assisted in distinguishing prehistoric/aboriginal from historical/Euro-American rock features.

Perhaps surprisingly, prehistoric/aboriginal cairn sites proved to be the most common kinds of sites in the study area, with 13 recorded. Three historic sites were also recorded, along with 13 isolated artifacts.

Three primary issues result from the archaeological concerns which originally motivated and informed this study and the nature of the results that were obtained. The first of these is chronology in the larger sense, and the possibility of Late/Terminal Pleistocene sites more specifically. The second involves the high-density of rock cairns and structures and their significance. The third is the general distribution and density of sites.

Chronology

Temporally diagnostic artifacts, first, were quite rare. Aside from a 1910 dime found at a historic camp, only two diagnostic artifacts were found. One of these is an isolated Humboldt projectile point fragment made of obsidian. Humboldts are poorly dated points. Consensus suggests that they ranged from about 6000 B.C. to A.D. 600, and perhaps even later in eastern California (Justice 2002:157), meaning that they have minimal chronological value. A second point fragment was found on a large lithic scatter. This is made of jasper and exhibits pot-lid flake scars from exposure to heat. It is a large atlatl point that generally (though not perfectly) resembles a Gypsum series of points. These date from roughly 2000 to 800 B.C. (ibid:294), during the Late Archaic period.

A small quantity of obsidian was also identified and collected from the sites, intended for hydration dating (in progress at the time of this writing). Hydration dating may prove useful relative to the ages of the sites where these specimens were found but, given the general rarity of obsidian in the project area (and in truth the unreliability of hydration dating), this may not resolve the chronological picture in any larger sense.

More important for temporal placement, even if any resulting inferences or interpretations are intrinsically provisional in nature, are the geomorphological context and nature of the sites. Three (related) factors are significant in this regard: (1) site location relative to the Tioga high-stand; (2) presence/absence of alluvial fan deposits and sand sheets masking Pleistocene/Holocene landforms; and (3) the nature of the soils on site. As is clear, potential Late Pleistocene sites can be identified (based on surface evidence) only on intact landforms at and above 690 m elevation. Terminal Pleistocene/Early Holocene sites could occur and be identified on intact ancient groundsurfaces anywhere above the playa bottom. In both of these first cases oxidized soils should be present on site as an indicator of stability and age. Sites on active alluvial fans or sand sheets, in contrast, are necessarily more recent in age; the soils at these locations lack stability and aren't oxidized, and these sites can occur at any (fan) elevation. Recent surface sites obviously also can occur anywhere, with the association between these sites and their context and soils coincidental. The contextual and soils evidence used in site age assessments is then necessary but not sufficient information for age determination, absent clear indications of the presence of a subsurface deposit (e.g., from an arroyo cut-bank).

Note that alluvial fan development proved to be occurring along almost the entirety of the northern side of the sub-basin and over much of the southwestern side, with a sand sheet on the playa bottom; that is, the majority of our study area was masked. Unlike the ACEC, the study area does not contain extensive exposures of preserved ancient landforms and then large numbers of exposed early sites. Intact preserved landforms instead proved to be isolated features that were the exception rather than the rule.

Four locations nonetheless were recorded that potentially may comprise early sites. These are as follows:

<u>Site #04-10</u>: This site is located on a large SW-NE-trending ballena/ridgeline on the southwest side of the sub-basin, about 2.5 km southeast of Teagle Wash. An ephemeral drainage trends W-E, curving along the upland (S) side of the ballena and site, exiting into the basin at the east end of the site. The site in other words sits on what would have been a small "peninsula" during the high-stand, backed by a stream.

This site is very large (~430 x 250 m) in size. It contains an extensive but low-density lithic scatter and a number of features, including five low rock cairns along the north brow of the landform; a rock alignment, a campstone (hearth) feature, and a bedrock grinding slick. An oxidized paleosol is exposed in rodent backdirt piles, indicating that the site area has been stable since (at least) the Terminal Pleistocene. The campstone feature appears to be eroding out of the groundsurface, suggesting that a subsurface deposit may be present. Although the site may have multiple temporal components, its location (immediately above the high-stand) and the soils present make it a candidate for high-stand occupation.

<u>Site #04-12</u>: This is a small lithic scatter located around a tufa tower on the north side of the sub-basin, about 675 m southwest of the Randsburg Wash (Christmas Canyon) Road. While most of the N side of the basin is covered (or being covered) by an alluvial fan, localized conditions here have left exposed a small portion of the Terminal Pleistocene/Early Holocene landscape, which is about 15 m below the Tioga high-stand elevation.

The site consists of a low- to moderate-density lithic scatter that is about 25 m in diameter and is predominated by heat-treated jasper, chalcedony, and chert flakes. Site soil is slightly oxidized sand that is most likely either Terminal Pleistocene/Early Holocene or Middle Holocene in age (given the development of the primary Mojave Desert sand sheet between about 7000 and 5000 YBP). The debitage appears to be eroding out of the sand, but testing is obviously required to verify the existence of a subsurface deposit. If such is present, site age is dated by the soils and likely falls somewhere between the Middle Holocene to Paleoindian periods.

Site #04-20: This site is a large but low-density lithic scatter located on a stream terrace on the south side of Teagle Wash at about 695 m elevation; that is, immediately above the mouth of Teagle Wash during the Tioga high-stand. Artifacts present at the site were limited to about 40 chalcedony biface thinning flakes and three obsidian flakes. All of the chalcedony appeared to be heat-treated, which is a characteristic of Paleoindian lithics. Site soil is a highly oxidized clayey silty sand that is reddish brown in color and is clearly quite ancient. A subsurface archaeological deposit here necessarily would be minimally Paleoindian/Early Archaic in age, given the soils. The site location also makes the site a candidate for high-stand age.

Site #04-21: This site is something of a match for the previous site inasmuch as it occurs on the opposite stream terrace on Teagle Wash (i.e., on the north side), in essentially the same elevational position and relationship to the high-stand. It consists of a series of campstone features that are eroding out of slightly oxidized silty gravely sand that is itself capped by lacustrine sands and gravels. The site appears to be a temporary camp/habitation dating to the Pleistocene or Early Holocene.

Sites #04-20 and -21, in fact, logically may represent an early-dating site complex, given the presence of oxidized soils and buried habitation remains and their location at the outlet of Teagle Wash, the largest drainage in the region. Subsurface testing is obviously required to evaluate this possibility, however.

In summary, geomorphological conditions in our study area proved poorly suited for the preservation of early sites, given the general absence of intact Late Pleistocene/Early Holocene landforms. This circumstance emphasizes all the more the significance and rarity of the conditions and sites within the nearby ACEC. And it also underscores the general difficulties in conducting Pleistocene/Early Holocene archaeological research in eastern California. As we have emphasized previously (Whitley and Dorn 1993), the transition from the Pleistocene to the Holocene involved more than just a change in climate and vegetation. It also represented a major shift in erosional versus depositional regimes worldwide. Accompanying this circumstance in the Mojave Desert were periods of sand sheet development during the Terminal Pleistocene and Mid-Holocene (Bach 1995). Both of these events had profound implications for site preservation and archaeological inference. Unless and until these facts and natural processes are adequately accommodated in our archaeological research, we cannot assume that the absence of evidence for occupation represents legitimate evidence for an absence of occupation.

Rock Cairns and Structures

A second issue is the presence, age, and significance of rock cairns and other structures. These features were seemingly unusually common in the ACEC, and for this reason their possible presence within our lakeshore study area warranted investigation. As noted above, what we generically refer to as rock cairns were the single most common site type that we recorded during the intensive survey project (13 out of 26 prehistoric sites, representing fully 50% of this total); indeed, rock cairns were significantly more common than lithic scatters (only 8 or 31%). Note however that individual rock cairns were also present on habitation sites and some lithic scatters, so even these figures slightly underestimate the importance of this aspect of the Christmas Canyon sub-basin archaeological record.

We have discussed above the characteristics that we used to distinguish prehistoric/aboriginal from historical/Euro-American cairns and have noted that Blair and Fuller-Murillo (1997) documented a series of uses for rock circles in the ethnographic record. These include: caches, dehydration palettes, planters, water-catchers, living structure outlines, sleeping circles, fish traps, hunting blinds, vision quest circles, leaching stations, basket/pot rests, ceremonial/initiation ritual components, and hearths. Jay von Werlhof (2004) has also outlined a series of surface features and their functions. His discussion concerns Yuman-speakers in the lower desert, in the Colorado River Valley area and, for this reason, it emphasizes modifications of desert pavement rather than the creation of rock structures, per se. Still, these are conceptually similar to rock features, and his list of five types of such sites is clearly important to consider. It includes vision quest rings (about one meter in diameter); two-to-three-meter clearings with a rock cairn in the center, representing the Yuman-speakers' sacred creation mountain, Avikwa'ame; a two-to-three-meter clearing with a single boulder functioning as a kind of altar; two concentric circles with a foot trail leading outward, again representing Avikwa'ame; and a wide foot path leading upslope to a large cleared circle (5 – 10 meters in diameter), sometimes with a central cairn, representing the Xam Kwitcam ritual pilgrimage path leading to Avikwa'ame.

Von Werlhof's (2004) list is obviously (and intentionally) specific to Yuman-speakers and, for this reason, it emphasizes the symbolic importance of the Avikwa'ame (Newberry Peak), which was the central feature in the sacred landscape for these Colorado River dwellers. The particular characteristics of the site types on von Werlhof's list are, for this reason, unlikely to be applicable to Christmas Canyon, which was part of Numic territory, at least ethnographically. But there is an important implication of his list that notably contrasts with the one compiled by Blair and Fuller-Murillo's (1997). This is the emphasis that von Werlhof places on ritual rather than mundane functions, an emphasis that, given the economic-technological-mundane bias of most archaeologists, might seem quite foreign. But, as Walker (2002) has emphasized, American archaeology has overemphasized the place of the mundane in archaeological interpretations—what Walker calls an overemphasis on "practical reason"—resulting in a misapprehension of traditional nonwestern cultures, where religious/ritual activities were both very common, and inextricably intertwined with everyday life. Von Werlhof's list may not be directly applicable to Christmas Canyon, but it is a cautionary bromide about interpretations that necessarily assume that mundane functions are somehow more likely than religious/ritual ones. Indeed, von Werlhof's ethnologically based typology suggests instead that the overemphasis on mundane functional interpretations reflects a presentist western bias. This should be an embarrassment for anthropologically trained researchers who have no excuse for failing to understand the range of variation found in cultures.

Note further that Blair and Fuller-Murillo's (1997) discussion is limited to *rock circles*, per se; that is, roughly circular or donut-shaped concentrations of cobbles. Our use of the term "rock cairns" generically is intended to include their "rock circles" and other rock constructions that they do not consider. These specifically include *cairns* or stacked piles of cobbles and small boulders; *alignments* (geoglyphs) or horizontally arranged arrays of rocks; *campstone features*, which are the most common expression of hearths; and miscellaneous rock constructions, such as deadfall Figure-4 rock traps.

Hunt (1960:189) illustrates an archaeological example of a Figure-4 trap. This has loosely piled cobble walls on either side of a lower flat rock and a flat upper deadfall which was supported by the stick Figure 4. One example of a rock cairn in the ACEC appears to be a collapsed deadfall trap.

Campstone features are particularly common in habitation sites and probably too frequently are ignored by archaeologists. We have discussed them in some detail elsewhere (W&S Consultants 2001), and we recorded a number of examples of them during this project. In summary form, these are concentrations of fire-affected rock resulting from stone boiling or from use as hearthstones. Hearthstones are fire-affected rocks that we believe were originally placed in fires as heat-sinks and then heat-radiators—a logical practice in areas where fuel is both relatively rare and small in size, and where a burning fire would die out relatively quickly. These also might result from sweat lodge use. Blair and Fuller-Murillo (1997) note that, where available, limestone was also added to campfires and earth ovens to extend cooking time and heat radiation.

Hearthstones typically are volcanic in origin (e.g., basalt), to prevent explosion upon intense heating (as occurs with certain igneous and sedimentary rocks). Stone boiling can employ almost any kind of rock because less prolonged heat is involved. Regardless of precise function, campstone features are clear indications of hearths and at relatively ephemeral camps (e.g., single night stays involving small groups) may be the most visible aspect of the archaeological record.

There is an archaeological implication of campstone features that warrants mention. The hollow, circular ring of rocks that is the contemporary Euro-American conceptual model of a "proper" campfire may be inappropriate as a model for the form and structure of prehistoric hearths. Indeed, where we have seen ethnographic examples of indigenous campfires (in Central American and sub-Saharan Africa) there are at best three rocks abutting the fire itself (thereby providing stable tripod-like support for cooking vessels), not a ring of rocks. And none of the archaeological hearths (as indicated by concentrations of charcoal and ash) that we have excavated in southern California, the Great Basin, Mesoamerica, and southern Africa has been encircled by cobbles. Campstone concentrations, in other words, are probably better hearth indicators than small circles of rock. Relatively small circles of rocks, in fact, are more likely vision quest structures than hearths, if they are truly aboriginal in origin, and about a meter or more in diameter, as suggested by von Werlhof (2004).

Stacked cairns and alignments (geoglyphs) had a few different uses. Perhaps the most common of these involved ritual activities. As we have noted previously (Whitley et al. 1999a; Whitley et al. 2004), stacked rock cairns were created in south-central California during shamans' vision quests (Gayton and Newman 1940:43; Applegate 1978:34), in the Colorado Desert by non-shamans seeking supernatural aid (Patencio 1943:73; Bean et al 1992:96), and during children's vision quests on the Columbia Plateau (during which they also painted rock art to portray their visionary imagery) as well as during Plateau rituals more generally (see Teit 1896:227, 1900:317, 320-321,1906:275, 1909:590, 1930:194, 283, n.d.; Anonymous 1916:3; Steward 1936:413; Cline 1938:138-139; Mandelbaum 1938:111; Malouf and White 1953:35; Leechman et al 1955:38; Caldwell and Carlson 1956; Chartkoff 1983; Buckley 1986; Jett 1986; Keyser 1992:47-48; York et al 1993; Winthrop et al 1995; Hann, Keyser and Minthorn n.d.). In both cases the creation of these structures was related to the perceived need for intense physical exertion (including running long distances) which, along with fasting, isolation and meditation, were common components of the Native American vision questing tradition—and which obviously promoted the induction of a visionary experience.

Rock cairns certainly could have mundane purposes. Earth ovens are the most obvious of these and, in regions where agave is common, these are typically found in sandy wash bottoms (where soil conditions allow easy digging). Earthoven form is also relatively distinctive, typically consisting of large mounds of fire-affected rock in a soil and ash matrix. Use of rock cairns as trail markers/directional aids is also a commonly cited function, but we know of no ethnographic data that support such an interpretation, and it seems somewhat implausible (inasmuch as trails are quite visible on the desert groundsurface). Cairns have also been suggested to have served as "dummy hunters" used to drive game. In the last case, location and viewshed would serve as functional indicators.

Stacked rock cairns and alignments were also created during rituals. Ritually produced examples are obviously distinctive morphologically from earth ovens, at least, and they should be common in areas where vision questing was regularly practiced. The Coso Range was probably the most important such location in the Great Basin, as indicated by the ethnographic record and by its massive concentration of petroglyphs (Whitley 1998a, 1998b, 2000; Whitley et al. 1999b). The Panamint and Searles Valley areas contain the largest known concentration of rock cairns and alignments/geoglyphs (von Werlhof 1987; cf. von Werlhof 1995, 2004)—as our own studies indicate. This does not appear to be a coincidence, for two reasons (Whitley et al. 2004). The ethnographic record concerning the creation of rock structures during vision questing is limited in California, as is the record on vision questing more generally. But if we use the nearby and very rich Columbia Plateau ethnographic record as a model, one pattern is quite clear: the process of conducting a vision quest was lengthy, it involved multiple expeditions, and it moved across the landscape. One result is that, on the plateau at least, the rock structures created during vision quests were never in the same locations as the rock art that was also made at the end of this ritual period and process. That a massive concentration of rock cairns and alignments is present in the valleys and ranges immediately east of the Cosos, where the petroglyphs are concentrated, fits the pattern suggested by the plateau data even though, unfortunately, there are no known ethnographic data of any kind (positive or negative) specifically identifying such a practice among Numic-speaking peoples in the Great Basin.

That the Searles/Panamint archaeological data fit the Plateau ethnographic pattern is encouraging. Perhaps more importantly, there is archaeological confirmation of the association of rock cairns with shamans and vision questing in this region. This has been found on the "Panamint Islands" in the Panamint Valley, north of Searles Lake proper. As we have illustrated previously (Whitley et al. 1999a; cf. Whitley et al. 2004), quartz crystals were almost universally associated with shamans, the breaking of quartz occurred during vision questing and is associated with vision quest activities and structures, and quartz hammerstones were used to create petroglyphs. Concentrations of broken quartz cobbles have been found in association with rock cairns on the Panamint Islands, and a large quartz crystal offering has been identified within one of the cairns. The implication here is obvious: the concentration of rock cairns and alignments at this location was associated with shamanic vision questing. Although we found no fractured quartz or quartz crystals associated with any of our Christmas Canyon cairns, at least some of them are likely to have been made during vision questing.

Our prehistoric/aboriginal cairn data are summarized in Table 2.

Table 2. Cairn Data

Site #	Association	Description of cairn(s)
04-1	w/ lithic scatter	horizontal concentration lg. cobbles
04-3	w/ lithic scatter	rock ring lg. cobbles
04-4	w/ lithic scatter	linear array, partly buried
04-5	isolated	collapsed (?) vertical lg. cobbles
04-7	isolated	vertical pile lg. cobbles
04-8	isolated	vertical pile lg. cobbles
04-9	w/ scatter	2 horizontal conc. small cobbles
04-10	w/ habitation	5 vertical piles lg. cobbles 1 alignment
04-13	isolated	collapsed (?) vertical lg. cobbles
04-14	isolated	collapsed (?) vertical lg. cobbles
04-18	isolated	horizontal concentration lg. cobbles
04-22	w/ lithic scatter	3 collapsed (?) vertical lg. cobbles
04-23	isolated	2 embedded circular concentrations
04-24	isolated	horizontal concentration lg. cobbles
04-25	isolated	lg. rock ring
04-27	isolated	vertical pile lg. cobbles
04-28	isolated	horizontal concentration lg. cobbles
04-29	isolated	vertical pile lg. cobbles

Vertical rock piles (Figure 7) are invariably made of large cobbles and small boulders. We suspect that a number of our recorded cairns represent collapsed or fallen examples of these vertical piles, based on the size and distribution of their constituents, but these may represent a separate type themselves. On the Panamint Islands, vertical rock piles are associated with broken quartz cobbles and a quartz crystal. At least in some cases, they therefore resulted from shamanic vision questing activities. A ritual origin for at least some of the Christmas Canyon examples seems likely based on location (e.g., on exposed alluvial flats where hunting would be unlikely) and associational context (e.g., where multiple examples were created in close proximity). Furthermore, although there are at least two aboriginal trails within the sub-basin, they do not have any associated vertical rock piles, to our knowledge.

There appear to be at least four different types of rock structures in our study area:

1 - Vertical rock piles
2 - Rock alignments
3 - Approximately circular concentrations of embedded cobbles
4 - Rock rings

Rock alignments (Figure 8) are patterned arrays of cobbles. As noted previously, these geoglyphs are particularly common in the Panamint/Searles region (Whitley 2000), and it is not surprising that a few have been identified in the Christmas Canyon sub-basin. Efforts have been made to identify and then decipher an iconography in these rock features (Davis 1981). But the simple fact is that very few of these features exhibit a clearly recognizable form and, when such a form is evident, it invariably involves the simplest of geometric patterns (such as our example from Site #04-10, a straight line).

Figure 7. Vertical rock pile from Site # 04-8.

Figure 8. Small rock alignment on Site # 04-10. Note GPS (pointing N) for scale.

Here the issue is straightforward: iconography exists only for purposes of graphic and visual communication, which is a social process. If there is no consistency or patterning in a body of graphic imagery (such as repeated motifs), then there is no basis for inferring that the graphic forms represent an iconographic corpus and that they served a social communicative purpose. A comparison with the Columbia Plateau ethnographic and archaeological records clarifies this point. Rock alignments were commonly created there during vision questing. Like rock cairns, these were built at some distance from the rockart sites where the primary visions were received and recorded by painting or pecking. And like rock cairns, the alignments were created as part of the physical exertion involved in the vision quest ordeal, with no intended iconographic meaning. Alignments were in this sense created for *instrumental* purposes in that they lacked *referential* meaning (Whitley 2005), and our use of the generic term "geoglyphs" to describe them is misleading. They are arrangements of rocks but they are not glyphs. The distinction between horizontal alignments and vertical stacks of rocks, then, is in some cases artificial and potentially very misleading.

A third type of rock structure is a roughly circular concentration of embedded stones or cobbles (Figure 9), a number of which were identified in the study area. These vary in overall size and in the size of the constituent rocks (perhaps simply as a function of what is immediately available). Blair and Fuller-Murillo (1997) include these in their study of rock circles and interpret them as the remnants of storage caches: cobbles placed over excavated cysts, buried baskets, and so on, to mark and protect them (e.g., from scavenging animals). Blair and Fuller-Murillo also interpret more or less donut-shaped concentrations of these embedded stones as "opened" or disturbed caches. There are certainly ample archaeological examples of caches in midden deposits and caves covered by rocks, and this interpretation of at least some of the Christmas Canyon examples seems quite plausible. (The excavation of certain of these features potentially could test this interpretation.) Moreover, it is possible that some of the features that we have identified as collapsed vertical rock piles may instead be opened caches; at this point more research is required to adequately distinguish possible different origins and functions for these features.

Figure 9. Circular concentration of embedded cobbles from Site # 04-9.

The final cairn type is the large rock ring, one example of which was recorded. Blair and Fuller-Murillo (1997) suggest that these types of structures may have served as leaching stations, pot rests, caches, or drying palettes. Such a structure also might have served as a hearth although, if so, it seems unlikely that this example would have been aboriginal or prehistoric, given its size. It may then have been a vision quest structure, not a feature created for mundane purposes.

Rock cairns and structures in other words proved to comprise a significant portion of the archaeological record in our study, extending the pattern seen in the ACEC. Though little studied by most archaeologists, their commonness in this area suggests that they would be a fruitful topic for additional research, as Jay von Werlhof's (2004) pioneering research has shown.

Site Distribution and Density

A few final comments need be made about site distribution and density. As should be clear, site density was significantly lower within our study area than in the nearby ACEC. This appears to be the result of the different geomorphological conditions in the two areas. The preserved ancient landforms within the ACEC contain within them a substantial even if dispersed quantity of raw lithic material in the form (especially) of float jasper, and the majority of ACEC sites are small chipping stations associated with this lithic resource. This resource is all but absent on the open alluvial flats and sand sheet that characterize the remainder of the sub-basin. Whether or not exposures of this resource (along with similar sites) are buried by alluvium and the sand sheet is unknown (though this is likely). What is known is that the character of the archaeological record differs dramatically between the two areas, yet this may not necessarily reflect significant differences in prehistoric landuse patterns and adaptation.

Conclusions

A few observations need to be made in light of the archaeological record in the Christmas Canyon area. The first concerns potential or putative early sites and their relationship to Late/Terminal Pleistocene shorelines. As our work within the ACEC proper illustrates, the potential for these kinds of sites and site environment relationships is obviously high in areas like the ACEC, with particularly good landform preservation. But possible early sites may be present even in regions having in gross terms poor Pleistocene landform preservation, as a couple of examples from our larger shoreline survey showed. Preservation of likely landforms in these cases may be fortuitous, but the archaeological discovery of the sites on them was much less than hit and miss. In fact, we predicted sites at these exact locations partly because of the preserved landforms, but also because of their setting at the mouths of major washes along the Pleistocene lake shore.

The Pleistocene lakes, as we have suggested above, were more like the coast than like modern western lakes, for a number of reasons. One of these is salinity which, in at least some of the lakes, would have restricted major settlements to fresh water sources—which is to say to stream or river mouths. But another potentially important factor would have been wave action. This was much more significant along the lake shorelines than one might initially think, given the sizes of these lakes and the kinds of winds that the desert region experiences. Unprotected steep shores, which are common on many of the Pleistocene lakes, would have been regularly pounded by waves. This in turn would have made them dangerous to access and use and suboptimal for the support of lacustrine resources (including lake margin vegetation and the fauna that this would attract).

Protected shoreline settings with a freshwater source, then, would have been the most favorable locations for Late Pleistocene villages. This is exactly the kind of setting and context present at Christmas Ridge, where we have recorded a complex of sites sitting on an ancient paleosol. A projectile point base from this site complex has been VML dated between 11,000 and 14,000 YBP. Obviously this is a provisional date, and it remains to be seen whether intact sub-surface deposits are present in this location and whether additional chronometric work will support this initial assay. But there is at least room for optimism concerning Late Pleistocene archaeology in this area.

The final point that we would like to emphasize concerns the rock cairns, structures and alignments that are relatively common in the Christmas Canyon sub-basin—and, as von Werlhof's (1987, 1995) work has shown, in the Panamint/Searles region more generally. Despite his efforts and those of a few additional archaeologists such as Blair and Fuller-Murillo (1997), the profession as a whole is surprisingly ignorant and apparently little interested in these kinds of features. It is not uncommon, for example, for archaeologists to almost immediately assume that cairns are necessarily historical mining claim markers, regardless of whether any mining activity ever occurred in the area in question. We hope that this attitude will itself soon disappear and that more effort will be exerted towards studying and understanding these kinds of rock features, enigmatic though some of them may still be. Jay von Werlhof's research on this topic has certainly provided a solid foundation and direction that California desert archaeologists would be wise to follow.

Acknowledgments

We thank Russ Kaldenberg and Judyth Reed for their enthusiasm and help with all of the Christmas Canyon projects, Susan Hector for her assistance with the Spangler Open Area study, and Luz Ramirez de Bryson for her contributions. We also thank our crew members for their dedication and efforts: Bruce Gothar, Michael Kuhn, Bob Sheets, Rob Snibley, Mike Hakim, and Tim Schmonk.

References

Anonymous
1916 Indian Tells Story of City Hall Rock. *Oregonian Newspaper*, Monday, January 24, 3.

Applegate, R.B.
1978 *?Atishwin: The Dream-Helper in South-Central California*. Socorro (NM): Ballena Press.

Bach, Andrew J.
1995 Climatic Controls on Aeolian Activity in the Mojave and Colorado Deserts, California. PhD dissertation, ASU, Tempe.

Bean, L.J., S.B. Vane and J. Young
1991 *The Cahuilla Landscape: The Santa Rosa and San Jacinto Mountains*. Menlo Park (CA): Ballena Press Anthropological Papers 75.

Bischoff, J. L. and K. Cummings
2001 Wisconsin glaciation of the Sierra Nevada (79,000 - 15,000 yr B.P.) as recorded by rock flour in sediments of Owens Lake, California. *Quaternary Research* 55:14-24.

Blair, L.M. and M. Fuller-Murillo
1997 *Rock Circles of Southern Nevada and Adjacent Portions of the Mojave Desert*. Nevada Department of Transportation, Cultural Resources Section, Environmental Division.

Buckley, T.
1986 Lexical Transcription and Archaeological Interpretation: "A Rock Feature Complex from Northwestern California." *American Antiquity* 51:617-618.

Caldwell, W.W. and R.L. Carlson
1954 Further documentation of "Stone Piling" during the Plateau Vision Quest. *American Anthropologist* 56: 441-442.

Campbell, E.W.C., W.H. Campbell, E. Antevs, C.A. Amsden, J.A. Barbieri and F.D. Bode
1937 *The Archaeology of Pleistocene Lake Mohave: A Symposium*. Southwest Museum Papers Number 11. Highland Park.

Cerveny, N., R. Kaldenberg, J. Reed, D.S. Whitley, J. Simon and R.I. Dorn
2006 A new strategy for analyzing the chronometry of constructed rock features in deserts. *Geoarchaeology* 21(3):281-303.

Cline, W.
1938 Religion and World View. In *The Sinkaietk or Southern Okanagan of Washington*, edited by L. Spier. General Series in Anthropology 6, Contributions from the Laboratory of Anthropology 2. Menasha, Wisconsin, 133-182.

Chartkoff, J.L.
1983 A Rock Feature Complex from Northwestern California. *American Antiquity* 48:745-760.

Davis, E.L., editor
1978 *The Ancient Californians: Rancholabrean hunters of the Mojave Lakes Country*. Los Angeles: Natural History Mueum of Los Angeles County, Science Series 29.
1981 Shamanism and Ikonography [sic] of Paleo-Indians. *ASA Journal* 5(1):43-49.

Dorn, R. I.
1998 Age determination of the Coso rock art. In E. Younkin, ed., pp. 69-96, *Coso Rock Art*. Maturango Press, Ridgecrest.

Dorn, R.I.
2001 Chronometric techniques: engravings. In D.S. Whitley, ed., pp. 167-189, *Handbook of Rock Art Research*. Walnut Creek (CA): AltaMira Press.

Gayton, A.H. and S.S. Newman
1940 Yokuts and Western Mono Myths. *University of California Anthropological Records* 5. Berkeley, 1-110.

Hann, D., J. Keyser and P. Minthorn
n.d. Columbia Plateau Rock Art: A Window to the Spirit World. In D.S. Whitley, ed., *Ethnography and North American Rock Art*. Walnut Creek: AltaMira Press (In Press).

Hunt, A.
1960 *Archeology of the Death Valley Salt Pan, California*. University of Utah, Anthropological Papers 47. Salt Lake City.

Jett, S.C.
1986 Observations Regarding Chartkoff's California "Rock Feature Complex." *American Antiquity* 51: 615-616.

Justice, N.D.
2002 *Stone Age Spear and Arrow Points of California and the Great Basin.* Bloomington: Indiana University Press.

Keyser, J.D.
1992 *Indian Rock Art of the Columbia Plateau.* Seattle: University of Washington Press.

Lajoie, K. P.
1968 Late Quaternary stratigraphy and geologic history of Mono Basin, eastern California. Ph.D. dissertation, University of California, Berkeley.

Leechman, D., M. Hess and R. Fowler
1955 Pictographs in southwestern Alberta. *National Museum of Canada, Bulletin* 136:36-53.

Liu, T.
2003 Blind testing of rock varnish microstratigraphy as a chronometric indicator: results on late Quaternary lava flows in the Mojave Desert, California. *Geomorphology* 53:209-234.

Malouf, C., and T. White
1953 The origin of pictographs. In G. Griswold, ed., pp. 30-35, *Archaeological sites in the Flathead Lake region, Montana: A symposium*. Anthropology and Sociology Papers 15. Missoula (MT): Montana State University.

Mandelbaum, M.
1938 The Individual Life Cycle. In *The Sinkaietk or Southern Okanagan of Washington*, edited by L. Spier. General Series in Anthropology 6, Contributions from the Laboratory of Anthropology 2. Menasha, Wisconsin, 103-130.

Marston, R.A.
2003 Editorial note. *Geomorphology* 53:197

Patencio, F.
1943 *Stories and Legends of the Palm Springs Indians*. Palm Springs (CA): Palm Springs Desert Museum.

Smith, G. I., V. Barczak, G. Moulton and C. Liddicoat
1983 Core KM-3, a surface to bedrock record of late Cenozoic sedimentation in Searles Valley, California. *U.S. Geological Survey Professional Paper* 1256:1-24.

Smith, G. I. and J. L. Bischoff
1993 Core OL-92 from Owens Lake, southeast California. *U.S. Geological Survey Open File Report* 93-683:1-398.

Smith, G. I. and F. A. Street-Perrott
1983 Lakes of the Western United States. In H. E. Wright, Jr., ed., *Late Quaternary environments of the United States*, in Porter, S. C., ed., p. 190-211, *Volume 1, The Late Pleistocene*. Minneapolis: University of Minnesota Press.

Steward, J.H.
1936 Petroglyphs of the United States. *Smithsonian Institute Annual Report for 1936.* Washington, D.C., 405-425.

Teit, J.
1896 A rock painting of the Thompson River Indians. *American Museum of Natural History, Bulletin 8*. New York, 227-230.
1900 The Thompson River Indians of British Columbia. *Memoirs of the American Museum of Natural History*, Vol. II, Part 4. New York.
1906 The Lillooet Indians. *Memoirs of the American Museum of Natural History*, Vol. II, Part 5. New York, 193-300.
1909 The Shushwap. *Memoirs of the American Museum of Natural History*, Vol. II, Part 8. New York.
1930 The Salishan Tribes of the Western Plateau. *Bureau of American Ethnology 45th Annual Report*. Washington, D.C.: Smithsonian Institution, 23-396.
n.d. Notes on rock painting in general, 1918. Manuscript, Glenbow Alberta Institute Archives.

von Werlhof, J.
1987 *Spirits of the Earth: Volume 1, The North Desert*. El Centro: Imperial Valley College Museum.
1995 Geoglyphs in Time and Space. *Proceedings of the Society for California Archaeology* 8:61-68.
2004 *That They May Know and Remember: Volume 2, Spirits of the Earth*. Ocotillo: Imperial Valley College Desert Museum Society.

W&S Consultants
2001 Phase II Test Excavations For The Life Of The Mine Project Area, Boron, Kern County, California. Manuscript on file, CSUB AIC.
2003 Report on Mapping 16 Sites in the Christmas Canyon ACEC. Manuscript on file, BLM, Ridgecrest.
2004 High-Stand Shoreline Survey in the Spangler Open Area, San Bernardino County, California. Manuscript on file, BLM, Ridgecrest.

Walker, W.H.
2002 Stratigraphy and Practical Reason. *American Anthropologist* 104:159-177.

Whitley, D.S.
1998a Finding rain in the desert: Landscape, gender, and far western North American rock art. In C. Chippindale & P.S.C. Taçon, eds., pp. 11-29, *The Archaeology of Rock-Art*. Cambridge: Cambridge University.
1998b Meaning and Metaphor in the Coso Petroglyphs: Understanding Great Basin Rock Art. In E. Younkin, ed., pp. 109-174, *Coso Rock Art: A New Perspective*. Ridgecrest: Maturango Museum.
2000 *The Art of the Shaman: Rock Art of California*. Salt Lake City: Univ. of Utah Press.
2005 *Introduction to Rock Art Research*. Walnut Creek: AltaMira Press (In Press).
n.d. The Archaeology of Religion. In H. Maeschner and A. Bentley, eds., *Handbook of Archaeological Theories*. Walnut Creek: AltaMira Press (In Press).

Whitley, D.S. and R.I. Dorn
1993 New Perspectives on the Clovis vs. Pre-Clovis Controversy. *American Antiquity* 58:626-647.

Whitley, D.S., R.I. Dorn, J.M. Simon, R. Rechtman and T.K. Whitley
1999a Sally's Rockshelter and the Archaeology of the Vision Quest. *Cambridge Archaeological Journal* 9:221-246.

Whitley, D.S., J. Simon & R. Dorn
1999b The vision quest in the Coso Range. *American Indian Rock* Art 25:1-32.

Whitley, D.S., J. Loubser and D. Hann
2004 Friends in Low Places: Rock art and landscape on the Modoc Plateau. In C. Chippindale and G. Nash, eds., pp. 217-238, *The Figured Landscapes of Rock Art: Looking at Pictures in Place*. Cambridge: Cambridge University.

Winthrop, R.H., K.R. Winthrop and D.J. Gray
1995 Rock Feature Sites on the Mt. Hood National Forest: An Inventory and Evaluation. Unpublished report on file, Mt. Hood National Forest, Gresham, Oregon.

York, A., R. Daly and C. Arnett
1993 *They Write Their Dreams on the Rock Forever: Rock Writings in the Stein River Valley of British Columbia*. Vancouver (BC): Talonbooks.

A FESTSCHRIFT HONORING THE CONTRIBUTIONS OF CALIFORNIA ARCHAEOLOGIST JAY VON WERLHOF

Edited by Russell L. Kaldenberg

Contributors:

Mark Allen – Mark earned his doctorate in anthropology at UCLA. His research interests include chiefdoms, hunter-gatherers, and prehistoric warfare. He is also committed to the teaching and practice of sound cultural resource management. While the Mojave Desert is his current main focus, he has conducted research in Kentucky, Illinois, Guatamala, Fiji, Hawaii, New Mexico, and New Zealand. He has previously taught or worked at UCLA, Pomona College, the Center for American Archeology (Kampsville, Illinois), and Fort Irwin in the Mojave Desert. He is presently Associate Professor of Anthropology at California State Polytechnic University. Pomona, and conducts a project or two each year for Epsilon Systems Solutions. Mark recently co edited *The Proceedings of the Millennium Conference*, published by Maturango Museum Press, and *The Archaeology of Warfare*, published by the University Press of Florida.

Rebecca Apple – Rebecca is a principal and senior archaeologist at EDAW in San Diego, California. She has over 20 years of experience in archaeology. Her field experience encompasses the southwestern United States, specifically desert areas of California. She received her M.A. in anthropology from San Diego State University. Her masters thesis dealt with the Coso volcanic field

Cheryl Bowdeen-Renna – Cheryl is an associate and staff archaeologist in San Diego California. She has over 15 years of experience in archaeology. Her work has been primarily in the Southwestern United States and Israel. She received her B.A. in anthropology from San Diego State University.

George Carter – George spent over 60 years researching early humans and their relationship with the environment. He first excavated with Malcolm Rogers at the Harris Site in San Diego County, and was a lifetime believer in the ancient antiquity of humans in California, particularly in San Diego and in the deserts such as at the Calico Early Man Site. George Carter died in 2004. He was professor emeritus at Texas A and M University.

Niccole Cerveny – Niccole's research interests are geomorphology and rock art conservation. She completed her Ph.D. in 2005 in the Geography Department at Arizona State University on rock weathering and its implications for rock art dating and conservation. She has also applied varnish dating techniques to stone structures in the Mojave Desert. She teaches at Mesa Community College, Arizona, and is working towards the integration of rock art conservation into the community college physical science curriculum.

Morlin Childers – Morlin was the curator of paleontology at the Imperial Valley College Desert Museum in El Centro. He had an acute interest in the peopling of the Americas and the artifacts left by them in what is now called the Yuha Desert. His interests were particularly in interrelationships between the early humans and big game animals such as elephants. He was a believer in a Malpais culture. He also believed that one could see the evolution of stone tool technology in the new as well as the old worlds. He passed away in 1988.

James W. Cleland – Jamie is a principal and senior archaeologist at EDAW in San Diego, California. He has enjoyed working in the California deserts for over 25 years. His association with Jay von Werlhof began when he first arrived in California. His other field experiences include coastal and montane California, the Virginia piedmont and the Blue Ridge Mountains, and the Indus Valley, Pakistan. His is a past president of the Society for California Archaeology and received a Ph.D. and an M.A. in anthropology from the University of Virginia. Jamie can be reached at EDAW corporate headquarters in San Diego.

C.W. Clewlow, Jr – "Billy" received his Ph. D. from the University of California, Berkeley, and is president of Ancient Enterprises, Inc. He has long been considered an authority on the Great Basin and the desert west, and his studies of rock art as well as his inquiries into the Early Holocene have set standards for the field. He is an experimental archaeologist having used electronic gadgets to define site boundaries years before ground-penetrating radar was popularized. Billy has explored the possibilities of pre-Clovis, early hominids in the New World, and the meaning of stone monuments throughout his study areas with vigor. His focused work has rewarded him with one of the most honored awards that can be given to a practicing archaeologist, the Society for California Archaeology's 2004, Thomas F. King

Award for Excellence in Cultural Resource Management. When not in the Black Rock Desert, Nevada, Billy can be reached at his corporate headquarters in Santa Monica, California.

John R. Cook – J.R. is a principal at ASM Affiliates and one of Jay's newer old friends. He first met Jay in 1974 during an SDSU-sponsored excavation of a Lake Cahuilla shoreline site, and has had the pleasure of working with him on numerous occasions over the years. Jay was instrumental in teaching J.R. in the finer points and subtleties of desert archaeology, knowledge that is crucial to a thorough understanding of the prehistory and has been incorporated into the many CRM studies authored by ASM staff. J.R.'s interests include the Paleoindians of the west and Late Prehistoric ceramic technology.

Courtney Ann Coyle – Courtney is an attorney whose private law firm's work includes representing American Indian tribal governments and entities in San Diego and in San Bernardino and Imperial Counties in planning, legislative, policy, and project initiatives. Her practice focuses on protecting tribal heritage resources including cultural landscapes and traditional cultural properties under CEQA, NEPA, and other local, state, and federal laws. On behalf of the Quechan Indian Nation, a bill sponsor, she helped to craft and worked to successfully gain passage of SB 18 and related bills over the last four years. Courtney was named *California Lawyer Magazine's* Environmental Lawyer of the Year for 2003 for her work on the passage of California's new mining reclamation and backfill measures the first of their kind adopted by any government. She also serves on several Boards advocating for cultural resources protection, including the Trust for Public Land's California Advisory board, and is an appointee to the City of San Diego's Commission for Arts and Culture. Courtney can be reached at her corporate headquarters in La Jolla, California.

Ronald Dorn – Ron is a Professor of Geography at Arizona State University. He received his Ph. D. from the University of California, Los Angeles, and he specializes in rock weathering, chronometrics and rock art research. He is the author of *Rock Coatings* (Elseevier, 1998) in addition to over a hundred research papers. He has been elected a Guggenheim Fellow and is the recipient of numerous prestigious awards in anthropology and geology. Ron can be reached at Arizona State University, Tempe, Arizona.

Diane M. Erwin – Diane is currently the Paleobotany Curator and Collection Manager of the Museum of Paleontology at the University of California, Berkeley. She received her B.S. and M. S. degrees in botany from Ohio University and a Ph.D. in paleobotany from the University of Alberta in Edmonton, Canada. Her current research interests include work on the descriptive systematics of Late Cenozoic leaf compression floras and petrified plant and wood remains from western North America to better understand this region's plant diversity and biogeographic histories during this time. This work has led also to an interest in ancient plant hosts—insect relationships such as those between oak and cynipid wasps and wooly aphids (adelgids) and spruce.

William Eckhardt – William is a California archaeologist with more than 25 years of experience and practice in the study, protection, and management of cultural resources. In 1976 William earned his bachelor of arts in anthropology with distinction in Archaeology from San Diego State University, then began his career in archaeology in the contract sector. From 1983-1996 he served in multiple capacities for the Department of Navy as archaeologist, contract technical representative, and historic preservation officer responsible for cultural resources located on 1.1 million acres of military testing range in the Mojave Desert. He is currently a senior archaeologist with Jones & Stokes, an environmental, planning, and resource management firm, and serves as a board member (2003-2006) to La Puerta Foundation, a California nonprofit public-benefit corporation.

Gail Egolf – Gail received his B.A. from Chico State University in anthropology. His interests in archaeology developed during his youth, growing up on a farm in Indiana which contained numerous archaeological sites and which the family had homesteaded in the second decade of the nineteenth century. Gail worked on forests in northern California including the Klamath National Forest. He found his way to southern California as area archaeologist for the El Centro Resource Area where he quickly made lifetime friends with Jay and Sherilee von Werlhof. Gail left El Centro to continue his family tradition of farming. He and his wife Judy keep in regular contact with the von Werlhofs, and now that they are retired they are winter snowbirds in Yuma, Arizona, where they can talk archaeology with Jay. Gail and Judy make their permanent residence in Indiana.

Tim Gross – Tim earned a B.A. in Anthropology from San Diego State University and an M.A. and a Ph.D. from Washington State University. He has been a student of Southern California's past since the 1960s. He began working in the Anza Borrego Desert State Park in the early 1970s and had the pleasure of meeting Jay von Werlhof at that time. Tim is currently engaged in the study of San Dieguito material from the C.W. Harris site in western San Diego County, and is interested in how the San Dieguito Complex figures in the larger picture of the early occupation of North America's west coast. Tim was instrumental in founding the San Diego Archaeology Center and in finding homes for orphaned archaeological collections.

Phil Hines – Phil is an archaeologist for the Off-Highway Motor Vehicle Recreation Division of California Department State Parks and Recreation. He has worked on a variety of projects ,including historic and prehistoric archaeological project throughout the state since 1975. His main interests are in prehistoric archaeology, particularly of arid lands. Between 1985 and 1990 he worked closely with Francis Riddell mapping and recording the ruins of Tambo Viejo in Acari, Peru. He began working for the Off-Highway Motor Vehicle Division in 1998 and has conducted cultural resources inventories on the Off Highway Vehicle Parks since then. Phil received both his bachelor's and Master's degrees in anthropology from California State University at Long Beach, California. He was a student of Franklin Fenenga. He can be reached at the Division of Off Highway Vehicles, in Sacramento, California.

Loretta Jackson – Loretta is the Cultural Resources Coordinator of the Hualapai Tribe in Peach Springs, Arizona. She regularly conducts government staff on tours of the archaeological resources along the Colorado River within Hualapai traditional territory. She has worked closely with the Bureau of Land Management and the Bureau of Reclamation archaeologists to protect cultural resources.

E. Henry James – We all miss Henry. He wrote his tribute to his friend Jay shortly before he passed away at the age of 86. Henry only had a high school diploma, but he had sharp eyes and sharper intuition as to where trails were to be found. Henry published two monographs on desert trails and was always quick to admonish public lands managers to do a better job of managing the lands because these trails, which he felt were hundreds and maybe thousands of years old, are very fragile, and it takes only a single bad managerial decision or a reckless individual to destroy something that should be protected and left for future studies.

Boma Johnson – Boma received a double master's degree in archaeology and ethnology from Brigham Young University in 1973. He worked from 1975 until 1999 for the Bureau of Land Management as a Field Archeologist/Cultural Resources Management Specialist. His research focus was on learning the origin and the purpose of petroglyphs and geoglyphs found through the Colorado River area. After retiring from the BLM, he moved to St. George Utah, where he is giving public lectures, teaching classes, and leading rock art tours to share his knowledge on the rock art of the ancient Native Americans of the Southwest. Boma can be reached at his new company "Archeology Plus," at 185 West Center St., Ivins, Utah 84738. Boma's phone number is 435 688-7325. He may also be reached at www.archaeology-plus.info.

Russell L. Kaldenberg – Russell received his B.A. and M. A. degrees in anthropology from California state universities. He has worked in cultural resources management since 1974 and in federal service since 1976. He has held positions such as Desert District Archaeologist and State of California Archaeologist for the Bureau of Land Management, as well as Field Manager for the BLM in Palm Springs. He served as Regional Archaeologist for the U.S. Forest Service, Milwaukee, Wisconsin, and Base Archaeologist for the China Lake Naval Air Weapons Station. He serves as a member of the Board of Directors for the Mojave Desert Heritage and Cultural Association and has served as an appointee of Governor Wilson on the State Historical Resources Commission. Russ has been honored by receiving the Mark R. Harrington Award for Conservation Archaeology, two Governor's Historic Preservation Awards, and most recently a Meritorious Civilian Service Award by the Department of the Navy. Russ lives in Trona, California, where he is a lifetime member of the Searles Valley Historical Society and where he studies the prehistory and history of Searles Lake and Valley environs.

Jeff LaFave – Jeff is a trial attorney in San Diego. He is also a rock art recorder. He has visited rock art sites throughout the U.S. and in 14 other countries, including France, South Africa, and Namibia. He received his J.D. from the University of Redlands in 1987. He has published articles about rock art including two recent rock art papers. He has also lectured about rock art topics and is a member of various rock societies including the American Rock Art Research Association and the Rock Art Foundation. Jeff is a California Archaeological Site Steward and regularly monitors the condition of its adopted archaeological site through the Bureau of Land Management. Jeff may be reached at jlafave1@cox.net.

Janie Loubser – Janie is a consulting archaeologist working for New South Associates in Atlanta, Georgia. He received his Ph.D. from the University of the Witwatersrand and specializes in rock art research and conservation. He is a graduate of the Getty Institute's yearlong rock art conservation training program that was conducted in 1987 in Australia. Janie has published widely on topics ranging from the Southern African Iron Age, to southeastern Woodlands archaeology, to rock art interpretation throughout the world, to rock art site management. He has worked throughout southern and eastern Africa, Australia, the Caribbean, and North and South America. He also writes comic books featuring archaeologists and their adventures, with *Archaeology: The Comic* (Alta Mira Press, 2003), his first effort in what is proving to be a useful (and humorous) approach to archaeological education.

Daniel McCarthy – Daniel received his B.S. and M.A. in anthropology from the University of California, Riverside. He has worked at Anza-Borrego State Park of Joshua Tree National Park, and throughout southern California compiling photographic inventories of rock art of these areas and throughout the region. Currently, he is Tribal Relations Program Manager for the San Bernardino National Forest.

Research interests include desert archaeology, aboriginal trail systems of the Sonoran, Mojave, and Colorado deserts; rock art distribution, documentation and interpretation; establishing conservation and protection guidelines; and recording string figures. He conducts southern California Native American cultural awareness classes providing information about native plant uses (particularly agave, acorns, and pinyon), rock art, and material culture. Daniel works with tribal governments (Serrano, Cahuilla, and Luiseno) and organizations such as the Southern California Indian Basketweavers Organization, where he is currently vice president.

Meg McDonald – Meg received her doctorate in anthropology at the University of California, Riverside. Her dissertation research focused on excavations at the Indian Hill Rockshelter in Anza-Borrego Desert State Park, a well-known site that contains deposits spanning 5,000 years of occupation. Since 1984, Meg has worked in the desert regions of California and Arizona and in the San Bernardino, Riverside, and San Diego areas of southern California. Currently she is the Cultural Resources Manager at the Marine Corps Air-Ground Combat Center, Twentynine Palms, California, and may be contacted there.

Judyth E. Reed – Judyth received her B.S. from the University of Iowa in classical archaeology and her M.A. in anthropology from the University of California, Riverside. Her specialization is dry lands environmental adaptation, particularly the southern California desert. Judyth has worked in federal service with the Bureau of Land Management and the US Forest Service since 1978. She served as Forest Archaeologist for the Los Padres National Forest, and worked for the Mendocino National Forest and the San Bernardino National Forest. Her BLM career began with the California Desert Plan. She then became the Indio Resource Area archaeologist. Later she worked as the Lead Archaeologist in the Ridgecrest Field Office and is presently staff liaison for the Wyoming State Office, Bureau of Land Management, Cheyenne, Wyoming, and the Wyoming SHPO. Judyth edited the Society for California Archaeology Proceedings for five years and is currently on the Editorial Board of the Archaeological Conservancy and the *Journal of California and Great Basin Anthropology*. She recently co-edited the *Proceedings of the Millennium Conference*, published by the Maturango Museum Press.

Howard Schorn – Howard has a B.A. degree in geology from Macalester College in St. Paul, Minnesota, and an M.A. in paleontology from UC Berkeley. He worked for 30 years as Collections Manager of paleobotany collections in the Museum of Paleontology at Berkeley. His main professional interests have centered on Neogene plants, especially the conifers of western North America. He has collected fossil plants of Devonian through Pleistocene age from the western states, including northern Alaska. His hobbies include black-and-white landscape photography and not too surprisingly, fossils. The history of just about anything is his favorite reading subject.

Joseph Simon – Joe is a consulting archaeologist and a partner in W&S Consultants, Simi Valley, California. His research specialties are geoarchaeology and the Pleistocene archaeology of the California Deserts. His recent publications include *The Archaeology of Ayers Rock, Inyo County, California* (Maturango Museum Press, 2005, with David S. and. T. K. Whitley). Joe lives in Simi Valley, California.

Joan S. Schneider – Joan received both M.S. and Ph.D. degrees in anthropology from the University of California, Riverside. She has worked in the deserts of California and Arizona for over 20 years. Currently, an Associate State Archaeologist in the Colorado Desert District, she continues her research interests both in the U.S. western desert and Israel. Her research focuses on the geoarchaeological aspects of millingstone quarries and the roles that women played in prehistoric hunter-gatherer societies. Joan can be reached at the State of California Department of Parks and Recreation, Colorado Desert District, 200 Palm Canyon Drive, Borrego Springs, California 92004.

Jackson Underwood – Jackson holds his M.A. and B.A. in anthropology from San Diego State University and his Ph.D. from the University of California, Los Angeles. He has worked in heritage resource management and applied anthropology since 1980. His ethnographic research includes work with the Navajo, Hopi, Tarahumara, Quechan, Mojave, Tohono O'odham, Maricopa, Kumeyaay, Luiseno, and Gabrieleno. He has conducted applied anthropology research in a San Diego beach community and among mentally ill homeless people in Los Angeles. While the vast majority of his archaeological work has been in southern California, he has also conducted archaeological survey and excavations in New Mexico, Arizona, Nevada, southern Oregon, and northern California. He also has extensive experience in oral history. He has written over 130 heritage resources reports and has presented approximately 20 professional papers and written 10 monographs and articles. His primary research interest is in archaeology and ethnography of the California deserts and western Arizona. Jackson is currently the corporate archaeologist and ethnographer for RECON in San Diego.

Michael R. Walsh – Michael received his Ph.D. from the University of California, Ls Angeles. He is a Research Associate at the Cotsen Institute of Archaeology, UCL, and Senior Archaeologist for Ancient Enterprises, Inc, a cultural resources management firm. His numerous publications reflect wide interests in both California and the North American Southwest. He may be reached at Ancient Enterprises, Santa Monica, California.

Claude N. Warren – Claude is Distinguished Professor Emeritus in the Department of Anthropology and Ethnic Studies, University of Nevada, Las Vegas, where he taught for 28 years. He has conducted archaeological research in southern California for 48 years and has published numerous articles and monographs on the archaeology of southern California deserts. He was the Barrick Distinguished Scholar at the University of Nevada in 1988, named Distinguished Lecturer by the Southwest Anthropological Association in 1994, and given the Society for California archaeology Lifetime Achievement Award in 1996. Recent articles by Warren include works on the Lake Mojave and Pinto Periods in the Mojave Desert, the Archaeopolitics of the Mojave Sink, and the Age of Clovis Points at China Lake. Claude may be reached at the University of Nevada, Department of Anthropology, Las Vegas, Nevada.

David S. Whitley – David is a consulting archaeologist, a partner in W&S Consultants and an Adjunct Professor of Geography at Arizona State University. His Ph.D. is from the Univrsity of California, Los Angeles, and its focus was on Coso Range rock art. David is a prolific publisher and one of the most renowned rock art specialists in the world. His recent books include *The Art of the Shaman: Rock Art of California* (University of Utah Press, 2000), *Handbook of Rock Art Research* (AltaMira Press, 2001), *Introduction to Rock Art Research* (Left Coast Press, 2005), *Discovering North American Rock Art* (University of Arizona Press, 2005, with L. Loendorf and C. Chippendale), and *The Archaeology of Ayers Rock, Inyo County, California* (Maturango Museum Press, 2006, with T.K. Whitley and J. Simon). He edits the Archaeology of Religion book series for AltaMira Press. David was honored by the Society for California Archaeology with the receipt of the Thomas F. King Award for Excellence in Cultural Resources Management. David lives in Fillmore, California.